THE TWO-BODY FORCE IN NUCLEI

THE TWO-BODY FORCE IN NUCLEI

Proceedings of the Symposium on the Two-Body Force in Nuclei
held at Gull Lake, Michigan, September 7-10, 1971

Edited by
S.M. Austin and G.M. Crawley

Cyclotron Laboratory, Department of Physics
Michigan State University
East Lansing, Michigan

ℙ PLENUM PRESS • NEW YORK-LONDON • 1972

Library of Congress Catalog Card Number 72-76009

ISBN 0-306-30598-4

© 1972 Plenum Press, New York
A Division of Plenum Publishing Corporation
227 West 17th Street, New York, N.Y. 10011

United Kingdom edition published by Plenum Press, London
A Division of Plenum Publishing Company, Ltd.
Davis House (4th Floor), 8 Scrubs Lane, Harlesden, London, NW10 6SE, England

Printed in the United States of America

Symposium Organization

under the sponsorship of

The National Science Foundation
The Office of Naval Research
The Physics Department, Michigan State University

Symposium Committee

S. M. Austin
G. M. Crawley
H. McManus

PREFACE

The idea of this symposium grew out of our discussions on the need to review the advances that had been made in the theoretical description of inelastic scattering reactions in the last few years. Since a microscopic description of inelastic scattering uses realistic effective interactions, we felt that it was appropriate to begin such a summary with a discussion of the free two-nucleon force.

However as we thought further about this review, it became increasingly apparent that a rather more ambitious program linking the free two-nucleon force and nuclear matter calculations both to shell model calculations and to reaction theory, would be appropriate and perhaps even necessary to do full justice to the subject. We hope that the symposium as it emerged did fulfill these aims, better perhaps than we expected.

There are some comments required concerning the presentation of the material. First the papers are grouped by session number for convenient gathering of the same topic in the same place. Secondly, because of the rather tight constraint on pages, it was possible to print only those contributions which were presented orally at the symposium. The remainder are included as abstracts. The full text of all the contributed papers is available as a Michigan State University Cyclotron Laboratory Report (MSUCL 39/1971). All invited papers are included except Professor Macfarlane's talk on "Nuclear Spectroscopy with High Energy Projectiles" which unfortunately was not available at the publication deadline. We did not record discussion at the symposium but left the inclusion of any comments to the discretion of the speaker.

Finally, we should like to thank the many people who helped make the symposium a success, including nearly all the faculty and many of the graduate students at the MSU Cyclotron Laboratory. In particular we should like to thank Hugh McManus, Gerry Brown and Malcolm Macfarlane for advice and comments on the program and all the distinguished physicists who willingly acted as chairmen; we should also like to thank Stan Fox, Duane Larson, Peter Miller, Bill Wagner and Andy Kaye who were all of great assistance in the practical aspects of helping the symposium run smoothly. Finally it is a great pleasure to thank Julie Perkins, our indefatigable secretary for all her labors both at Gull Lake and in the following months in cheerfully typing a good deal of the final manuscript.

S. M. Austin
G. M. Crawley

PROGRAM OF
GULL LAKE SYMPOSIUM ON THE TWO-BODY FORCE IN NUCLEI

Session I - Tuesday Morning

Symposium Overview   M. H. Macfarlane
(ANL)

The Free Two-Nucleon Force

Phase-Shift and Potential-Model   P. Signell
   Analysis of Experimental Information   (Michigan State
University)

Elementary-Particle Models of the   G. E. Brown
   Two-Nucleon Force   (Stony Brook
& Nordita)

Contributions

Session II - Tuesday Evening

Off-Energy-Shell Effects

Review of Experiments on Nucleon-   M. L. Halbert
   Nucleon Bremsstrahlung   (ORNL)

Some Basic Questions in Nucleon-   L. Heller
   Nucleon Bremsstrahlung   (LASL)

Off-Energy-Shell Effects in Many-   F. Tabakin
   Nucleon Systems   (University of
Pittsburgh)

Session III - Wednesday Morning

Calculations of the Effective Reaction Matrix
in Nuclear Matter and Finite Nuclei

An Effective Reaction Matrix Derived   H. A. Bethe
   from Nuclear Matter Calculations   (Cornell University)

The Effective Two-Body Interaction   B. R. Barrett
   in Finite Nuclei and its Calculation (University of
Arizona)

Contributions

Session IV - Wednesday Evening

   <u>Effective Interaction from Empirical</u>
   <u>Studies of Finite Nuclei</u>

   The Two-Body Interaction in Nuclear   J. McGrory
      Shell Model Calculations   (ORNL)

   The Effective Nucleon-Nucleon Interaction   J. P. Schiffer
      Deduced from Nuclear Spectra   (ANL)

   Contributions

Session V - Thursday Morning

   <u>The Effective Interaction for Inelastic</u>
   <u>Scattering and Charge Exchange Reactions</u>

   The Description of Inelastic   R. Schaeffer
      (p-p') Scattering   (Berkeley, Saclay)

   Core Polarization Effects in   H. McManus
      (p-p') Reactions   (Michigan State
                                                    University)

   The Effective Two-Nucleon Interaction   S. M. Austin
      from Inelastic Proton Scattering   (Michigan State
                                                    University)

   Contributions

Session VI - Thursday Evening

   <u>The Effective Interaction from Studies</u>
   <u>with Complex Projectiles</u>

   The Microscopic Description of   V. A. Madsen
      Inelastic Scattering and Charge   (Oregon State
      Exchange with Complex Projectiles   University)

   Experimental Review of the (^{3}He, ^{3}He'),   P. G. Roos
      (^{3}He,t) and (α,α') Reactions   (University of
                                                    Maryland)

   Contributions

Session VII - Friday Morning

   Nuclear Spectroscopy with High   M. H. Macfarlane
      Energy Projectiles   (ANL)

   Symposium Summary   M. Baranger
                                                     (MIT)

CONTENTS

SESSION VI: The Effective Interaction from Studies with
              Complex Projectiles

Chairman: A. Bernstein, MIT

SESSION VII: Summary

Chairman: J. S. Blair, University of Washington, Seattle

SYMPOSIUM OVERVIEW

Malcolm H. Macfarlane

Argonne National Laboratory and University of Chicago

I must confess to uncertainity as to precisely what an overview is, particularly an overview of a conference given before the
conference. It smacks of summarizing what speakers have said
before they have said it, which reminds me of a jingle that Scottish
natives use to pacify tourists who complain about the quality of
Scotland's mountain highways:

> If you'd seen these roads before they were made
> You'd throw up your hands and bless General Wade.

It is not clear to me which of the various parts I should play--
General Wade, the native or the tourist.

This symposium has, not one, but two topics--the interaction
between free nucleons and the interaction between nucleons in nuclei.
Both are effective interactions in the usual sense; it is necessary
to work in a restricted subspace of the complete Hilbert space of
the system, truncation effects being taken into account by replace-
ment of the true interaction in the complete space by an effective
interaction in the restricted subspace. The free-nucleon interaction is effective since mesonic degrees of freedom and excited
states of the nucleon are excluded; the interaction between nucleons
in nuclei is effective since not only are mesonic degrees of freedom
excluded from the active Hilbert space but also all nuclear configurations except a small number that involve the degrees of
freedom of valence nucleons.

In both cases, detailed treatment of the effective interaction
proceeds in three distinct stages.

1

1) Extraction of a phenomenological effective interaction from the experimental data. For the free-nucleon interaction, this involves phase-shift and potential parameterizations of nucleon-nucleon scattering data; for the effective interaction in nuclei we deal with shell-model studies of nuclear level-structure and microscopic analyses of inelastic-scattering data.

2) Calculation of the effective interaction from more fundamental considerations. The free-nucleon interaction is to be computed from a meson theory of nuclear forces; the effective interaction in nuclei is to be related to the free-nucleon interaction.

3) Assessment of the quality of agreement between phenomenological and calculated interactions. In this comparison it is found that not only do the computed interactions fail to account for some of the data but also the models of the interactions underlying the computations contain much information not tested by the available data.

Table 1 shows how the symposium has been organized around the above three aspects of the two effective-interaction problems. There is a hole in this Table under the heading "Gaps and Limitations in the Comparison of Phenomenological and Calculated Interactions between Nucleons in Nuclei." My "overview" talk will in fact be an attempt to fill this hole in the table of contents of the symposium; in so doing I will concentrate on the effective interaction involved in bound-state shell-model calculations.

TABLE 1.--Topics covered in the various sessions (Roman numerals) of the Symposium.

	Free-Nucleon Interaction	Interaction between Nucleons in Nuclei
Extraction of Phenomenological Interactions from Experimental Data	I,II	IV,V,VI
A priori Calculation of Effective Interactions	I	III
Gaps and Limitations in Comparison of Phenomenological and Calculated Interactions	II	?

Let us first recall how phenomenological shell-model interactions are determined from nuclear level data. First, a finite
model vector-space is constructed by filling particles into an
underlying set of single-particle orbitals; all of these orbitals
except for a few near the nuclear Fermi surface are assumed to be
either completely filled or empty. This is illustrated for 1p-shell
nuclei (Li^6 to O^{15}) in Figure 1. Here the model space for a given
nucleus is spanned by construction of all antisymmetric states with
good total angular momentum in the configurations $(1p_{3/2}, 1p_{1/2})^n$,
where n runs from 2 (Li^6) to 11 (O^{15}) for 1p-shell nuclei. The
adaptation to any set of valence orbitals $(1p_{3/2}, 1p_{1/2})^n \rightarrow$
$(j_1, j_2, \ldots j_\mu)^n$ is obvious.

The effective interaction for use in this model space is
<u>assumed</u> to be a two-body interaction and is therefore completely

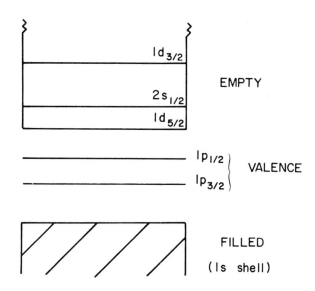

Figure 1.--Active single-particle orbitals for 1p-shell nuclei
(Li^6 to O^{15}).

characterized by the finite set

$$\{x_i\} = \{<j_1j_2JT|V|j_1'j_2'JT>\} \tag{1}$$

of two-body matrix elements, where j_1, j_2, j_1', j_2' run over the valence orbitals. "Best-fit" two-body matrix elements are then determined by numerical least-squares fit to observed level-energies E_α^{exp} in the range of nuclei under consideration. (In the example mentioned above, the nuclei involved are Li^6, Li^7, Be^8, Be^9... N^{14}, O^{15}.) The function

$$F(\{x_i\}) = \sum_{\alpha=1}^{N} \omega_\alpha (E_\alpha^{exp} - E_\alpha^{calc.}(\{x_i\}))^2 \tag{2}$$

is therefore minimized as a function of the two-body matrix elements $\{x_i\}$. The values $\{x_{i0}\}$ of these parameters at minimum are the "best-fit" two-body matrix elements.

The appropriate calculated interaction is derived from a suitable free-nucleon interaction potential, possibly the Hamada-Johnston potential or one of its more up-to-date relatives. In all quantitative studies to date, a two-body Brueckner reaction matrix is first computed for the nucleus with two nucleons in the appropriate active orbitals (Li^6 in the example mentioned above.) This two-body reaction matrix, corrected for core-polarization and perhaps for a few other simple truncation effects, is taken to be the calculated effective two-body interaction for the range of nuclei under consideration.

Comparison of phenomenological and calculated interactions is now a matter of comparing two long lists of two-body matrix elements—one obtained from a least-squares fit to nuclear level-energies, the other calculated using Brueckner theory from the force between free nucleons. The trouble with this sort of comparison is that, as is clear from the above description, the two effective interactions are rather different in character, each involves assumptions and practical restrictions not shared by the other. We shall discuss here two of the main sources of mismatch.

The first point is that the list of phenomenological two-body matrix elements is likely to contain much less information than it appears to. Certain linear combinations of the two-body matrix elements are much better determined by the data than others. Consider the Taylor expansion of the least-squares function F of eq. (2) about the minimum $x_i = x_{i0}$ (where $\partial F/\partial x_i = 0$):

$$F(x_i) \approx F(x_{i0}) + \sum_{ij} \left[\frac{\partial^2 F}{\partial x_i \partial x_j}\right]_0 (x_i - x_{i0})(x_j - x_{j0}) \qquad (3)$$

It is clear that the second-derivative matrix determines how rapidly $F(x_i)$ changes when the parameters x_i are displaced from their values at minimum; the larger this change, the better-determined the parameter. The error matrix

$$H_{ij} = \left[\frac{\partial^2 F}{\partial x_i \partial x_j}\right]_0^{-1} \qquad (4)$$

contains the desired information in quantitative form; its eigenvectors are the uncorrelated linear combinations of two-body matrix elements, the corresponding eigenvalues h_k being the mean-square uncertainties in these linear combinations.

The eigenvalues of the error matrix for the 1p-shell calculation of Cohen and Kurath[1] are shown in Figure 2. There are 11 independent

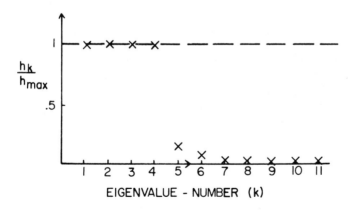

Figure 2.--Eigenvalues of the error matrix in the 1p-shell calculation of Cohen and Kurath.

two-body parameters[†]; of these seven are very well determined,
four are very poorly determined. Similar results are found in
other shell-model least-squares fits. As the number of two-body
matrix elements increases the proportion of poorly-determined linear
combinations increases. In other words, as the shell-model Hilbert
space gets larger, more work is done for less proportional output.

In comparing phenomenological and calculated effective interactions, it is obviously appropriate to focus on the linear
combinations of matrix elements that are well-determined in the
least-squares fit. When this is done, it is found that the deviations
between calculated and fitted matrix elements are systematically
smaller for the well-determined linear combinations. This is
illustrated in Figure 3, taken from a shell-model study[2] by Arima,
Cohen, Lawson and Macfarlane which treats the nuclei O^{18}, F^{18}, O^{19},
F^{19} and Ne^{20} within the configurations $(1d_{5/2}, 2s_{1/2})^n$. The
eigenvalues of the error matrix and the deviations Δ_k between fitted
and calculated values of the corresponding linear combinations of
matrix elements are plotted. The calculated matrix elements are
those computed by Kuo and Brown[3] from the Hamada-Johnston interaction between free nucleons but other calculations give similar
results. The much greater errors in the poorly-determined parameters are strikingly clear. That agreement should be better
when we separate those aspects of the phenomenological interaction
that arise from the properties of nuclei from those that are
introduced by what amounts to a random-number generator is at
least pleasing and may suggest that the degree of agreement achieved
between phenomenology and more basic theory has some real content.

The physical distinction between well-determined and poorly-
determined combinations of matrix elements involves approximate
symmetries of the system. In the 1p-shell, space--(supermultiplet)
symmetry is a fairly good quantum number; the energies of states of
the highest space-symmetry depend predominantly on the effective
interaction in even states. The poorly-determined parameter-
combinations in Figure 2 are dominated by the odd-state interactions.
In similar fashion, seniority is a good approximate symmetry of
identical-nucleon systems; the well-determined parameters are then
predominantly those that determine the energies of states of lowest
seniority. Finally, the salient qualitative feature of early
ds-shell nuclei is their willingness to assume a spheroidal
equilibrium shape; this deformability is closely related to SU_3
symmetry. The well-determined parameters in Figure 3 are predominantly those that determine the effective two-body interaction
between particles in the lowest deformed single-particle (Nilsson)
orbitals. The general qualitative point is clear; whenever there

---

[†]There are fifteen independent two-body matrix elements for a
general interaction. The assumption that the interaction is local
reduces the number of independent parameters to 11.

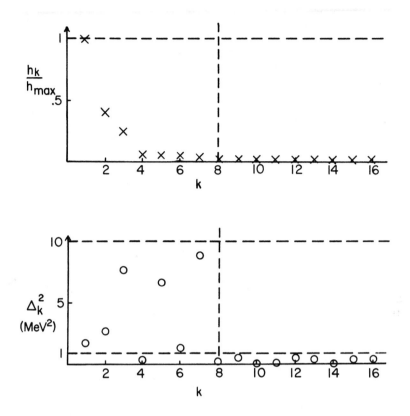

Figure 3.--Eigenvalues of the error matrix and deviations Δ_k between phenomenological and computed two-body parameters for ds-shell calculations (16 parameters).

is a good approximate symmetry, the only linear combinations of interaction parameters that can be determined with any precision are those that specify the energies of states of the highest symmetry.

There are other respects in which the process of extracting phenomenological two-body interactions is basically different from that leading to computed interactions.

1) The various theoretical schemes that lead to a computed interaction with a reaction matrix as leading term all treat the systems with two, three, four . . . valence nucleons on a separate

footing. The few calculations done for more-than-two-particle
systems clearly show that many-body effects are sizeable. In the
phenomenological fitting procedure, in contrast, a two-body interaction is assumed and many-body effects are minimized in some
fashion by spreading them over a range of different nuclei with
different numbers of valence nucleons.

2) In the phenomenological fitting procedure, only one or
two of the lowest states of each energy matrix are used in the
fitting procedure, since only for such states can an appropriate
term in the least-squares function of eq. (2) be identified. All
other eigenvalues are simply thrown away; in large shell-model
calculations more than 90% of the computed eigenvalues are ignored.
The theoretical schemes for computing an effective interaction from
the force between free nucleons, on the other hand, are based on
retaining the entire model subspace, all eigenvalues of each energy
matrix. It is quite possible that convergence problems, many-body
effects and other troublesome imponderables can best be handled by
concentrating their effects in the part of the shell-model spectra
to be thrown away in comparing theory and experiment.

My conclusion, then, is that further progress in computing
nuclear level spectra from the force between free nucleons will
require attention to a problem that has so far been ignored. How
can we impose on the calculation of effective interactions from
the force between free nucleons, the practical constraints that are
introduced in phenomenological fitting procedures by averaging
over many different nuclei and by throwing away most of the computed
eigenvalues in comparing theory and experiment?

REFERENCES

1. S. Cohen and D. Kurath, Nuclear Physics 73,1(1965).

2. A. Arima, S. Cohen, R. D. Lawson and M. H. Macfarlane, Nuclear
    Physics 108,94(1968).

3. T. T. S. Kuo and G. E. Brown, Nuclear Physics 85,40(1966).

PHASE-SHIFT AND POTENTIAL-MODEL ANALYSIS OF EXPERIMENTAL

INFORMATION *

P. Signell

Michigan State University, East Lansing, Michigan

First I would like to make a few remarks about potentials and
then go on to phase shift analyses and how some of the popular
potentials compare to them. Figure 1 shows the sort of progress
that has been made in recent years. The 1962 $^1S_0$ Hamada-Johnston[1]
potential descended to tremendous depths at around 0.5F, just outside an infinitely positive core. Now we have[2] much more sensible
hard and soft core potentials and they are much easier to use in
computer calculations since they do not require such small radial
increments for a tabular specification. Finite-core alternatives[3,4]
are shown in Figure 2: all three of these give almost identical
phase shifts up to 300 MeV, but beyond that their predicted phases
diverge. Finally, in Figure 3 we see some momentum-dependent
alternatives[5]. Each of the p^2-dependent potentials gives identical
phase shifts for all energies, identical also to those from the
0.4F hard-core potential shown. The phases for the 0.1F hard
("soft") core and finite core potentials again diverge from these
and from each other above about 300 MeV. That's all I have to say
about recent developments in phenomenological potentials. Now let
us turn our attention to phase shifts and see how well the potentials
do.

We[6] have recently reanalyzed much of the two-nucleon elastic
data below 350 MeV, in what are called "single-energy analyses".
I would suggest that you never use results from what are called
"multi-energy analyses" unless you have convinced yourself that
you clearly understand the meanings of the errors quoted for such
analyses. For our single-energy analyses, the quoted errors are
derived from an error matrix. A little later on we will indicate
how we have tried to minimize correlations between the phases in
order to make the quoted errors as close as possible to standard

deviations with the usual statistical meaning. We have revised the
latest data compilation of MacGregor, Arndt, and Wright[7], making
a number of corrections, additions, and deletions. In some cases
these changes have decreased the uncertainties in the deduced phase
shifts, in some cases increased them, in some cases only changed
the values. We hope that our final data set accurately represents
the experimentalists' accomplishments. The models we have chosen
to compare to the "derived-data" phase shifts in the rest of the
Figures are:

*hard core: Hamada-Johnston[1] labeled HJ, YALE[8], and REID(HC)[9]

*boundary condition: Lomon-Feshbach[10], LF

*one-boson-exchange: Bryan-Scott[11], BSIII

*separable: Tabakin[12], TABAKIN

*soft core: Reid[9], REID(SC)

*energy-dependent phase shift representation, MAW-X[13]

*finite core: Bressel-Kerman-Rouben[4]-BK and BKR

We first show the phase shifts for the isospin one (I=1)
states, which are mainly determined by the usually-accurate proton-
proton scattering data. Almost all models reproduce the lowest
angular momentum state, the $^1S_0$, so we show in Fig. 4 only the
experimentally-determined phase shifts, the HJ prediction as
representative of most models, and the corresponding values for the
one-boson-exchange BSIII potential. Note that all values shown are
with respect to the energy-dependent phase shift representation
MAW-X. One should obviously be somewhat cautious about using the
BSIII potential in other calculations. Figures 5 and 6 show the
$^1D_2$, with a fair number of potentials being too attractive and
the BSIII being much too repulsive.

In Fig. 7-12 we have plotted the so-called C, T, and LS
phases[14] which are straight linear combinations of the usual ^3P_J
phases with J=0, 1, 2. The linear combination coefficients are
defined to be the matrix elements of the central, tensor, and
spin-orbit operators. The reasons for using such linear combinations are twofold: first, it is an attempt to decrease correlations
between the phases in the data analyses; second, such phases will
be related to integrals over the corresponding potentials. This is
particularly true for the higher angular momentum states, least
true for ^3P_C. For the latter, one notices in Fig. 7-8 that the
BSIII result is somewhat extreme, and that the 210 MeV data point
seems low. The "tensor" L=1 phase shifts ^3P_T in Fig. 9-10 seem
fairly well matched by all of the models and the same can be said

for the "spin orbit" phases $^3P_{LS}$ in Fig. 11-12. Note that for
several energies the error bars are smaller than the central circles
so can not be shown. The final I=1 phase shown is the L=3 $^3F_{LS}$ in
Fig. 13, which has a bearing[15] on our understanding of spin-orbit
splitting in nuclei. The predictions of the HJ potential and of
the realistic one-boson-exchange ("ρ+ω") models[15] are grossly
incorrect. The recent elementary-particle-physics calculation of
Chemtob and Riska[16], labeled CR(THEORY) in Fig. 13, is a substantial
improvement over the ρ+ω model and is also superior to HJ. We will
hear more about the CR calculation in Prof. Brown's talk, so I will
just remark that one can imagine both the experimental and theoretical values changing somewhat as techniques in both areas undergo
further refinement[17].

Now for the isospin-zero phases, which are determined solely
by the less-accurate neutron-proton data. First, let me remark
that we have recently found[18] that there are two solutions for the
I=0 phases near 330 MeV. These are shown in Fig. 14 as data bars
in the space of two phase parameters. The IX and X labels show
the solutions published by MacGregor, Arndt, and Wright in their
papers IX[14] and X[13]. The experimentalists will have to determine
the true solution. We have labeled them α and β and show both in
the I=0 graphs.

In Fig. 15-16 one finds very strange behavior for the LF and
BSIII $^1P_1$ phases. The open circle at 50 MeV is what results from
using the old Harwell[20] NP angular distributions. The diamond
results from substituting the new Oak Ridge[21] backward-hemisphere
data for the corresponding Harwell one, and the solid circle results
from using an altered version[22] of the Harwell forward-hemisphere
data. The latter alteration was made in an attempt to undo a
correction which had been applied to their data by the Harwell
group.

Virtually all models satisfy the $^3\bar{S}_1$ phase shifts: some
representative predictions are shown in Fig. 17. In Fig. 18-19
we show the $^3\bar{S}_1$-$^3\bar{D}_1$ coupling parameter $\bar{\varepsilon}_1$. The one-boson-exchange
models, including those labeled GREEN[23] and INGBER[24], are all seen
to be low. The potential labeled INGBER is the one reported by
Brueckner[25] at the Montreal Conference. Its increased binding in
nuclear matter can mostly be ascribed to this very sizable difference in the $\bar{\varepsilon}_1$ parameter, which is a measure of the central/tensor
ratio. If the $\bar{\varepsilon}_1$ data points hold up, one would have to rule out
the INGBER potential. One should also note that the TABAKIN and
("unconstrained") MAW-X ε_1 curves do not go through a derived
datum[26] at the negative-energy point corresponding to the deuteron.
Neither does the other ("constrained") solution which is given in the
MAW-X paper. The ^3D_C, ^3D_T, and $^3D_{LS}$ phases shown in Fig. 20-25
seem to contain only one striking anomaly- the $^3D_{LS}$ TABAKIN curve.

The REID curves are not shown because they are only defined for the $^3D_1$ and $^3D_2$ states. The latter are shown in Fig. 26-29, with the $^3D_3$ in Fig. 30-31. Finally, the $^3D_3$-$^3G_3$ coupling parameter ε_3 is shown in Fig. 32. All values except MAW-X and the β solution are close to the one-pion-exchange value, labeled 1π. Note especially the theoretical curve of Chemtob and Riska[16]: it would be wiped out by the β-type of 330 MeV solution quoted by MAW-X. The $\bar{\varepsilon}_3$ parameter is set equal to zero in the Tabakin potential, but that is probably a very good approximation for most nuclear calculations.

REFERENCES

1. T. Hamada and I. D. Johnston, Nucl. Phys. <u>34</u>, 382 (1962).

2. P. Signell, Proc. 1969 Midwest Theory Conference, Physics Dept., Univ. of Iowa, 1970.

3. M. S. Sher, P. Signell, and L. Heller, Ann. Phys. <u>58</u>, 1 (1970).

4. C. B. Bressel, A. K. Kerman and B. Rouben, Nucl. Phys. <u>A124</u>, 624 (1969).

5. M. Miller, M. Sher, P. Signell, N. Yoder and D. Marker, Phys. Lett. <u>30B</u>, 157 (1969).

6. J. Holdeman and P. Signell (to be published).

7. M. H. MacGregor, R. A. Arndt, and R. M. Wright, Univ. of Cal. Rad. Lab Report No. UCRL-50426 (1968), unpublished.

8. K. E. Lassila, M. H. Hull, Jr., H. M. Ruppel, R. A. McDonald, and G. Breit, Phys. Rev. <u>126</u>, 881 (1962).

9. R. V. Reid, Jr., Ann. Phys. <u>50</u>, 411 (1968).

10. E. L. Lomon and H. F. Feshbach, Ann. Phys. <u>48</u>, 94 (1968).

11. R. Bryan and B. L. Scott, Phys. Rev. <u>177</u>, 1435 (1969).

12. F. Tabakin, Ann. Phys. <u>30</u>, 51 (1964).

13. M. H. MacGregor, R. A. Arndt, and R. M. Wright, Phys. Rev. <u>182</u>, 1714 (1969).

14. L. Heller and M. S. Sher, Phys. Rev. <u>182</u>, 1031 (1969).

15. P. Signell, in <u>Polarization Phenomena in Nuclear Reactions</u>, Ed. H. H. Barschall and W. Haeberli, U. of Wisc. Press, Madison, Wisc. (1971).

16. M. Chemtob and D. O. Riska, Phys. Lett. 35B, 115 (1971).

17. P. Signell, Mich. State University preprint No. COO-2061-13
     (1971), to be published.

18. P. Signell and J. Holdeman, Mich. State Univeristy preprint
     No. COO-2061-14 (1971), to be published.

19. M. H. MacGregor, R. A. Arndt, and R. M. Wright, Phys. Rev.
     173, 1272 (1968).

20. J. P. Scanlon, G. H. Stafford, J. J. Thresher, P. H. Bowen
     and A. Langsford, Nucl. Phys. 41, 401 (1963).

21. M. J. Saltmarsh, M. L. Halbert, C. R. Bingham, and A. van
     der Woude, ORNL Report No. 4649, p. 3 (May, 1971).

22. The published values had been altered in an angle–dependent
     way in order to force agreement with a particular total cross
     section measurement.

23. R. W. Stagat, F. Riewe, and A. E. S. Green, Phys. Rev. C3,
     552 (1971).

24. L. Ingber and Potenza, Phys. Rev. C1, 112 (1970).

25. K. Brueckner, Proc. Int. Conf. on Properties of Nuclear
     States, Press of University of Montreal, Montreal (1969).

26. P. Signell, Phys. Rev. C2, 1171 (1970).

*Supported in part by the U. S. Atomic Energy Commission.

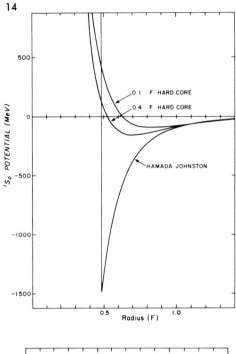

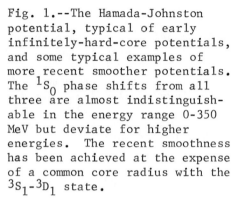

Fig. 1.--The Hamada-Johnston potential, typical of early infinitely-hard-core potentials, and some typical examples of more recent smoother potentials. The $^1S_0$ phase shifts from all three are almost indistinguishable in the energy range 0-350 MeV but deviate for higher energies. The recent smoothness has been achieved at the expense of a common core radius with the $^3S_1$-$^3D_1$ state.

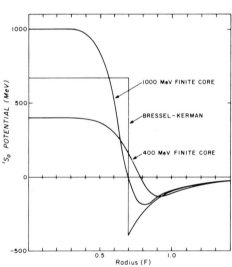

Fig. 2.--Several recent finite-height potentials which give almost indistinguishable $^1S_0$ phase shifts up to 350 MeV. Lowering of the core height below 400 MeV results in progressive worsening of the fit to the 0-350 MeV $^1S_0$ phase.

Fig. 3--Several of the potentials from Figs. 1 and 2, and two p^2-dependent potentials which give $^1S_0$ phase shifts identical to those of 0.4 F Hard Core one at all energies. This was achieved by carrying out an exact mathematical transformation on the 0.4 F HC potential.

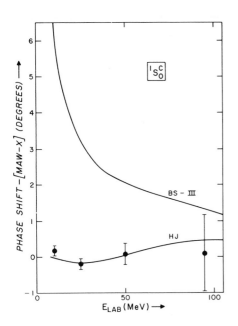

Fig. 4.--Coulomb-type nuclear $^1S_0$ phase shifts from analyses of the data, along with the predictions of three models. For display convenience the data and two of the models have been shifted by the predictions of the third model. The quality of fit of the Bryan-Scott one-boson-exchange-plus-parameters potential is seen to be poorer than that of the Hamada-Johnston potential.

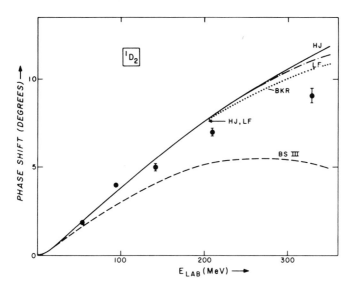

Fig. 5.--The predictions of 4 models for the nuclear bar $^1D_2$ phase shift. Where error bars are not shown on the data points, they are the size of the circle or smaller. Note that all except BS-III are too attractive at the higher energies. Strengthening of the quadratic spin-orbit interaction would correct this.

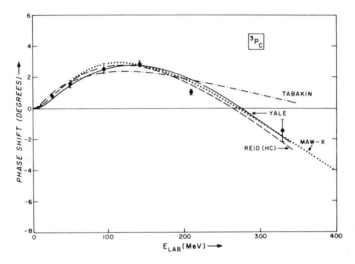

Fig. 6.--As in Fig. 5, with 5 model predictions. The Reid Soft
Core (SC) phases are shown only where they differ significantly
from those of the Reid Hard Core (HC) potential. Note that the
Reid (SC) curve goes through all data bars except the one at
95 MeV. This casts some suspicion on the data and or analysis
at that energy.

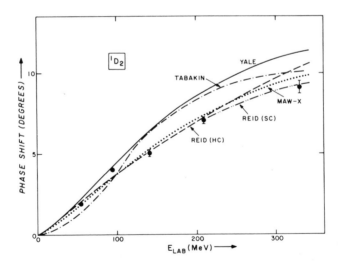

Fig. 7.--The "Central" linear combination of the nuclear bar $L=1$
phase shifts, $^3P_c = (^3P_0 + 3\,^3P_1 + 5\,^3P_2)/9$, along with the predictions
of four models.

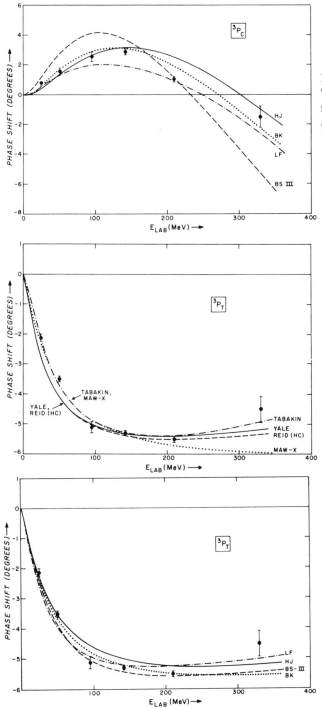

Fig. 8.--As in Fig. 7. The Reid(SC) values are not shown because they are almost identical to those shown for Reid(HC).

Fig. 9.--The "Tensor" linear combination; $^3P_T = (-5/72)(2\,^3P_0 - 3\,^3P_1 + \,^3P_2)$.

Fig. 10.--As in Fig. 9. Note that not one of the curves in Fig.'s 9 and 10 matches all of the data points.

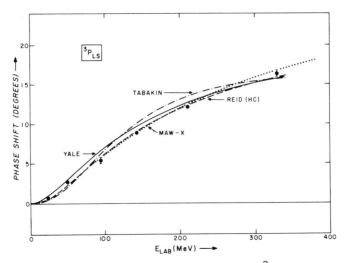

Fig. 11.--The "Spin-Orbit" linear combination $^{3}P_{LS} = (-1/12)$ $(2^{3}P_{0}-3^{3}P_{1}+5^{3}P_{2})$.

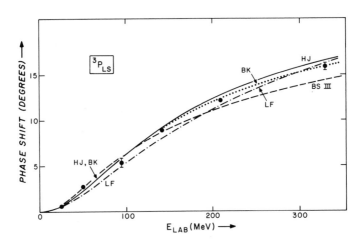

Fig. 12.--As in Fig. 11. Note the smallness of the error bars on the data. These result from using the complete $^{3}P_{0}$-$^{3}P_{1}$-$^{3}P_{2}$ error matrix. Much larger error bars would result if one (erroneously) computed them directly from the error bars on the $^{3}P_{0}$, $^{3}P_{1}$, and $^{3}P_{2}$ phases.

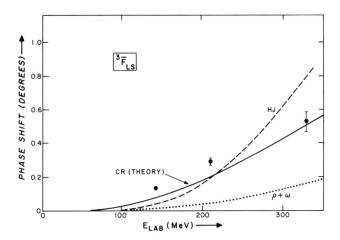

Fig. 13.--The "Spin-Orbit" linear combination of the L=3 nuclear bar phases, $^3F_{LS} = (-1/168)(20\,^3F_2 + 7\,^3F_3 - 27\,^3F_4)$. If one matched the $^3F_2$, $^3F_3$, and $^3F_4$ nuclear bar phases with central, tensor, and spin orbit potentials, the $^3F_{LS}$ phases would presumably be very close to the phase shifts produced by the spin orbit potential alone. The reason for the strong discrepancy between the theoretical CR curve and the data at 142 and 210 MeV is not known, but both the theoretical derivation and the phase shift analyses at these two energies are suspect. The uncertainty in the analyses should be cleared up when analytic approximation theory is used to reanalyze the data.

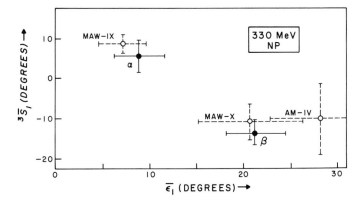

Fig. 14.--The "α" and "β" solutions produced by analysis of the 330 MeV NP+PP data, shown here in the combined space of the $^3S_1$ nuclear bar phase shift and the $^3S_1$-$^3D_1$ nuclear bar coupling parameter ε_1.

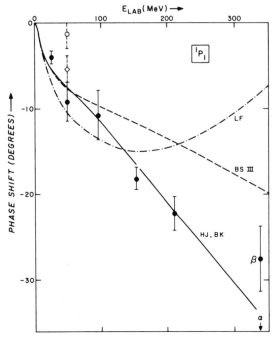

Fig. 15.--The α solution is
at −43.3°±2.1°.

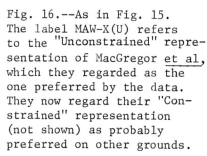

Fig. 16.--As in Fig. 15.
The label MAW-X(U) refers
to the "Unconstrained" representation of MacGregor et al,
which they regarded as the
one preferred by the data.
They now regard their "Constrained" representation
(not shown) as probably
preferred on other grounds.

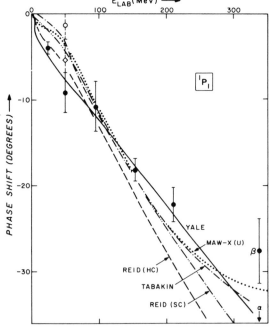

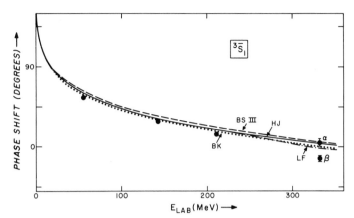

Fig. 17.--Note that the phase shift is represented as going to 180° at zero energy, a reflection of the existence of the deuteron bound state. All models are in close agreement on this phase parameter.

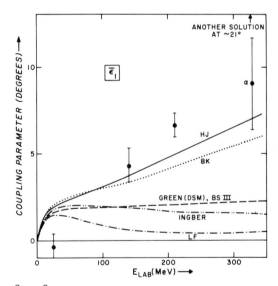

Fig. 18.--The $^3S_1$-$^3D_1$ nuclear bar coupling parameter ε_1, which is closely related to the asymptotic $^3D_1/^3S_1$ wave function ratio in the deuteron. The latter is at E_{LAB}=-4.4 MeV. Actually, the most direct connection is between the Blatt-Biedenharn parameter ε_1 and the asymptotic ratio, but the Blatt-Biedenharn ε_1 is readily computable from the nuclear bar $^3S_1$ and $^3D_1$ phases and the nuclear bar ε_1.

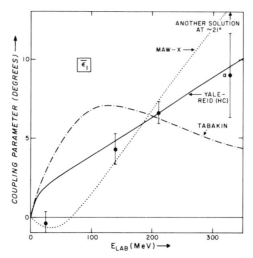

Fig. 19.--As in Fig. 18.   The β solution is at 21.2°±3.1°.   The
MAW-X representation shown is the "Unconstrained" one.   It gives
the wrong sign for the asymptotic ratio at E_{LAB}=-4.4 MeV.

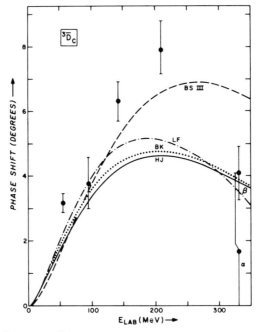

Fig. 20.--The "Central" linear combination of L=2 nuclear bar
phases: $^3D_C=(1/15)(3\,^3D_1+5\,^3D_2+7\,^3D_3)$.

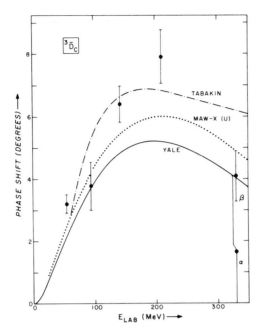

Fig. 21.--As in Fig. 20. Note that none of the models seems able to reproduce the data.

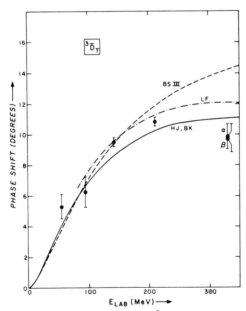

Fig. 22.--The "Tensor" combination: $^3\bar{D}_T = (-7/120)(3\,^3D_1 - 5\,^3D_2 + 2\,^3D_3)$.

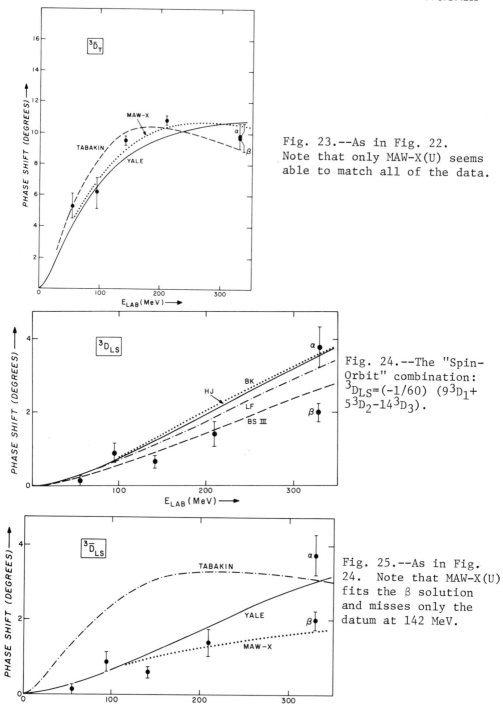

Fig. 23.--As in Fig. 22.
Note that only MAW-X(U) seems
able to match all of the data.

Fig. 24.--The "Spin-
Orbit" combination:
$^3D_{LS} = (-1/60) (9^3D_1 +
5^3D_2 - 14^3D_3)$.

Fig. 25.--As in Fig.
24. Note that MAW-X(U)
fits the β solution
and misses only the
datum at 142 MeV.

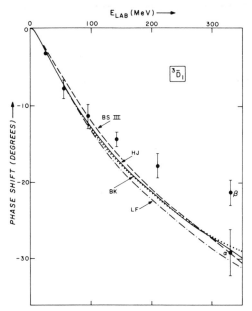

Fig. 26.--The $^3D_1$ nuclear bar phase shift. A number of models show good agreement with the α solution for this deuteron state. Notice, however, that the trend of the data at lower energies seems to favor the β solution.

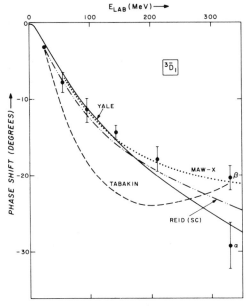

Fig. 27.--As in Fig. 26. Perhaps the true value of $^3D_1$ at 330 MeV lies between the α and β solutions, as with Reid(SC).

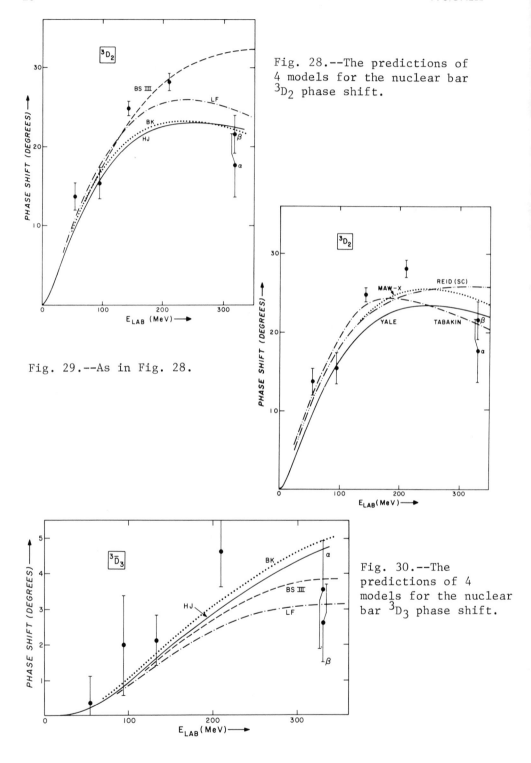

Fig. 28.--The predictions of 4 models for the nuclear bar $^3D_2$ phase shift.

Fig. 29.--As in Fig. 28.

Fig. 30.--The predictions of 4 models for the nuclear bar $^3D_3$ phase shift.

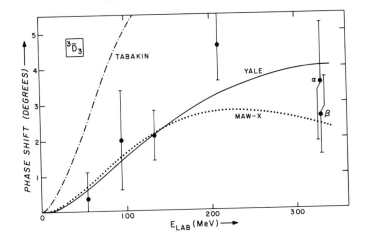

Fig. 31.--As in Fig. 30. Note the departure of the Tabakin phase shift from those of the other six models at low energies.

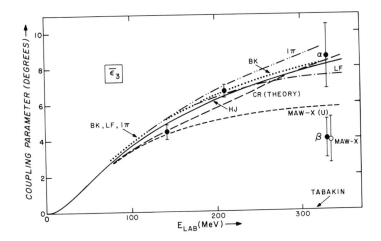

Fig. 32.--The $^3D_3$-$^3G_3$ nuclear bar coupling parameter ε_3. The one-pion-exchange curve is indicated by the symbol "1π". Any reascnable unitarization scheme will give the values shown, because the 1π $^3D_3$-$^3G_3$ partial-wave amplitudes are very close to being unitary. Note that the α solution is favored; by the theoretical calculation, by the trend of the data at lower energies, and by the 1π values.

ELEMENTARY-PARTICLE MODELS OF THE TWO-NUCLEON FORCE

G. E. Brown

State University of New York, Stony Brook and NORDITA

I. INTRODUCTION

Let me begin my talk by saying that little progress was made
in understanding the origin of the nucleon-nucleon force from the
time of great activity in the early fifties until recently. The
activity in 1950-1953 was concerned with incorporating Yukawa's
idea of meson exchange into half-way respectable formalism. Names
like Levy, Klein, Brueckner and Watson are associated with these
developments. The one-pion-exchange term, held responsible for
the long-range part of the nucleon-nucleon interaction, was incorporated into the phase-shift analyses, by the Livermore and Yale
groups, and found to enable significantly better fits of the data.

In intervening years, certain technical progress was made,
especially in the development of dispersion-theory techniques by
Amati, Leader and Vitale[1] and in the incorporation of certain
relativistic effects by Lomon and Partovi[2]. These relativistic
effects turned out to be absolutely essential in separating off the
interated one-pion-exchange term from the two-pion exchange process
so as to be able to define a two-pion-exchange potential. John
Durso will talk about this in a contributed paper.

These important technical developments were, however, largely
buried in the euphoric ecstasy of one-boson-exchange potentialists.

The idea behind the one-boson exchange potentials was that,
just as exchange of the pion gives the one-pion exchange part of the
potential, exchange of heavier known bosons, the ρ, η and ω, should
give other contributions. A mysterious, and largely fictitious,
σ-particle was also added.

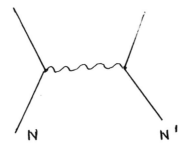

Fig. 1.   The one-pion-exchange potential. Here the lines N
and N´ represent nucleons; the wavy line, the pion.

One-boson-exchange potentials are clearly inadequate, and it
is silly to believe that a description of the nucleon-nucleon force
can be given in terms of them.

The hope of the one-boson-exchange theorists was that two-pion-
exchange processes could somehow be summarized in terms of single
bosons. For example, when the two exchanged pions were coupled to
isospin T=1, they could be replaced by a ρ-meson.

Let us look at the two-pion exchange process

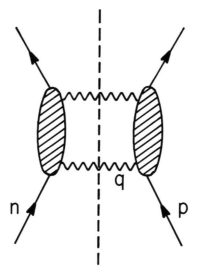

Fig. 2.   Two-pion exchange interaction. The shaded areas
indicate that essentially anything can happen.

Now, one can consider this as arising from the product of processes

and

in a well-defined (within the framework of dispersion relations) sense. The shaded area represents the amplitude for emitting two pions; if there are strong correlations between the two pions, such as go into making up the ρ-meson, these will go into this shaded area.

Now, the two pions have total energy $\sqrt{t}$ given by[*]

$$\sqrt{t} = \sqrt{k^2+\mu^2} + \sqrt{q^2+\mu^2} \tag{1}$$

which is $\sqrt{t} = 2\sqrt{q^2+\mu^2}$ in the C.M.S. of the two pions.

Let us classify the two-pion states according to J, the total angular momenta of the two pions, and the isospin of the two pions. Thus, f_+^0 represents the amplitude for emitting two pions in a J=0, isospin symmetric state[**]. That is,

J=0, T=0

$$f_+^0(t) =$$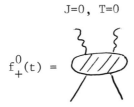

represents the amplitude for emitting two pions in a J=0, T=0 state.

---

[*]I use the usual variables of dispersion theory here.

[**]I am actually cheating in the notation; the lower suffix + here refers to helicity, which I don't wish to go into here.

From the familiar argument based on the uncertainty principle, if a system of energy $\sqrt{t}$ is exchanged between two nucleons, this leads to a Yukawa potential of range $\hbar/\sqrt{t}$.

Multiplying amplitudes for emission and absorption of the two pions, one finds the resulting potential to be essentially

$$V(r) = \int |f_0^+(t)|^2 \frac{e^{-\sqrt{t}\,r}}{r}\, dt \qquad\qquad (2)$$

to within uninteresting numerical and kinematic factors.

Thus, the two-pion-exchange potential can be represented as a superposition of Yukawas of different ranges corresponding to differing masses. The question now is to determine what the $|f(t)|^2$ look like. If they look like

$$|f(t)|^2 = C\,\delta(t-t_0) \qquad\qquad (3)$$

then the potential $V(r)$ can be represented by a one-boson exchange, the mass of the boson being $\sqrt{t_0}$.

We shall see that they do not.

Let us first look at the case which should be best for the one-boson-exchange model, where the ρ-meson is exchanged. Here the angular momentum of the pions is J=1. Signell[3] pointed out some time ago that the spin-orbit splitting in, say, the F-wave nucleon-nucleon scattering, was not well fitted by even the best nucleon-nucleon potentials. Let us look at Fig. 3, where the quantity describing the spin-orbit splitting

$$\Delta_{LS} = \frac{1}{168}\,(27\,\delta_{J=4} - 7\,\delta_{J=3} - 20\,\delta_{J=2}) \qquad\qquad (4)$$

as obtained from the nucleon-nucleon phase-shift analyses is compared with predictions from the Hamada-Johnston potential. The latter is similar to the one-boson-exchange models in that the spin-orbit force is of short range, as would arise from the exchange of the heavy ρ- and ω-mesons.

The point to be made--the point I believe Signell made--is that the energy dependence demands something like a distribution of exchanged masses, the longer-range (lower-mass) parts of the interaction being investigated at low energies of the nucleon-nucleon scattering, the shorter-range parts being investigated at high energies.

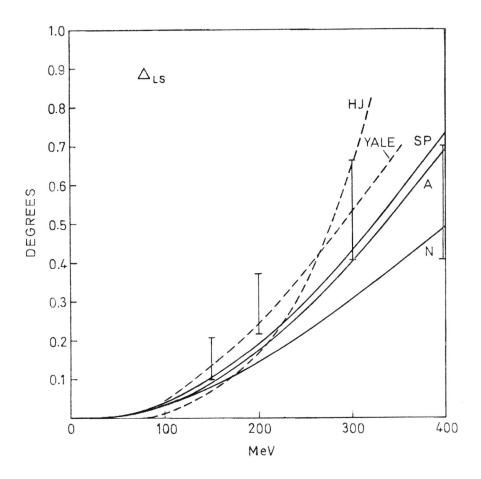

Fig. 3.   Comparison of various predictions of Δ_{LS} to the corresponding values obtained from the Livermore energy-independent solutions[4].   HJ stands for Δ_{LS} from the Hamada–Johnston potential[5], SP for Δ_{LS} from the best fit of Chemtob and Riska[6] and A from Brown and Durso [7]; both SP and A include the two-pion continuum.   YALE stands for the prediction of the Yale phase shifts.

But I don't want to base my arguments for the need to do
properly the two-pion exchange simply on the inadequacy of the one-
boson-exchange potentials to explain experiment, rather, it can be
shown theoretically that the continuum of two-pion exchange cannot
be represented by a single meson.

II.   THE TWO-PION EXCHANGE CONTRIBUTION

After saying so many negative things, I should now be more
positive.   I wish to describe the first part of work with Marc
Chemtob, John Durso and Dan-Olof Riska.   This part is concerned with
obtaining the two-pion-exchange contribution to the scattering matrix.
John Durso will describe the reduction to a potential in a contributed paper.

Let us consider the evaluation of the $f_+^0(t)$.   From f_+^0 we can
describe the force arising from the exchange of scalar, $T=0$ objects
with a continuum of masses.

First of all, we could consider the $f_+^0(t)$ which would arise
with only the nucleon as intermediate state, i.e.,

$$\left[f_+^0(t) \right]_{Box} = \quad \text{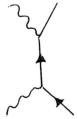}$$

I use the lower suffix "Box" because using this as outlined earlier,
we obtain results equivalent to those from evaluating the Feynman
box diagram, as done by Lomon and Partovi[2].

It is known, however, that for the process of pion-nucleon
scattering, use of the Feynman graph

plus crossed graph gives very bad results.   The $\left[f_+^0(t) \right]_{Box}$ is a piece (projected with respect to J) of the same amplitude, but in a different range of t; that is, it is negative for pion–nucleon scattering, positive for our case where we want to calculate the nucleon–nucleon force. In fact, to be more precise, our $f_+^0(t)$ comes from

$$N + \bar{N} \rightarrow \pi + \pi,$$

but for energies t of the $N + \bar{N}$ system far below physical threshold.

We shall discuss two approaches to find $f_+^0(t)$ for small t. In the first, we determine the common amplitude for

$$\pi + N \rightarrow \pi + N$$

and

$$N + \bar{N} \rightarrow \pi + \pi$$

from the soft–pion developments. We state these in the following way. The common amplitude for the above process is

$$T(s,t,u) = - A(s,t,u) + i \, \gamma \cdot Q \, B(s,t,u) \tag{5}$$

where $Q = \frac{1}{2}(q-k)$. (See Fig. 2.) Also, the Mandelstam variable

$$s = (p+q)^2 \tag{5.1}$$

and s,t,u are related by

$$s + t + u = 2M^2 + 2\mu^2 \tag{5.2}$$

where μ is the pion mass. Now, it is easy to show that in the limit of pion four–momentum going to zero, the second term in T becomes

$$i \, \gamma \cdot Q \, B(s,t,u) \underset{q_\mu \Rightarrow 0}{\Longrightarrow} \frac{G^2}{M} \tag{6}$$

aside from a pion–nucleon form factor, which we suppress here. In old–fashioned terms, one can say that this term comes completely from virtual pion pairs as intermediate states in the Born–approximation expression for B.

Adler[8] and Weinberg[9] have shown that (aside from the ubiquitous σ–term, which we shall not discuss here), the term A goes to just the same limit, so that

$$T(s,t,u) \underset{q_\mu \Rightarrow 0}{\Longrightarrow} 0 \tag{7}$$

That is to say, the amplitude for pion-nucleon scattering goes to zero as pion four-momentum goes to zero. This is nothing more than an expression of the old "pair suppression" which one has known about for 20 years.

In order to enforce the soft-pion condition, we make a subtraction in the dispersion theoretic expression for A.

$$A^+(s,t,u) = A^+(M^2,t,u(M^2)) \tag{8}$$

$$+(s-M^2)\frac{1}{\pi}\int\frac{\sigma_A^+(s',t)ds'}{(s'-s)(s'-M^2)}$$

$$+(u-u(M^2))\frac{1}{\pi}\int\frac{\sigma_A^+(s',t)ds'}{(s'-u)(s'-u(M^2))}$$

where (See eq. (5.2))

$$u(M^2) = M^2+2\mu^2-t \tag{8.1}$$

We know from the work of Adler and Weinberg that

$$A^+(M^2,t,u(M^2))$$

is equal to G^2/M for zero pion mass (implied by zero pion four-momentum). Within the spirit of soft-pioneering*, we assume this not to vary much going from zero pion mass to the physical mass.

Thus,

$$A^+(s,t,u) = \frac{G^2}{M} + (s-M^2)\frac{1}{\pi}\int\frac{\sigma_A^+(s',t)ds'}{(s'-s)(s'-M^2)}$$

$$+(u-u(M^2))\frac{1}{\pi}\int\frac{\sigma_A^+(s',t)ds'}{(s'-u)(s'-u(M^2))} \tag{9}$$

Now, the contribution lowest in energy (that is, in s') to the $\sigma_A^+(s',t)$ will come from the region of the (3,3) isobar, where $s' \sim (M+2\mu)^2 \cong M^2+4M\mu$.

---

* In fact, one can view the subtraction constant as coming from the exchange between nucleon and pion of a heavy sigma meson. The statement here is that the sigma mass must be large compared with the pion mass.

To order μ^2, $u(M^2)$--which is given by eq. (8.1)--is equal to M^2, so that in the region of $s \sim M^2$ (which corresponds to the region of low pion momentum q) the two integrands are equal to order μ/M, and one has

$$A^+(s,t,u) \cong \frac{G^2}{M} + (s+u-M^2-u(M^2)) \frac{1}{\pi} \int \frac{\sigma_A^+(s',t)ds'}{(s'-s)(s'-M^2)} \tag{10}$$

$$= \frac{G^2}{M} \, ,$$

the latter equality holding since $s+u = 2M^2 + 2\mu^2 - t$ and $M^2 + u(M^2) = 2M^2 + 2\mu^2 - t$.

In the case of B, we propose to use only the Born terms

$$B^\pm = G^2 \left(\frac{1}{M^2-s} - \frac{1}{M^2-u} \right) \tag{11}$$

because, for small s and t, one is quite close to the nucleon pole.

The calculation of the f^J's is now easy[7] and we find

$$f_+^0 = \frac{G^2}{4\pi} \{ M(h \arctan \frac{1}{h}) \tag{12}$$

$$- \frac{t}{2M} \}$$

where

$$h = \frac{M^2 + \vec{p}^2 + \vec{q}^2}{2i|\vec{p}||\vec{q}|} \tag{12.1}$$

In Fig. 4 we show the f_+^0, calculated from eq. (12) as a function of t. For comparison, the f_+^0 obtained by Nielsen et al.[10] by analytic continuation from pion-nucleon scattering without our assumptions of small t, etc., is shown. The agreement of our f_+^0 with their Ref_+^0 is good over the entire region plotted.

Our treatment does not determine $Im f_+^0$. However, in the range $4\mu^2 < t \leq 16\mu^2$,

$$f_+^0 = e^{i\delta_0} |f_+^0| \tag{13}$$

Eq. (13) is usually assumed to hold for values of t well above $16\mu^2$. In any case, in the low-t range, one can generate an imaginary part as

$$Im f_+^0 = \tan \delta_0 \, Re \, f_+^0 \tag{14}$$

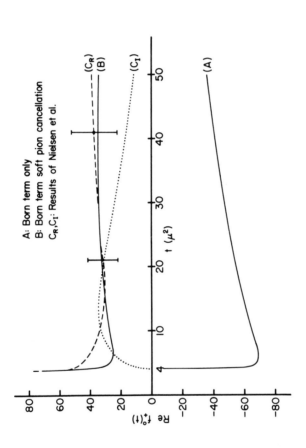

Fig. 4. The f_+^0 obtained from eq. (12) is shown as the solid line ——B. The Re f_+^0 and Im f_+^0 obtained in Ref. 10 are shown by the dashed --- and dotted ... lines, respectively. Also shown is the $[f_+^0]_{Box}$ obtained from Born approximation (the lower curve A). All amplitudes are plotted in units of the pion Compton wavelength.

Taking Weinberg's result[9] for the scattering length, a=0.2 in units of the pion Compton wavelength, one sees that this Im f_+^0 is small, in the region we consider, compared with Re f_+^0, especially since it is the squares of each which count in $|f_+^0|^2$.

In Fig. 5 we show the Born amplitude, and the contribution of the ρ-meson to the $f_-^1(t)$ amplitude. We are, of course, unable to determine the ρ- contribution, which comes at $t \sim 29\mu^2$, but this can be taken directly from experiments, such as the Novosibirsk colliding e^+-e^- beam experiments. It is, however, clear that the low-t behavior is determined by the Born term here. It should be noted here that OBEP models would neglect this Born term.

The nucleon-nucleon scattering in partial waves of high angular momentum is determined by the small-t behavior of the amplitudes, as Signell has emphasized[11]. For example, contributions from the region $t=4\mu^2$ are weighted by a factor of ~150 at $r = 2\hbar/\mu c$ in 300 MeV nucleon-nucleon scattering as compared with contributions at $t=28\mu^2$. In order to test our weighting functions, we supplemented our Re f_+^0 by a small imaginary part, manufactured from eq. (14) by taking

$$\tan \delta_0 = \frac{q}{(\frac{1}{a})(1-\frac{q^2}{q_x^2})+q \cot \delta_x} \tag{15}$$

with a=0.2$\hbar/\mu$c, qx = 2.5. Here δ_x is arbitrarily chosen so as to give a maximum in δ_0 of 45° at an energy of 700 MeV in the pion-pion c.m.s.

Results for the F-wave nucleon-nucleon scattering using the one-pion-exchange amplitude plus the two-pion-exchange amplitude calculated as described above are shown in Fig. 6.

As noted before, the $f_+^0(t)^2$ gives a weighting function describing the interaction between nucleons which is scalar in all respects; i.e., the sort of interaction that would result from the exchange of spin-zero isoscalar mesons of various masses $\sqrt{t}$. Note that the square of our f_+^0 curve B in Fig. 4 is only about 1/5 of that of the Born term, curve A, in the region $t \sim 4$ to $10\mu^2$. This is the pair suppression discussed above and indicates that the nucleon box diagram without pair suppression gives much too much inter-mediate-range attraction in this channel.

In this work described above, only the S-matrix entered. We did not go back to a potential, with the attendant ambiguities. The reduction to a potential will be discussed in a contributed paper by John Durso.

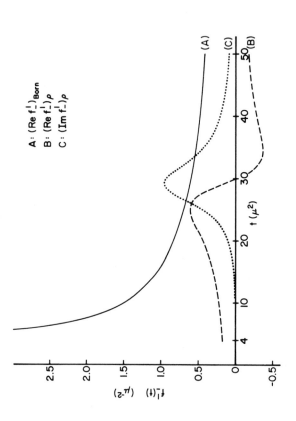

Fig. 5 The solid line shows the Born amplitude, the dashed line, the real part of the amplitude from the ρ-meson, the dotted line, the imaginary part; parameters for the ρ-meson contribution are $\Gamma_\rho = 120$ MeV, $E_\rho = 29\mu^2$.

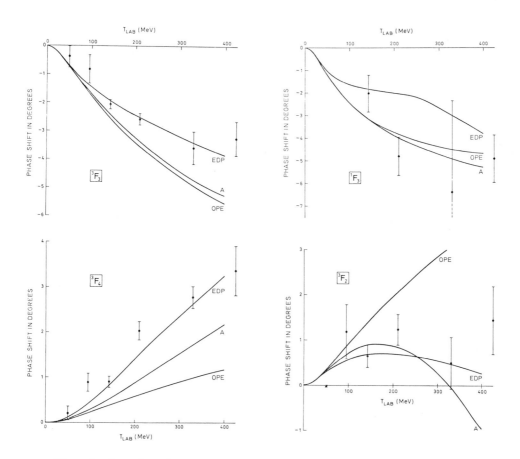

Fig. 6   Results from our weighting functions for F-wave nucleon-nucleon scattering compared with the Livermore[12] energy-independent phase shifts and energy-dependent phase-shift analysis (curves labelled EDP).   The curves labelled OPE correspond to predictions from one-pion exchange; curves labelled A are our theoretical results.

I would like to thank Marc Chemtob, John Durso and Dan-Olof Riska, who not only participated in, but actually did most of the work I am reporting on here.

REFERENCES

1. D. Amati, E. Leader and B. Vitale, Nuovo Cimento 17(1960)68; 18(1960)409; 18(1960)458; Phys. Rev. 130(1963)750.

2. E. Lomon and H. Partovi, Phys. Rev. D2(1970)1999.

3. P. Signell, Proceedings of the Third International Symposium on Polarization Phenomena in Nuclear Reactions, University of Wisconsin Press.

4. M. H. MacGregor, R. A. Arndt, R. M. Wright, Phys. Rev. 182 (1969)1714.

5. T. Hamada and I. D. Johnston, Nuclear Physics 34(1962)382.

6. M. Chemtob and D. O. Riska, Phys. Letts. 35B(1971)115.

7. G. E. Brown and J. W. Durso, Phys. Letts. 35B(1971)120.

8. S. Adler, Phys. Rev. 137(1965)B1022.

9. S. Weinberg, Phys. Rev. Letters 17(1966)616.

10. H. Nielsen, J. Lyng Petersen and E. Pietavinen, Nuclear Physics B22(1970)525.

11. P. Signell, Advances in Nuclear Physics 2(1969)223.

DEVELOPMENT OF A SUPER SOFT CORE POTENTIAL MODEL OF THE NUCLEON-

NUCLEON INTERACTION

R. de Tourreil and D. W. L. Sprung

ORSAY and McMaster University

It is by now well known that the elastic scattering data do not require the nucleon-nucleon potential to contain infinite or even strong finite repulsion at short distance. This was shown in the $^1S_0$ state by Srivastava and Sprung[1] who constructed a number of 'super soft core' potentials whose maximum repulsion ranged from 70 MeV to 260 MeV. A rough criterion for an SSC potential is that in nuclear matter the second order perturbation theory terms are 10 to 20% of the corresponding first order ones. At about the same time, Pires, de Tourreil and Gogny (PDG)[2] presented a complete potential, choosing as radial form a sum of gaussians. Their intention was to fit not only the elastic scattering data, but also nuclear matter saturation properties and the radii of finite nuclei in the Hartree Fock approximation. Modest agreement with all three criteria was obtained, but at the expense of a high quality fit to phase shifts, especially in the singlet odd state.

The present work is directed to producing a complete SSC potential which provides an excellent fit to phase shifts and other two body data, and includes the OPEP tail. The forms employed are essentially Yukawas modified by gaussian or similar cut-off functions to keep the potential soft at short distances. The potential is not yet in its final form, but acceptable forces have been found in three of the four spin, isospin (S,T) subspaces and the remaining singlet odd state presents no difficulty. In this note we discuss some characteristics of the potential which are noteworthy. Some preliminary potentials are included to show the type of forms employed.

SOME CONSIDERATIONS IN FITTING THE POTENTIAL

The potential is fitted to reproduce the single energy phase shift analyses of MacGregor, Arndt, and Wright.[3] In addition to

43

the central, tensor and spin orbit components, an L^2 dependent
force is used in each (S,T) subspace. This operator is preferable
to the so called quadratic spin-orbit force $\sigma^1 \cdot L \, \sigma^2 \cdot L$ or equivalently
the operator L_{12} adopted by Hamada and Johnston,[4] on grounds of
simplicity and in fitting the phase shifts. In singlet states, any
of these more complicated operators reduces to an L^2 force. In
triplet states the operator L_{12} is equivalent to the sum of L^2 and
an operator

$$Q_{12} = 3 \, \sigma^1 \cdot L \, \sigma^2 \cdot L - \sigma^1 \cdot \sigma^2 L^2$$

which causes a splitting of the three states of different J for
each L. Although in the triplet even states there is a need for
an L^2 force to make the mean D-wave interaction weaker then the
S-wave, there appears to be no evidence for an additional splitting
of the J-values beyond that given by the LS and tensor forces. We
have not found it necessary to invoke a Q_{12} force.

Our potential for the singlet even states has the form

$$V_C(r) = P_1 \exp(-r^4/p_2^2) + \{p_3 \frac{\exp(-p_4 x)}{x} + OPEP\}\{1-\exp(-r^4)\}$$

$$V_{L2} = \{p_5 \, Z(y) + p_7 \, Z(x)\} \frac{1}{x^3} \{1-\exp(-r^8)\}$$

$$V = V_C(r) + V_{L2}(r) \, L^2, \quad x = \mu r, \quad \mu = 0.7 \text{ fm}^{-1}$$

$$OPEP = -10.463 \exp(-x)/x, \quad Z(y) = (1+3/y + 3/y^2) \exp(-y)$$

$$y = p_6 x$$

$$P_1 \cdots P_7 = 378.5, \quad .4787, \quad -1003., \quad 3,600, \quad 6.403, \quad 2.026, \quad -.4090$$

The forms adopted are highly arbitrary. In particular the L^2 force
requires a very high order cut-off to compensate for the singular
radial form taken from Hamada and Johnston. Further improvements
are planned which will simplify the functional form.

One problem in fitting the triplet even state phase parameters
to MAW is to reproduce the steady rise with energy of the coupling
constant ε^1. For a weak potential this can be studied qualitatively
in terms of the Born approximation;

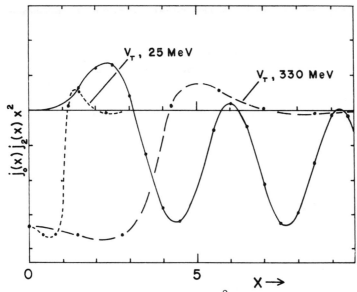

Fig. 1.--The Bessel function product $x^2 j_0(x) j_2(x)$ (solid line) and a schematic soft tensor force (dashed lines).

$$\varepsilon^1 = - \frac{1}{k^2} \frac{m}{\hbar^2} \int_0^\infty j_0(x)\, j_2(x)\, V_T(r)\, x^2\, dx, \quad x=kr$$

In Fig. 1 we show the Bessel function product $x^2 j_0(x) j_2(x)$, and superimposed on it a schematic soft tensor force. The radial scale is chosen to correspond to 25 MeV (dashed line) and 330 MeV (long dash line). The oscillations of the Bessel function product are mostly on the negative side of the axis. If the tensor force were always attractive and monotonic, as is the OPEP force, it is clear that as the energy increases, the tail of the tensor force integrated over the first negative loop of the Bessel function product would give a substantial negative contribution to ε^1, severely reducing its value when E=330 MeV. In the PDG potential one finds as a consequence that ε^1 either heels over or becomes negative at 200 or 300 MeV.

This tendency can be combatted if $V_T(r)$ takes on severely reduced or even small positive values in the region near 1 fm. An extreme form of such behavior is also illustrated in Fig. 1.

In this case the damage done by the negative loop of the Bessel function product is reduced, or even turned to advantage, and a monotonic increase with energy of ε^1 is obtained. Such a behavior of the radial form of $V_T(r)$ may not be completely artificial, since it is found in the potential of Ingber and Potenza[5] where repulsive contributions from the vector mesons combined with the π, σ and η meson forces give a $V_T(r)$ which is nearly zero at about 0.5 fm. We agree with Ingber that previous phenomenological potentials did not include such a possibility because they were guided too closely by the OPEP form, supplying only the required regularization of the $1/x^3$ singularity at the origin.

Because the other components of the force are weak in our potential, even the small r part of V_T plays a role in determining ε^1; it is not completely masked by a strong repulsion. In order to fit the deuteron quadrupole moment we require an attractive contribution to V_T of about two pion range.

A triplet even state potential has the form:

$$V = V_C(r) + V_{L2}(r)L^2 + V_{LS}(r)\ \underset{\sim}{L}\cdot\underset{\sim}{S} + V_T(r)\ S_{12}$$

$$V_{L2} = \{P_5\ Z(y) + P_7\ Z(x)\}\ \frac{1}{x^3}\ \{1 - \exp(-r^8)\}$$

$$V_{LS} = -P_8\ (1+P_9\ x)\exp(-P_9 x)\ \frac{1}{x^3}\ \{1-\exp(-r^6)\}$$

$$V_T = P_{10}\ \exp(\frac{-r^4}{P_{11}^2}) + \{P_{12}\ Z(y')/x - 10.463\ Z(x)/x\}\{1-\exp(-r^6)\}$$

$$V_C = \text{same as singlet even.}$$

$$Z(y) = (1 + 3/y + 3/y^2)\exp(-y) \qquad y = P_6\ x,\ x = 0.7r,$$

$$y' = P_{13}\ x$$

$$P_1 \ldots P_{13} = \begin{matrix} 460.0, & 0.67, & -1110.0, & 3.600, \\ 10.20, & 2.215, & -.3699, & \\ -18.50, & 4.20, & & \\ -178.4, & 0.92, & +500.7, & 4.352 \end{matrix}$$

Deuteron properties: $E_D = -2.215$ MeV, $Q = .264$ fm^2,

$$\mu = .857\mu_N,\quad P_D = 4.55\%$$

low energy properties: $a = 5.50$ fm, $r_0 = 1.847$ fm.

In nuclear matter the $^3S_1$ state gains about 9 MeV potential energy as compared to the Reid force, at normal density. There is not a strong saturating effect due to the tensor force, indicating that it is quite weak. Saturation relies upon odd state repulsion in this case.

Further work is in progress and it is hoped to publish a more detailed report in a few months time.

REFERENCES

1. D. W. L. Sprung and M. K. Srivastava, Nucl. Phys. A139 (1969) 605.

2. D. Gogny, P. Pires, and R. de Tourreil, Phys. Lett. 32B (1970) 591.

3. M. H. MacGregor, R. A. Arndt and R. M. Wright, Phys. Rev. 182 (1969)1714.

4. T. Hamada and I. D. Johnston, Nucl. Phys. 34 (1962) 382.

5. L. Ingber and R. M. Potenza, Phys. Rev. C1 (1970) 112.

TWO-PION-EXCHANGE NUCLEON-NUCLEON POTENTIAL

M. Chemtob, J. W. Durso and D. O. Riska

Niels Bohr Institute and NORDITA

The work reported on here may be regarded as an extension of earlier theoretical derivations [1,2] of the nucleon-nucleon potential with two important modifications. In the first place we include as much information from $\pi N \to \pi N$ and $N\bar{N} \to \pi\pi$ processes as is currently available from theoretical descriptions. The second modification is the use of the Blankenbecker - Sugar - Logunov - Tavkhelidze (BSLT) equation [3,4] to define the potential.

The first modification is illustrative of the advantages of using a dispersion relations approach to calculate the full two-pion-exchange amplitude rather than the field-theoretic approach. [5] In this way, information on πN rescattering effects (nucleon isobars in the intermediate states) as well as on $N\bar{N} \to \pi\pi$ helicity amplitudes can be accounted for in a simple way. The price one pays for this is the restriction of the amplitude calculated to on-energy-shell processes.

The second modification is important in several respects. Since the two-particle Green's function in the BSLT equation has the same elastic part as the free, relativistic two-particle Green's function, the resulting amplitude will satisfy a relativistic unitarity condition. [5,6] This means that the elastic unitarity cut of the field-theoretic amplitude will be identical with that of the iterated one-pion-exchange potential for an energy region extending from the NN elastic threshold to the pion production threshold. Removal of the iterated OPEP then results in a potential which is extremely energy independent. Furthermore, the iterated OPEP calculated in this way is finite, so that it is unnecessary to introduce infinite subtraction constants in order to calculate the potential.

We have calculated the potentials resulting from several assumed models for $\pi N \to \pi N$ and $N\bar{N} \to \pi\pi$ processes.[7,8] Some of the results are shown in Figs. 1-4. The curves labeled R show the potential calculated by taking the πN amplitude to be given by a sum of pole terms corresponding to the nucleon and the $P_{33}(1236)$, $P_{11}(1400)$ and $D_{13}(1520)$ isobars. Replacing the $N\bar{N} \to \pi\pi$ s-wave contribution due to isobar exchange by the contribution taken from the results of Nielsen, Petersen and Pietarinen[9] produces the potential S'. The addition of a finite-width ρ-meson to the nucleon isobar exchange contributions in the $N\bar{N} \to \pi\pi$ p-wave helicity amplitudes then yields the curves SP'. Finally, inclusion of the w- and η-exchange contributions gives the results labeled by T' which correspond to our most complete description of πN and $N\bar{N} \to \pi\pi$ processes. The potentials calculated are, as stated, very energy-independent and quite insensitive to the cut-off.

In most cases the calculated potential is close to the phenomenological Hamada-Johnston potential (curves HJ), particularly at distances ≥ 1 fermi. Significant is the fact that the corrections made by including $\pi\pi$ correlations via the s- and p-wave $N\bar{N} \to \pi\pi$ helicity amplitudes almost invariably improve the agreement. It is hoped that further refinements in the description of πN and $N\bar{N} \to \pi\pi$ processes will improve the potential still further.

REFERENCES

1. J. M. Charap and M. Tausner, Nuovo Cimento 18 (1960) 316.

2. W. N. Cottingham and R. Vinh Mau, Phys. Rev. 130 (1963) 735.

3. A. A. Logunov and A. N. Tavkhelidze, Nuovo Cimento 29 (1963) 380.

4. R. Blankenbecker and R. Sugar, Phys. Rev. 142 (1966) 1051.

5. M. H. Partovi and E. L. Lomon, Phys. Rev. D2 (1970) 1999.

6. G. E. Brown, A. D. Jackson, and T. Kuo, Nucl. Phys. A133 (1969) 481.

7. M. Chemtob and D. O. Riska, Phys. Letters 35B (1971) 115.

8. G. E. Brown and J. W. Durso, Phys. Letters 35B (1971) 120.

9. H. Nielsen, J. L. Petersen, and E. Pietarinen, Nucl. Phys. B22 (1970) 525.

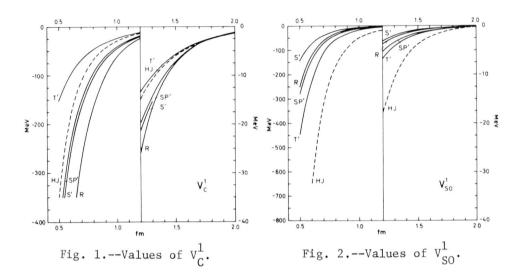

Fig. 1.—Values of V_C^1.

Fig. 2.—Values of V_{SO}^1.

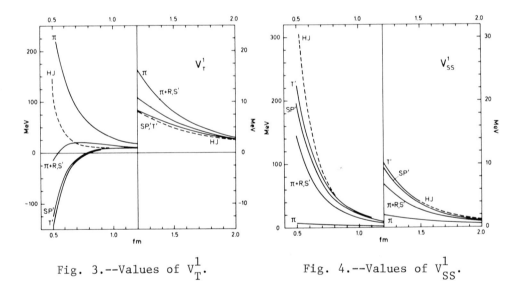

Fig. 3.—Values of V_T^1.

Fig. 4.—Values of V_{SS}^1.

THE TWO-PION EXCHANGE CONTRIBUTION TO THE NUCLEON-NUCLEON POTENTIAL

Geoffery N. Epstein and Bruce H. J. McKellar
University of Sydney
Sydney, N.S.W. 2006, Australia

The 2π exchange contribution to the nucleon-nucleon potential
is calculated following the Mandelstam representation method of
Cottingham and Vinh Mau. The $N\bar{N} \to \pi\pi$ amplitude which is central to
the calculation is obtained by analytic continuation of the low
energy πN scattering amplitude which is constructed from nucleon
and N* pole terms in a way consistent with the πN scattering data
and soft pion theorems. $\pi\pi$ interactions are treated using the
up-down solution for the S wave and ρ meson dominance for the P
wave. The uncorrelated 2π contribution is obtained from the work
of Partovi and Lomon.

It is to be emphasized that no arbitrary parameters or cut
offs are used in the calculations. The potentials obtained are
in good agreement with the phenomological potentials of the Yale
group and Hamada and Johnston. It is interesting to note that the
N* contribution is negligible and that the potential is essentially
independent of the notorious constant C_A^+ of Cottingham and Vinh
Mau.

ON THE DEFINITION OF A BOSON EXCHANGE POTENTIAL

C. J. Noble and K. C. Richards
Bartol Research Foundation
Swarthmore, Pennsylvania

Four one-boson exchange potential definitions have been used to
calculate the one pion exchange part of this potential. The four
sets of phase shifts calculated show a spread of 10% at 300 MeV for
most uncoupled channels. A few exhibit greater variation. This
uncertainty demonstrates a limitation of boson exchange potentials.

REVIEW OF EXPERIMENTS ON NUCLEON–NUCLEON BREMSSTRAHLUNG[*]

M. L. Halbert

Oak Ridge National Laboratory

CONTENTS

1. Introduction

2. Proton–Proton Bremsstrahlung (ppγ)

    A. Geometry
    B. Experimental Problems and Their Effects on the Data
    C. Experimental Results
    D. Theoretical Cross Sections
    E. Comparison of Theory and Experiment
    F. Gamma-Ray Angular Distributions

3. Neutron-Proton Bremsstrahlung (npγ)

    A. Experimental Results
    B. Comparison of Theory and Experiment

4. Conclusions and Speculations

1.   INTRODUCTION

It is now only about 6-1/2 years since the publication of the first experiments[1] on nucleon-nucleon bremsstrahlung. During that time we have come a long way. From conception and birth, with the initial great hope that here at last was a practical way of

---

[*] Work sponsored by USAEC under contract with Union Carbide Corp.

selecting the correct potential for the nucleon-nucleon force,
we passed through the youthful era of wild disagreements with
theoretical predictions, and then into comfortable middle age when
all experiments and theories seem to agree reasonably well. Recently
interest has tended to wane somewhat, particularly since it was
revealed that although off-shell effects are certainly being observed,
the bulk of the difference between the existing ppγ predictions
of various potential models can be traced to already-known differences
in their on-shell behavior.[2] Thus the original great promise has
failed to materialize. Nevertheless, the field is still vigorous
and has not yet reached senility, particularly with respect to
npγ experiments.

While in this introductory vein I should like to pay particular
tribute to the Harvard group - now dispersed - of experimenters[1]
and theorists[3] that recognized the feasibility of making these
measurements, conceived the symmetric geometry used in nearly all
subsequent experiments, made the first modern calculations, and
brilliantly executed the first measurements. The scatter diagrams
published by Gottschalk, Shlaer, and Wang in 1966[4] and 1967[5] remain
the prettiest that have been obtained. Moreover, of all the experiments done in the geometry they originated, theirs is the only one to
have demonstrated that there are, in fact, high-energy photons
accompanying the proton pairs.[4] And their cleverly done measurement
of the noncoplanar cross section[5] was the only one in existence
until very recently.

The emphasis in this review is on several questions. Are
the experimental results really in agreement with theory? If so,
which theory? What experimental problems might be causing trouble?
Are comparisons of theory with experiment being done properly?
What is the status of the more exotic kinds of measurements - for
example, those with unequal nucleon angles? What experiments are
now in progress and what types of results can be expected from
them?

2.   PROTON-PROTON BREMSSTRAHLUNG (ppγ)

A.   Geometry

The term "Harvard geometry" originally meant detection of
the two inelastic protons in a coplanar, symmetric arrangement.
In most experiments the coplanar ideal is not approached very closely
because the maximum noncoplanarity between the protons is a few
degrees at most, and counter heights comparable with this usually
must be used. In some experiments the proton polar angles have
been unequal. So nowadays the term is frequently applied to any
arrangement in which the energy and direction of the gamma ray are

not measured, but are inferred from the measured energy and direction of the two protons.

One of the pioneering series of measurements[6,7] was made in the "Rochester geometry", which involves detection of all three final particles. Each event is kinematically overdetermined several times and therefore this geometry discriminates well against spurious events. A similar experiment is in progress at Orsay at 156 MeV.[8] However, most experimenters have preferred to defeat the background by being clever with their proton counters rather than fight the difficult battle of gamma-ray detection.

Several coordinate systems are in use; the symbols θ and ϕ may mean different things in different papers. Careful discussions of the different systems have been presented in three theoretical papers.[9,10,11] These references also provide useful information on phase space considerations.

B. Experimental Problems and Their Effects on the Data

The experiments are time-consuming. Typical event rates are between 1 and 10 per hour. The experiment at McGill University[12] has the best statistics at particular angles, 4000 events spread over four angles; this took six weeks of cyclotron time for data collection. Experimenters are obliged to use counters subtending appreciable solid angles to obtain even these very low counting rates. In the past this has caused problems of interpretation because of the averaging over angles, principally the azimuthal angle. As will be shown in the next section, this problem is now under control for $pp\gamma$.

The new spark-chamber experiment at Manitoba[13] is in a class by itself - its event rate is over 100/hour. This is achieved by using counters that cover a very large solid angle. Although the number of events per hour for any particular pair of angles may not be any greater than with conventional equipment, they obtain data simultaneously at many angles.

The experiments are very tricky. Basically this is because the bremsstrahlung yield is about 10^6 times weaker than the elastic counting rate. A few percent of these elastics give substandard detector signals that simulate low-energy particles. Also, there may be low-energy protons that pass through slit edges or are otherwise degraded. Random coincidences between such occurrences are the main source of background.

To minimize the troublesome substandard pulses, many different methods have been used: dE/dx pulse-height selection in a transmission counter,[5,12,14-18] veto of long-range protons by a counter beyond

the one that stops the inelastic protons,[1,4,5,18] veto of elastic
events by detection of the conjugate nucleon,[12,19] measurement
of flight time between elements of a telescope,[1,4,5,12] and
reconstruction of the proton trajectories.[13] For rejection of
slit-scattered particles one experiment introduced "live" slit
edges.[17] If the protons have sufficient energy, it is preferable
to replace the front slits by a transmission counter.[4,5,12] In
the spark-chamber experiment[13] this problem does not arise.

The unwanted elastics may be prevented from entering the detector
telescopes by use of a magnet or other deflection system, but thus
far none have been used. Such a deflection system is probably
a necessity for experiments at very small proton angles; these would
be of great theoretical interest because they are further off the
energy shell.

Most of the above methods are aimed at reducing the number
of accidental coincidences that simulate ppγ. This does not,
however, eliminate _real_ coincidences of low-energy protons that
are not bremsstrahlung events. Such spurious events may result
from slit scattering. In Figure 1 we have a typical example of
the Harvard geometry used in one of the Oak Ridge experiments.[15]
The front slits of the two telescopes have to be fairly thick to
stop elastic protons. Since the entire chamber is filled with
hydrogen gas, many elastic-scattering events are occurring. You will
remember from your billiard-parlor experience that after an elastic
collision, equal-mass particles come off with a right angle between
them. Thus if the vertex of each right angle shown by the dashed lines
represents a point where an elastic collision has occurred, the
adjacent edges show us possible trajectories for the outgoing proton
pair. If both protons graze the slit edges they may be scattered
into the detectors with reduced energy. One thus has a real coincidence between two low-energy protons which is not a bremsstrahlung
event.

There are three ways to avoid these spurious events. One
is to insert baffles to stop these protons before they hit the
slit edges. This was what we did, as shown by the cross-hatched
region. Another is to confine the hydrogen gas so that no scattering
can occur at the vertex of either right angle. This was done for
the 25° data of Ref. 18. For their 30° data only one of the right
angles was eliminated, although they minimized slit scattering
by careful shaping of the slit edges. The third way, mentioned
earlier, is to dispense with the front slit entirely.[4,5,6,12,13]

If nothing is done to eliminate these events, one will have
an additional yield that will make the apparent ppγ cross section
too large. How serious can this be? Calculation of the effect
would be difficult and probably unreliable; I believe that an experimental determination would be the only safe way to evaluate it.

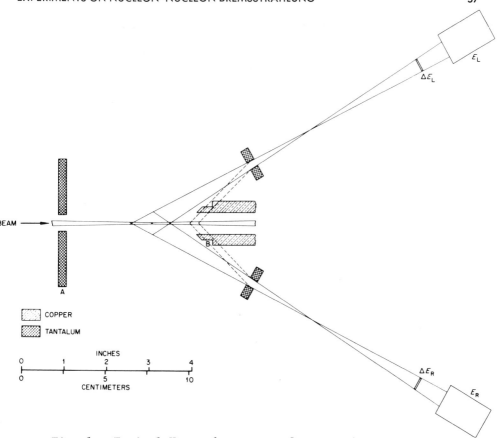

BEAM →

COPPER

TANTALUM

INCHES

0   1   2   3   4
0   5   10
CENTIMETERS

Fig. 1. Typical Harvard geometry for experiments at low energy. The dashed right angles show trajectories of elastic protons that might result in coincidences simulating ppγ events. The baffles B were inserted to stop such protons.

In our preliminary experiments, without the baffles, it appeared that at 35°, 35° the yield in the ppγ region was roughly twice what was expected for genuine events, and at 40°, 40° perhaps three or four times the expected rate. At 30° the effect should have been smaller (and it appeared so), but we did not run long enough to evaluate it. We simply inserted the baffles and stopped worrying.

I see no dependable way to evaluate this effect in experiments now dismantled that did not have protective measures built in. Neither would a measurement of the effect in one experiment be relevant to the effect in another since it depends on slit arrangements and proton energies.

Consequently, I believe that we should be prepared to regard the results from experiments subject to this problem as upper limits

only. This includes Refs. 19-25 and perhaps the 30° data of Ref. 18.
Since the published results of Refs. 20-22 are in effect already
only upper limits, this caution does not change their status. Part
of the data in Ref. 16 were obtained without fully adequate baffles,
but for the reasons outlined in that paper the effect is believed
to have been negligible.

C. Experimental Results

Most ppγ experiments have had large acceptance for noncoplanar
events. Thus the measured values are averages over Φ, the angle of
proton noncoplanarity. To compare with theoretical calculations of
the coplanar cross section, the data must be corrected. Perhaps
the most widely used Φ distributions are those calculated by
Drechsel and Maximon[9],[26] for the Hamada--Johnston potential. This
correction is now on a very firm basis. Almost identical results
for the Φ distribution were obtained when corrections for relativistic spin effects were included.[11],[27] The shape was also
practically the same for the Reid potential[9] and for a one-boson-
exchange model.[10] If the curves are plotted as a function of
Φ/Φ max, the shape is found to vary very little either with beam
energy or proton polar angle. The measurements of the Φ distribution at 42 MeV,[13] 64.4 MeV,[17] and 157 MeV[5] are consistent with this
apparently universal shape.

Marker and Signell[2] made a tabulation of all coplanar symmetric
cross sections for 30° and 35° experiments. In some cases published
results were updated by private communications. A consistent correction for acceptance of noncoplanar events was applied to all the
data by use of the Drechsel-Maximon Φ distributions.[9],[26] Table I
is an extension of their tabulation to include all data now available,
including nonsymmetric geometries. In a few cases the numbers
given here differ slightly from those of the original authors or
Ref. 2; these are explained in the footnotes.

Additional ppγ measurements are in progress or are planned for
the near future. As I mentioned earlier, a group at Orsay[8] is
making measurements at 156 MeV at very forward angles (15° on one
side, 18°, 21°, 24°, and 27° on the other). And the Manitoba group
has plans for next month to obtain 10 times as much data as they
published before.[13]

D. Theoretical Cross Sections

Four wide-ranging sets of integrated cross sections have been
published by theorists. Three of them are for potential models.
We direct our attention to results for the Hamada--Johnston potential.
The calculations of Virginia Brown[29] include processes in which

Table I. Experimental data for ppγ in Harvard geometry with proton angles θ_1, θ_2.

Beam Energy (MeV)	θ_1	θ_2	Institution	Distributions Measured θ_γ	Φ	Integrated Coplanar Cross section ($\mu b/sr^2$)	Ref.
3.5	30°	30°	Case Western Reserve	No	No	$0.15^a + 0.17, -0.15$	20
10	30°	30°	Rice Univ.	No	No	$<0.42^a$	21
10.5	30°	30°	Canberra	No	No	$0.5^b + 0.7^b, -0.5^b$	22
20	25°	25°	Univ. Washington	Yes	No	0.68 ± 0.07	18
20	30°	30°	Univ. Washington	Yes	No	0.69 ± 0.07	18
20	35°	35°	Univ. Rochester	No	Yes	$1.2^c \pm 0.5$	19
30	35°	35°	Manitoba/Oberlin	No	No	$2.10^d \pm 0.28^d$	25
33.5	30°	30°	UCLA	No	No	$3.6^b \pm 1.1^b$	23
42	e	e	Univ. Manitoba	Yes	Yes	f	13
46	30°	30°	Univ. Manitoba	Yes	Yes	$3.8^b \pm 0.7^b$	23
47.1	30°	30°	Oak Ridge	Yes	No	1.37 ± 0.29	15
48	30°	30°	Manitoba/Oberlin	No	No	$2.68^d \pm 0.45^d$	24
48	35°	35°	Manitoba/Oberlin	No	No	$3.93^d \pm 0.57^d$	24
52.3	33°	33°	Tokyo	No	No	$2.9^g \pm 1.1$	28

Beam Energy (MeV)	θ_1	θ_2	Institution	Distributions Measured θ_γ	Φ	Integrated Coplanar Cross section (μb/sr^2)	Ref.
61.7	30°	30°	Oak Ridge	Yes[h]	No	2.04 ± 0.24	14
61.7	30°	30°	Oak Ridge	No	No	2.27 ± 0.73	14
64.4	30°	30°	Oak Ridge	No	Yes	2.32 ± 0.20	17
65.0	20°	20°	Oak Ridge	Yes	No	1.49 ± 0.78	16
65.0	20°	26°	Oak Ridge	No	No	1.53 ± 0.54	16
65.0	23.3°	30°	Oak Ridge	No	No	1.50 ± 0.50	16
65.0	25°	25°	Oak Ridge	Yes	No	2.21 ± 0.35	16
65.0	25°	35°	Oak Ridge	Yes	No	1.98 ± 0.23	16
65.0	26°	35°	Oak Ridge	No	No	2.11 ± 0.32	16
65.0	26.5°	26.5°	Oak Ridge	No	No	2.30 ± 0.44	16
65.0	30°	30°	Oak Ridge	Yes	No	2.34 ± 0.38	16
99.0	25°	25°	McGill Univ.	Yes	No	3.77 ± 0.24[i]	12
99.0	30°	30°	McGill Univ.	Yes	No	5.14 ± 0.24[i]	12
99.0	35°	35°	McGill Univ.	Yes	No	9.01 ± 0.37[i]	12
99.0	40°	40°	McGill Univ.	Yes	No	18.83 ± 1.21[i]	12
157	30°	30°	Harvard Univ.	Yes[j]	Yes	10.2[b] ± 1.7[b]	5
157	35°	35°	Harvard Univ.	Yes[j]	Yes	14.7[b] ± 2.5[b]	5

Table I (continued)

157.8	30°	30°	Harvard Univ.	Yes	No	7.9 ± 20%	4
157.8	32.5°	32.5°	Harvard Univ.	Yes	No	10.1 ± 20%	4
157.8	35°	35°	Harvard Univ.	Yes	No	12.4 ± 20%	4
157.8	37.5°	37.5°	Harvard Univ.	Yes	No	13.3 ± 20%	4
157.8	40°	40°	Harvard Univ.	Yes	No	23.8 ± 20%	4
204[k]	30°	30°	Univ. Rochester	Yes[ℓ]	Yes[ℓ]	13.0[m] ± 2.4	6
204[k]	35°	35°	Univ. Rochester	Yes[ℓ]	Yes[ℓ]	14.0[m] ± 2.7	6
204[k]	40°	40°	Univ. Rochester	Yes[ℓ]	Yes[ℓ]	29.0[m] ± 6.0	6

a. Parabolic correction for noncoplanarity. This probably overestimates coplanar cross section.
b. From Ref. 2.
c. From Ref. 19, corrected for measured noncoplanarity.
d. From Ref. 26.
e. This experiment presents results for equal and unequal polar angles in 4° bins from 18° to 34°.
f. Consult Ref. 13. Cross sections are given graphically in terms of $d\sigma/d\theta_1 d\theta_2 (\mu b/rad^2)$.
g. Method of correction for noncoplanarity not explained by authors.
h. An improved measurement at 65 MeV of the 30°, 30° θ_γ distribution is found in Ref. 16.
i. Error estimates include quadratic addition of 2% absolute uncertainty.
j. Results presented in terms of ψ_γ.
k. Proton beam 90% polarized.
ℓ. Not Harvard geometry.
m. For equivalent Harvard geometry.

the nucleons scatter a second time after the photon is emitted.
These rescattering effects were found to increase the ppγ cross
section slightly ($\leq$ 0.2% for 62 MeV or lower, 4% for 158 MeV at
30°). Marker and Signell[2] made an approximate correction for Coulomb
effects. They found that the cross sections were reduced. The
upper part of Figure 2 shows the correction factor I extracted
from their work for various angles and energies. My purpose was
to devise some method, however crude, for estimating the Coulomb
effect in situations not treated explicitly in Ref. 2. It is my
understanding that Heller and Rich[30] have developed a method for
treating Coulomb effects exactly; I think their results agree with
the Marker-Signell approximate values for the angles of interest
here. Heller and Rich[30] also treated rescattering effects.
Dr. Heller may discuss these matters in the next talk.

The calculations of Liou and Cho[11,27] include a relativistic
spin correction, first pointed out by McGuire.[31] Their calculation
includes internal scattering effects and satisfies the gauge condition.
(Internal radiation is that produced during the nucleon-nucleon
scattering. It includes rescattering and radiation from the exchange
of charged mesons.) The lower part of Figure 2 shows the ratio
of their results with the correction to their results without it.[27]
Again, the cross section is decreased, this time the effect being
more important for higher energies.

I should perhaps mention here that the uncorrected Hamada-
Johnston results from Refs. 2, 9, 11, 26, 27, 29, and 30 are all
in good agreement. However, what we need for comparison with experiment
are results including all the significant corrections. I have
taken the liberty of multiplying the Marker-Signell Coulomb-corrected
cross sections by the relativity correction derived from the lower
part of Figure 2. This procedure is questionable, but I believe
we will probably come closer to realistic numbers this way than
by ignoring one or another of these corrections. Please bear in
mind, however, that I did not include rescattering effects. This
neglect is of no consequence for the results up to 65 MeV, but
may cause an underestimate of the theoretical predictions by a
few percent for energies $\geq$ 100 MeV.[29]

The fourth set of theoretical results is for a very different
type of calculation, one based on meson theory rather than a potential
model. This calculation, by Baier, Kühnelt, and Urban,[10] uses
a one-boson-exchange (OBE) model and takes into account exchanges
of π, ρ, ω, η, and σ mesons. It is relativistically correct, gauge-
invariant, and takes into account, at least partially, internal
radiation and Coulomb effects between the protons. However, it
does not satisfy unitarity.

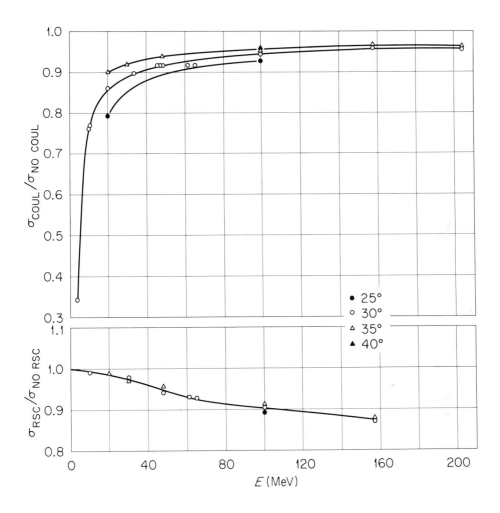

Fig. 2. Corrections to Hamada-Johnston calculations for Harvard
geometry at various angles and energies.
Upper part: Coulomb effect, points from Ref. 2.
Lower part: Relativistic spin effect, points from Ref. 27.

E.   Comparison of Theory and Experiment

In this section I consider only $d\sigma/d\Omega_1 d\Omega_2$, the cross section for selected proton angles, integrated over all gamma-ray angles.

For comparison with theory, I calculated the deviation of each experimental value from the (corrected) prediction and divided by the experimenter's error estimate. Figure 3 shows a plot of deviations from the Hamada-Johnston potential-model predictions corrected for Coulomb effects and relativity. The open points are to be regarded as upper limits because of the possibility of spurious events that I mentioned in Section B. In a few cases, I had to use Hamada-Johnston calculations from other authors[9,13,29] or interpolations, and then apply corrections obtained from Fig. 2.

If the theory really fits the data (or vice versa if you prefer), the distribution of points should approximate a Gaussian with unit standard deviation. Below 100 MeV the trend of the solid points is reasonably close to this, particularly for the 42-MeV data.[13] Near 65 MeV the theory may be somewhat too high (or the experimental errors underestimated), but it is nevertheless a believable representation of the data. This is the opposite of a statement we published last year;[17] the difference in my point of view comes from including both the Coulomb and relativistic corrections, and in looking at an ensemble of points rather than just one.

At the highest energies the theory tends to be low by about one standard deviation, which is not serious, especially if you recall that rescattering effects will increase the prediction a few percent. The situation at 99 MeV is more disturbing. Either the theory dips too low by about 20%, or the accuracy claimed for these points (4 to 6%) has been incorrectly estimated, or some combination of the two has occurred. Rescattering effects are probably no more than 2% here, and are insufficient to account for the discrepancies. Of course, on this kind of plot you are putting a spotlight on the most accurate points - a 15% discrepancy with theory will stand out prominently with 5% data, while with 30% data it would look quite acceptable. However, the 35° point, the one with the largest deviation, is very far out of line.

Figure 4 shows a similar plot for comparison with the OBE model. Some of the data shown in Figure 3, particularly most of those at 42 MeV, are not here because the theoretical calculations are not available. On the other hand, there are a few unequal-angle points at 65 MeV here that were not in Figure 3. The general trend is for the theory to be somewhat too high near 65 MeV, while the 99-MeV data look quite reasonable except again for the 35° point. The theory is in good agreement with most of the points for 158 and 204 MeV.

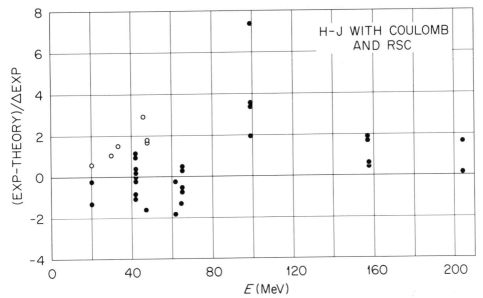

Fig. 3. Comparison of ppγ experiments with Hamada–Johnston
calculations with Coulomb[2] and relativistic[27] corrections.
The deviation from theory is expressed in units of the
experimental error.

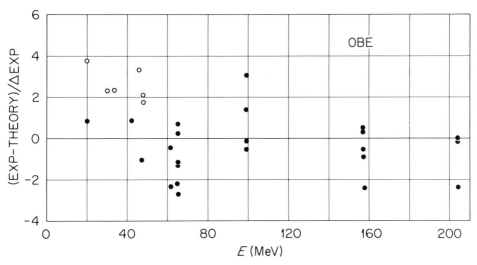

Fig. 4. Comparison of ppγ experiments with OBE calculations.[10]
The deviation from theory is expressed in units of the
experimental error.

F.   Gamma-Ray Angular Distributions

The measurement of the two proton energies and angles in the
Harvard geometry makes it possible to infer the gamma ray angles
θ_γ and ϕ_γ for each event. Figure 5 gives the kinematics for 30°,
30° events at 61.7 MeV. The axes show the proton energies; any
particular event will appear as one point on this diagram. The
outermost ring gives the locus of proton energies for coplanar
events. The gamma-ray polar angle goes from 0° to 180° as shown.
All noncoplanar events occur inside this ring. The locus for proton
noncoplanarity of 3°, 6°, and 9° are shown by the three full rings
within the outermost one. The dotted rings within these are for
10° and 11°. The dot near the center is for 11.86°, the maximum
angle of noncoplanarity. The important thing to note is that for
noncoplanar events, kinematics restricts the range of possible
θ_γ - there is a forbidden region near 0° and 180°.

Experiments that accept appreciable numbers of noncoplanar
events can thus not be compared directly with theoretical coplanar
distributions. To illustrate this point, we have in Figure 6 the
θ_γ distribution for 30°, 30° at 65 MeV (Ref. 16). The solid histogram
shows the experimental results with a few typical error bars. The
full curve is the coplanar calculation for 62 MeV by Drechsel and
Maximon.[9] The curve has been normalized to the same integral as
the experimental results. The shape is more or less suggestive
of the data, but there are a number of differences in detail. The
theory is much too high near 0° and 180°. Also, the back-angle
peak is higher than the forward peak, in contradiction with the
experiment, and the peaks occur at different angles.

We were able to obtain from Dieter Drechsel a complete set
of predictions for various angles of noncoplanarity. We then
weighted these with the relative efficiency in our geometry for
detecting various noncoplanar angles and folded them together.
The results are shown by the dashed histogram, again normalized
to the same area as the experiment. Notice that the fit is excellent.
It correctly shows the features such as the low yield near 0° and
180° and the lower peak height at back angles, and it shows better
agreement with the peak positions.

Unfortunately, theorists rarely publish their noncoplanar
θ_γ distributions, so it is usually impossible for the experimenter
to construct such a dashed histogram for his data. On the other
hand experimenters never publish their detection efficiency as
a function of noncoplanarity, which makes it impossible for theorists
who want to do the necessary folding-in for a particular geometry.
It seems that everyone is satisfied with crude comparisons of noncoplanar
data with coplanar predictions. For the original Harvard data,[4]
the comparison is almost acceptable because the noncoplanar acceptance
was not large, but this is not generally true. Perhaps this is

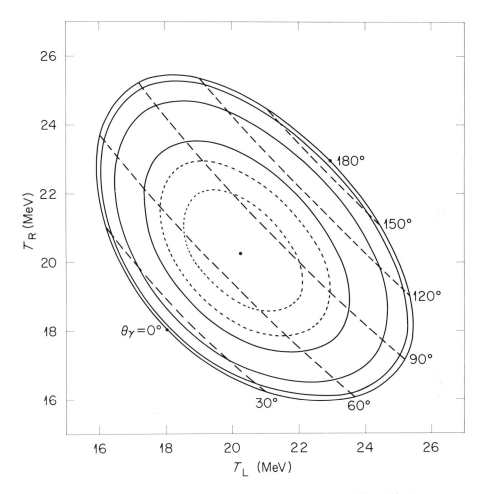

Fig. 5.   Kinematics for ppγ at 61.7 MeV in the 30°, 30° Harvard
          geometry.

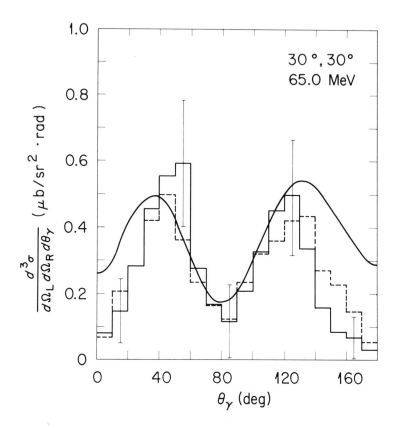

Fig. 6.   Gamma-ray angular distribution for 30°, 30° ppγ at 65.0 MeV.
The full histogram represents the experimental data.[16]
The experiment accepted noncoplanar events. The full line
is a (normalized) potential-model calculation[9] for coplanar
events. The dashed histogram is the result of folding
together predicted θ_γ distributions obtained from the same
model for various angles of noncoplanarity, weighted
according to the geometrical acceptance in the experiment.

merely a comment that no one expects to learn very much from the
θ_γ distributions - the shapes, at least for symmetric $pp\gamma$ geometry,
seem to be insensitive to the choice of model. The various
corrections mentioned earlier influence the predicted shapes mainly
near $\theta_\gamma = 0°$ and $180°$, where the data are probably the least reliable.

Some data are available for unequal proton angles,[13,16] and
here there appear to be differences between models. Figure 7 shows
the measured θ_γ distribution at 65 MeV for $25°$, $35°$. The left
half of the histogram refers to emission of the gamma ray on the
left side of the beam, where the 25° telescope was. Two coplanar
predictions are shown, the full curve from the OBE model[10] and
the dashed curve from McGuire and Pearce[32] who used a one-pion-
exchange off-shell parameterization of the quasiphases. Although
the data include appreciable noncoplanar contributions and the
warning I just gave will apply here, it nevertheless seems clear
that the OBE curve is a better representation of the data.

3. NEUTRON–PROTON BREMSSTRAHLUNG (npγ)

A. Experimental Results

These experiments are even more difficult than $pp\gamma$ and only
a few have been carried out. Each one used a different geometry.
Table II gives a summary of published results thus far. New results
from Harwell near 130 MeV will soon be forthcoming.[36] This experiment
covered n and p angles of $\pm 23°$, $\pm 26°$, $\pm 29°$, $\pm 38°$ and $\pm 20°$,
$\pm 32°$, respectively. It included coplanar and noncoplanar measurements.

The results from Ref. 35, measured in the Harvard geometry,
are given without correction for noncoplanar acceptance. There
is an indication from theory[10] that the cross section increases
as one goes out of the plane, in contrast with the $pp\gamma$ situation.

Two of the experiments in Table II were done with protons
on deuterium. In these cases there are six processes that can give
gamma rays,[7] and the npγ result for free neutrons and protons can
be obtained only after auxiliary calculations based on some model.
The two results are incompatible, even allowing for the difference
in beam energy.[7]

The agreement between the 197- and 208-MeV experiments is
mixed. The Rochester group[7] concludes that for $E_\gamma > 40$ MeV,
$\sigma_{np\gamma}/\sigma_{pp\gamma} = 50 \pm 20$, but the Davis group finds a ratio of about 3
in their more restricted range of angles.[35,37] On the other hand,
by using a theoretical ratio for the angle-integrated cross section
relative to the differential cross section, the Davis group

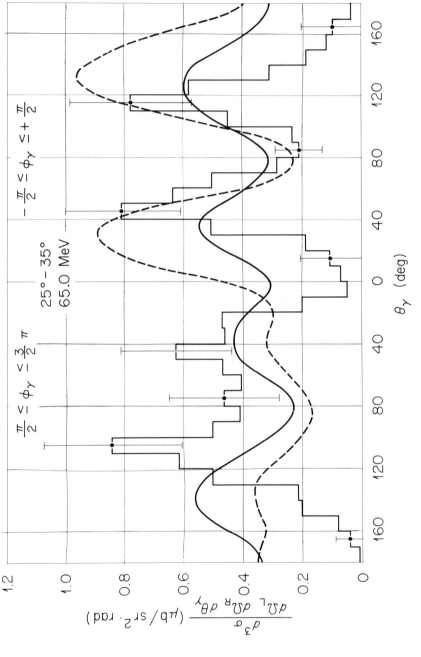

Fig. 7. Gamma-ray angular distributions for $\theta_L = 25°$, $\theta_R = 35°$ at 65.0 MeV. The histogram is the experimental result from Ref. 16. The experiment accepted noncoplanar events. The full curve is a coplanar OBE prediction[10] while the dashed curve is a coplanar prediction for a one-pion-exchange extrapolation of the quasiphases.[32]

Table II. Experimental data for npγ.

Beam	Energy (MeV)	Institution	Target	Observed particle(s)					Cross section	Ref.
				1	θ_1	2	θ_2	3		
n	14	U.C.L.A.	Plastic	n	30°	p	all	–	$<170\mu b/sr^2$	33
p[b]	140	Harwell	D_2O-H_2O	γ	30°–126°	–	–	–	$\sim 8\mu b$[a]	34
p[b]	197	Rochester	D_2	p	all	γ	all	p or d	$35\pm 12\mu b$	7
p[b]	197	Rochester	D_2	p	all	γ	60°	p or d	$3.4^a\pm 1.0\mu b/sr$	7
p[b]	197	Rochester	D_2	p	all	γ	108°	p or d	$2.5^a\pm 0.8\mu b/sr$	7
p[b]	197	Rochester	D_2	p	all	γ	147°	p or d	$1.8^a\pm 0.5\mu b/sr$	7
n	208±23	U.C.,Davis	H_2	n	30°	p	30°	–	$35\pm 14\mu b/sr^2$	35
n	208±23	U.C.,Davis	H_2	n	35°	p	35°	–	$57\pm 13\mu b/sr^2$	35
n	208±23	U.C.,Davis	H_2	n	38°	p	38°	–	$116\pm 20\mu b/sr^2$	35
n	208±23	U.C.,Davis	H_2	n	40°	p	30°	–	$114\pm 44\mu b/sr^2$	35
n	208±23	U.C.,Davis	H_2	n	45°	p	30°	–	$132\pm 53\mu b/sr^2$	35

a. For Eγ>40 MeV.

b. Proton beam 90% polarized.

extrapolated from their measurement a result[37] in agreement with
that from Rochester.[7]

B. Comparison of Theory and Experiment

(1). Cross sections integrated over θ_γ. Figure 8 shows
the current situation for the symmetric Davis results.[35] The
agreement with the coplanar OBE predictions[10] is satisfactory
for the symmetric cases. Not shown in Fig. 8 are the two results
for nonsymmetric geometry. The theoretical OBE values quoted
in Ref. 35 are a factor of two below the experimental results;
perhaps some of the discrepancy will disappear when enough information
is available to correct the experimental results for noncoplanar
acceptance. The symmetric Davis results[35] are also in satisfactory
agreement with the potential-model calculations by V. R. Brown,[38]
and even closer to the revised, gauge-invariant calculations
including exchange effects.[39]

It is interesting to note that for npγ the effect of the
internal radiation is very large.[38] The PGD point in
Fig. 8 is the result of an early calculation without rescattering.[40]
The calculations by McGuire[41] and McGuire and Pearce[42] are based
on a simple off-shell extrapolation of the quasiphases given
by one-pion exchange. Their results agree well with the 30° datum,
but are somewhat below the 35° and 38° data. (The 30° and 38°
data shown in Fig. 1 of Ref. 42 have been superseded; the correct
values are given in Ref. 35.)

(2). Cross sections differential in θ_γ. The Rochester
results are in reasonable agreement with the OBE calculations.[10]

4.   CONCLUSIONS AND SPECULATIONS

Viewed in a broad sense, theory and experiment for nucleon-
nucleon bremsstrahlung are in satisfactory agreement. In detail,
there are discrepancies, but the data now available do not permit
us to select one model of the nucleon-nucleon force over the
others.

On the assumption that the proposed thoroughgoing Manitoba
experiments are successful, one wonders whether further experiments
on ppγ are worthwhile. It seems to me that additional experiments
near 30°, 30° are probably not. But radical departures from
well-trodden paths might be quite important. Heller and Rich[30]
have found that at 100 MeV, significant model dependence may
be expected in the symmetric Harvard geometry for angles ∿15°,
and some similar comments have been made by McGuire and Pearce[32]
regarding some exploratory calculations for 5°. I know of no

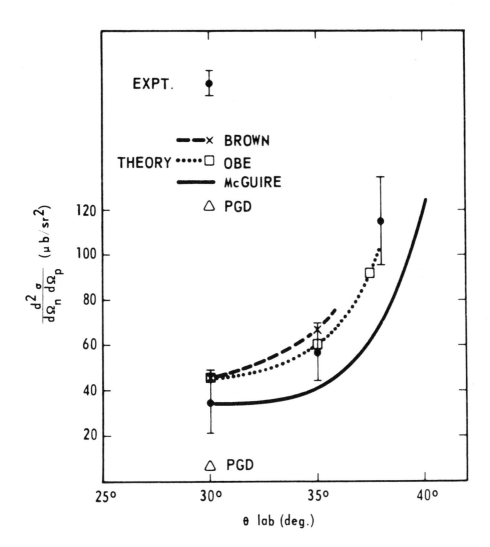

Fig. 8.   Comparison of symmetric npγ measurements with various
          calculations. The experimental data (and this figure)
          were taken from Ref. 35. The references for the
          theoretical curves are as follows: Brown, Ref. 38;
          OBE, Ref. 10; McGuire, Ref. 41; PGD, Ref. 40.

experimenters seriously considering such things at present. They promise
to be orders of magnitude more difficult than the already difficult
conventional experiments. It is my impression that firm statements by
theorists could stimulate experimenters to try some of these difficult
things.

With regard to npγ, the meeting between theory and experiment
has been amiable thus far. Real confrontation for integrated cross
sections cannot occur unless more accurate data become available. In
this respect, npγ offers rich diggings for the experimenter. More-
ever, it was recently pointed out by Brady[43] that the predicted
θ_γ distributions for coplanar symmetric npγ at 200 MeV given
in Ref. 10 and Ref. 38 are quite different. (The latest potential-
model calculations[39] give essentially the same θ_γ distribution
as the older ones.[38]) Thus there is promise that if the Harvard-
geometry θ_γ distribution were to be measured even crudely, real
progress could be made.

Finally, I should mention an interesting result which lies
outside the strict limitation of my title. I refer to an experiment
on proton-alpha bremsstrahlung published very recently.[44] The
interesting feature here is that rescattering plays
a very important role. In this case it appears to be connected
with the 2.9-MeV p-wave resonance in ^{5}Li.

REFERENCES

1. B. Gottschalk, W. J. Shlaer, and K. H. Wang, Phys. Lett.
    16, 294 (1965).

2. D. Marker and P. S. Signell, Phys. Rev. 185, 1286 (1969).

3. M. I. Sobel and A. H. Cromer, Phys. Rev. 132, 2698 (1963).

4. B. Gottschalk, W. J. Shlaer, and K. H. Wang, Nuc. Phys. 75,
    549 (1966).

5. B. Gottschalk, W. J. Shlaer, and K. H. Wang, Nuc. Phys. A94,
    491 (1967).

6. K. W. Rothe, P. F. M. Koehler, and E. H. Thorndike, Phys.
    Rev. 157, 1247 (1967).

7. P. F. M. Koehler, K. W. Rothe, and E. H. Thorndike, Phys.
    Rev. Letters 18, 933 (1967).

8. A. Willis, N. Osenda-Willis, R. Frascaria, N. Marty, and
    M. Morlet, 1970 Annual Report of I.P.N., Orsay (unpublished).

9. D. Drechsel and L. C. Maximon, Ann. Phys. (N.Y.) $\underline{49}$, 403 (1968).

10. R. Baier, H. Kühnelt, and P. Urban, Nucl. Phys. $\underline{B11}$, 675 (1969).

11. M. K. Liou and M. I. Sobel, to be published.

12. F. Sannes, J. Trischuk, and D. G. Stairs, Nuc. Phys. $\underline{A146}$, 438 (1970); Phys. Rev. Letters $\underline{21}$, 1474 (1968).

13. J. V. Jovanovich, L. G. Greeniaus, J. McKeown, T. W. Millar, D. G. Peterson, W. F. Prickett, K. F. Suen, and J. C. Thompson, Phys. Rev. Letters $\underline{26}$, 277 (1971).

14. M. L. Halbert, D. L. Mason, and L. C. Northcliffe, Phys. Rev. $\underline{168}$. 1130 (1968).

15. D. L. Mason, M. L. Halbert, and L. C. Northcliffe, Phys. Rev. $\underline{176}$, 1159 (1968).

16. D. L. Mason, M. L. Halbert, A. van der Woude, and L. C. Northcliffe, Phys. Rev. $\underline{179}$, 940 (1969).

17. D. O. Galde, M. L. Halbert, C. A. Ludemann, and A. van der Woude, Phys. Rev. Letters $\underline{25}$, 1581 (1970).

18. D. W. Storm and R. Heffner, to be published.

19. A. Bahnsen and R. L. Burman, Phys. Letters $\underline{26B}$, 585 (1968).

20. E. A. Silverstein and K. G. Kibler, Phys. Rev. Letters $\underline{21}$, 922 (1968).

21. A. Niiler, C. Joseph, V. Valkovic, R. Spiger, T. Canada, S. T. Emerson, J. Sandler, and G. D. Phillips, Phys. Rev. $\underline{178}$, 1621 (1969).

22. G. M. Crawley, D. L. Powell, and B. V. Narasimha Rao, Phys. Letters $\underline{26B}$, 576 (1968).

23. I. Slaus, J. W. Verba, J. R. Richardson, R. F. Carlson, W. T. H. van Oers, and L. S. August, Phys. Rev. Letters $\underline{17}$, 536 (1966).

24. R. E. Warner, Can. J. Phys. $\underline{44}$, 1225 (1966).

25. J. C. Thompson, S. I. H. Naqvi, and R. E. Warner, Phys. Rev. $\underline{156}$, 1156 (1967).

26. D. Drechsel, L. C. Maximon, and R. E. Warner, Phys. Rev. $\underline{181}$, 1720 (1969).

27.   M. K. Liou and K. S. Cho, Nuc. Phys. A160, 417 (1971).

28.   J. Sanada, M. Yamanouchi, Y. Tagishi, Y. Nojiri, K. Kondo,
      S. Kobayashi, K. Nagamine, N. Ryu, H. Hasai, M. Nishi, M. Seki,
      and D. C. Worth, Prog. Theor. Phys. 39, 853 (1968).

29.   V. R. Brown, Phys. Rev. 177, 1498 (1969) and private communication.
      The Bryan-Scott results in this reference have been superseded.

30.   L. Heller and M. Rich, Bull. A.P.S. 16, 559 (1971).

31.   J. H. McGuire, Phys. Letters 32B, 73 (1970).

32.   J. H. McGuire and W. A. Pearce, Nuc. Phys. A162, 561 (1971) and
      private communication.

33.   J. W. Verba, I. Slaus, J. R. Richardson, L. S. August,
      W. T. H. van Oers, and R. F. Carlson, in International
      Nuclear Physics Conference, Gatlinburg, 1966, ed. by Becker,
      Goodman, Stelson, and Zucker (Academic Press, N.Y., 1967),
      p. 619.

34.   J. A. Edgington and B. Rose, Nuc. Phys. 89, 523 (1966).

35.   F. P. Brady and J. C. Young, Phys. Rev. C 2, 1579 (1970).

36.   J. A. Edgington, V. J. Howard, S. S. das Gupta, F. P. Brady,
      I. M. Blair, C. A. Baker, and B. E. Bonner, Symposium on the
      Nuclear Three-Body Problem and Related Topics, Budapest,
      July 8-11, 1971, paper B-1.

37.   F. P. Brady, J. C. Young, and C. Badrinathan, Phys. Rev.
      Letters 20, 750 (1968).

38.   V. R. Brown, Phys. Letters 32B, 259 (1970).

39.   V. R. Brown and J. Franklin, Bull. A.P.S. 16, 560 (1971), and
      this Symposium .

40.   W. A. Pearce, W. A. Gale, and I. M. Duck, Nuc. Phys. B3,
      241 (1967).

41.   J. H. McGuire, Phys. Rev. C 1, 371 (1970).

42.   J. H. McGuire and W. A. Pearce, Nuc. Phys. A162, 573 (1971).

43.   F. P. Brady, in preparation.

44.   W. Wölfli, J. Hall, and R. Müller, Phys. Rev. Letters 27,
      271 (1971).

COMMENTS

Leon Heller: The relativistic spin correction of Liou and Cho which Halbert referred to should be used with some caution. They have omitted the term $-(\vec{p} - e\,\vec{A}/c)^4/8M^3$ coming from the expansion of $[(\vec{p} - e\,\vec{A}/c)^2 + M^2]^{\frac{1}{2}}$ even though it is the same order of magnitude as the spin orbit term which they have included. Furthermore the magnitude of the effect of the spin orbit term which they calculate is surprisingly large. For example at 100 MeV, with the protons at $25°$ to the beam and the photon going forward, inclusion of the spin orbit term reduces their cross section by a factor of more than 2. At backward photon angles their effect goes the other way and there is cancellation in the integrated cross section.

SOME BASIC QUESTIONS IN NUCLEON-NUCLEON BREMSSTRAHLUNG[*]

Leon Heller

Los Alamos Scientific Laboratory, University of California

Los Alamos, New Mexico 87544

The problem of the correct form of the electromagnetic current operator is considered when the Schrödinger equation has a nonlocal potential. By looking at the mesonic origin of exchange potentials, and assuming the mesons are 'minimally' coupled to the electromagnetic field, it is concluded that the minimal prescription for the nonlocal potential is generally wrong. Doing nothing about the momentum dependence of the potential corresponds to choosing a 'maximal' current; this violates the soft photon theorem but is gauge invariant. The correct current to associate with the exchange of a charged scalar meson is written down, and it is intermediate between the minimal and maximal choices. The implications for nucleon-nucleon bremsstrahlung calculations are discussed.

I. INTRODUCTION

Nucleon-nucleon bremsstrahlung has been considered a tool for distinguishing between different potentials which fit the elastic scattering data equally well. Implied is the assumption that the electromagnetic charge and current densities are known operators, so that when the electromagnetic interaction is treated as a perturbation in calculating the bremsstrahlung matrix element

$$\delta(\vec{P}_i - \vec{P}_f - \vec{k})(\varepsilon_0 M_0 - \vec{\varepsilon}\cdot\vec{M}) = \langle \Psi_f^{(-)} | \int d^3x e^{-i\vec{k}\cdot\vec{x}} [\varepsilon_0 \rho(\vec{x}) - \vec{\varepsilon}\cdot\vec{j}(\vec{x})] | \Psi_i^{(+)} \rangle \tag{1}$$

the only unknowns are the strongly interacting states Ψ. $\vec{P}_i$ and $\vec{P}_f$ are the total initial and final momenta of the particles, $\vec{k}$ is the photon momentum and $(\varepsilon_0, \vec{\varepsilon})$ is the photon polarization. In all the calculations of nucleon-nucleon bremsstrahlung which treat the strong interaction via a potential in the Schrödinger equation, the Schrödinger current due to point nucleons with charge and magnetic moment has been used, although this current is not conserved for the realistic potentials because they have nonlocal terms. In the course of our calculations using a variety of potentials we have reexamined this question and have learned some interesting things about the electromagnetic current, including some results which are not generally thought to be possible.

The three theoretical guidelines which should be followed in choosing the electromagnetic current are (i) relativity, (ii) current conservation, and (iii) meson theory. While it might appear that there is no hope of satisfying Lorentz invariance in any calculation which uses the Schrödinger equation, it is known[1] that the operators of position, momentum, and spin for a system of n interacting particles can be used to generate representations of the Poincaré group, and relativistic corrections to the electromagnetic interaction have been written down.[2] We shall not be concerned with this question any further in this paper.

In discussing the conservation of the current, and for the rest of the paper, we will consider the simplest possible case consisting of two spinless particles only one of which is charged. This system is adequate to demonstrate the interesting questions, but must be augmented for the nucleon-nucleon problem. The Schrödinger prescription for the charge and current densities is

$$\langle \Psi_f | \rho_1(\vec{x}) | \Psi_i \rangle = e_1 \int d^3 r_2 \, \Psi_f^*(\vec{x}, \vec{r}_2) \Psi_i(\vec{x}, \vec{r}_2)$$

$$\langle \Psi_f | \vec{j}_1(\vec{x}) | \Psi_i \rangle = -\frac{ie_1}{2m_1} \int d^3 r_2 \, \Psi_f^*(\vec{x}, \vec{r}_2) \overset{\leftrightarrow}{\nabla}_x \Psi_i(\vec{x}, \vec{r}_2)$$

(2)

where $\phi \overset{\leftrightarrow}{\nabla} \chi \equiv \phi \nabla \chi - \chi \nabla \phi$. It is well known that if Ψ_i and Ψ_f are solutions of the Schrödinger equation $H\Psi = i\partial\Psi/\partial t$ where H is the sum of the kinetic energies H_0 plus a local potential, then the current is conserved. For any more general potential

$$\langle \vec{r}_1, \vec{r}_2 | v | \vec{r}_1', \vec{r}_2' \rangle = \delta(\vec{R} - \vec{R}')\langle \vec{r} | v | \vec{r}' \rangle$$

(3)

where $\langle \vec{r} | v | \vec{r}' \rangle$ is not simply[3] $f(r)\delta(\vec{r} - \vec{r}')$, some addition to Eq.(2) is required to insure conservation. We have introduced center of mass and relative coordinates $\vec{R} \equiv (m_1\vec{r}_1 + m_2\vec{r}_2)/M$, $\vec{r} \equiv \vec{r}_1 - \vec{r}_2$,

with identical definitions for the primed coordinates, and
$M \equiv m_1 + m_2$. The form of Eq. (3) is determined by translation and
Galilean invariance; rotational invariance requires that v be a
(3-dimensional) scalar. Hermiticity requires that $\langle \vec{r}' | \dot{v} | \vec{r} \rangle^* =$
$\langle \vec{r} | v | \vec{r}' \rangle$. Although time reversal invariance is not essential to
the discussion we shall assume it in which case $\langle \vec{r} | v | \vec{r}' \rangle$ must be
real and symmetric in $\vec{r}$ and $\vec{r}'$.

II. TREATMENT OF NONLOCAL POTENTIALS

We first look at modifications of the current which leave the
charge density ρ unaltered. The general solution of this problem
was given by Osborn and Foldy.[4] A current density $\vec{j}_2(\vec{x})$ (which
is a two body operator) is added on to $\vec{j}_1(\vec{x})$, and is defined by

$$\langle \Psi_f | \vec{j}_2(\vec{x}) | \Psi_i \rangle \equiv i \iiiint d^3r_1 d^3r_2 d^3r_1' d^3r_2' \Psi_f^*(\vec{r}_1, \vec{r}_2)\langle \vec{r}_1, \vec{r}_2 | V | \vec{r}_1', \vec{r}_2' \rangle$$

$$\times e_1 \vec{\xi}(\vec{x} - \vec{r}_1, \vec{x} - \vec{r}_1') \Psi_i(\vec{r}_1', \vec{r}_2') \tag{4}$$

[If particle 2 is also charged, there is an additional term
$e_2 \vec{\xi}(\vec{x} - \vec{r}_2, \vec{x} - \vec{r}_2')$ in the integrand.] The condition that

$$\text{div}_x \, \vec{\xi}(\vec{x} - \vec{r}_1, \vec{x} - \vec{r}_1') = \delta(\vec{x} - \vec{r}_1) - \delta(\vec{x} - \vec{r}_1') \tag{5}$$

guarantees that the total charge and current densities obtained by
adding the contributions from Eq's. (2) and (4) is conserved. Indeed one verifies directly that with any $\vec{\xi}$ satisfying Eq. (5)

$$\frac{\partial}{\partial t} \langle \Psi_f | \rho_1(\vec{x}) | \Psi_i \rangle + \nabla \cdot \langle \Psi_f | \vec{j}_1(\vec{x}) + \vec{j}_2(x) | \Psi_i \rangle = 0$$

provided Ψ_i and Ψ_f are solutions of the Schrödinger equation
$(H_0 + V)\Psi = i\partial\Psi/\partial t$. $\vec{\xi}$ is otherwise arbitrary except that Hermi-
ticity requires $\vec{\xi}^*(\vec{x} - \vec{r}_1, \vec{x} - \vec{r}_1') = - \vec{\xi}(\vec{x} - \vec{r}_1', \vec{x} - \vec{r}_1)$, and the additional requirement of time reversal invariance for the electromag-
netic interaction would make $\vec{\xi}$ real and antisymmetric under interchange of $\vec{r}_1$ and $\vec{r}_1'$. Note that if time reversal is assumed
then $\vec{j}_2(x)$ is identically zero for any local potential and one
is right back to the normal current.

Writing each wave function as a product of a plane wave for
the center of mass motion and a function ϕ of the relative
coordinate, gives

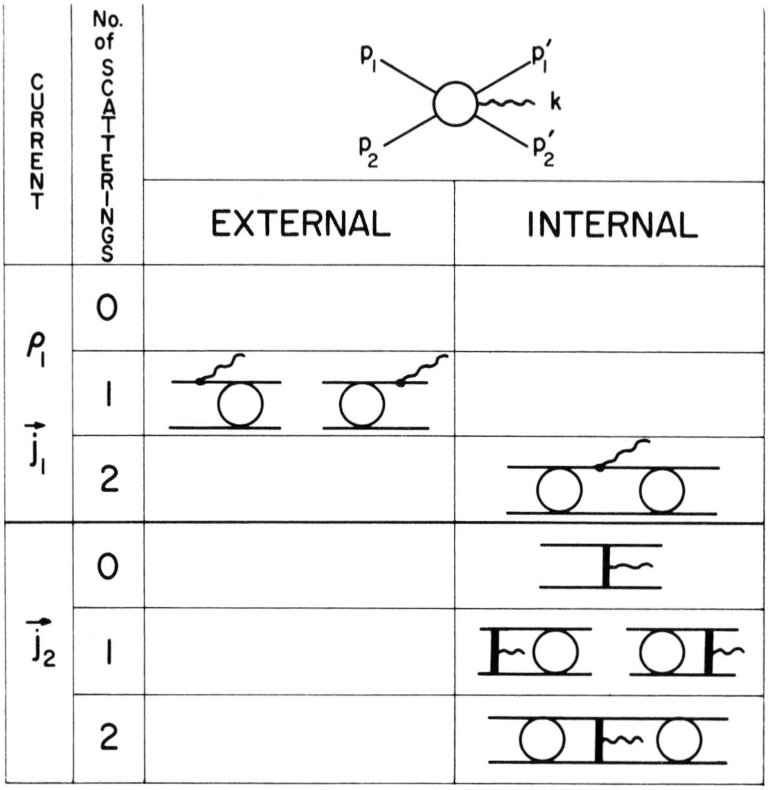

Fig. 1.   External and internal contributions to the bremsstrahlung
          matrix element coming from the ordinary current $\rho_1, \vec{j}_1$,
          and the extra current $\vec{j}_2$ required by the presence of a
          nonlocal potential. The circles represent scattered waves
          (or, equivalently, off-shell T-matrices). ρ_1 and $\vec{j}_1$ are
          represented by a photon attached to the charged nucleon.
          $\vec{j}_2$ is represented by a photon attached to a region of
          strong interaction.

$$M_0 = e_1 \int d^3 r e^{-i\frac{m_2}{M}\vec{k}\cdot\vec{r}} \phi_f^{(-)*}(\vec{r})\phi_i^{(+)}(\vec{r})$$

$$\tag{6}$$

$$\vec{M}_1 = \frac{e_1}{2m_1 i}\int d^3 r e^{-i\frac{m_2}{M}\vec{k}\cdot\vec{r}}\left[i\frac{m_1}{M}(\vec{P}_i+\vec{P}_f)\phi_f^{(-)*}(\vec{r})\phi_i^{(+)}(\vec{r}) + \phi_f^{(-)*}(\vec{r})\overset{\leftrightarrow}{\nabla}_r \phi_i^{(+)}(\vec{r}) \right]$$

$$\vec{M}_2 = ie_1 \int\int d^3 r d^3 r' \phi_f^{(-)*}(\vec{r})\langle\vec{r}|v|\vec{r}'\rangle\phi_i^{(+)}(\vec{r}')\int d^3 u e^{-i\vec{k}\cdot\vec{u}}\vec{\xi}(\vec{u}-\frac{m_2}{M}\vec{r},\vec{u}-\frac{m_2}{M}\vec{r}')$$

with $\vec{M} = \vec{M}_1 + \vec{M}_2$. The strong interaction and the electromagnetic radiation occur simultaneously in $\vec{M}_2$. If ϕ_i and ϕ_f are each written as a plane wave plus a scattered wave, then there are four distinct contributions to M_0 and $\vec{M}$. These are called zero scattering, single scattering (two of these), and double scattering, according to whether the number of scattered waves is zero, one, or two. Figure 1 shows the breakdown by type of current and number of scatterings, into external and internal emission.

What can be said about $\vec{\xi}(\vec{x}-\vec{r}_1,\vec{x}-\vec{r}_1')$? In the time reversal invariant case if a wire carrying unit current extends from $\vec{r}_1$ to $\vec{r}_1'$ and has a shape which is symmetric about the median plane, then it is an elementary solution of Eq. (5). That is, for any point $\vec{x}$ not on the wire $\vec{\xi} = 0$, and for any point on the wire $\vec{\xi}$ is a unit vector in the direction of the current multiplied by a two dimensional delta function in the orthogonal plane. (Axial symmetry requires that the curve be rotated about the segment from $\vec{r}_1$ to $\vec{r}_1'$ and that the current be uniformly distributed over the resulting surface.) The general solution of Eq. (5) consists of an arbitrary superposition of such surfaces with one unit of total current.

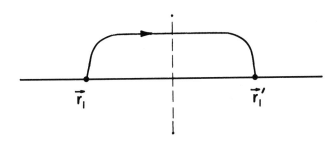

Fig. 2. An elementary solution of the equation
$\text{div}_x \vec{\xi}(\vec{x}-\vec{r}_1,\vec{x}-\vec{r}_1') = \delta(\vec{x}-\vec{r}_1) - \delta(\vec{x}-\vec{r}_1')$. The wire must be rotated about the line from $\vec{r}_1$ to $\vec{r}_1'$.

   It is straightforward to show that if the effect of an arbi-
trary nonlocal potential $\langle \vec{r}_1, \vec{r}_2 | V | \vec{r}_1', \vec{r}_2' \rangle$ is expressed as a super-
position of displacement operators,[5] and if each ∇_{r_1} is then re-
placed by $\nabla_{r_1} - ie_1 \vec{A}(\vec{r}_1)$, the resulting electromagnetic current is
precisely the one where $\vec{\xi}$ follows the straight line path from $\vec{r}_1$
to $\vec{r}_1'$. The proof is the same as that given by Sachs[6] for an exchange potential. This provides a very graphical description of
the 'minimal' electromagnetic current as the one which takes the
shortest path; see Fig. 3. Evaluating $\int d^3x e^{-ik \cdot x} \vec{\xi}(\vec{x}-\vec{r}_1, \vec{x}-\vec{r}_1')$
with $\vec{\xi}$ a unit vector between $\vec{r}_1$ and $\vec{r}_1'$ gives

$$i(\vec{r}_1' - \vec{r}_1) \frac{e^{-ik \cdot \vec{r}_1'} - e^{-ik \cdot \vec{r}_1}}{\vec{k} \cdot (\vec{r}_1' - \vec{r}_1)} \qquad \text{(MINIMAL)}$$

and in the limit $k \to 0$ this becomes $\vec{r}_1' - \vec{r}_1$.

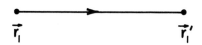

Fig. 3. The Minimal Current. For points on the straight line seg-
         ment connecting $\vec{r}_1$ to $\vec{r}_1'$, $\vec{\xi}$ is the unit vector
         $(\vec{r}_1' - \vec{r}_1)/|\vec{r}_1' - \vec{r}_1|$ multiplied by a two-dimensional δ-
         function in the orthogonal plane. For all other points
         $\vec{\xi} = 0$.

   It is now possible to define the 'maximal' electromagnetic
current. In reference 4, $\vec{\xi}$ was expressed as the sum of an ir-
rotational and a solenoidal part, $\vec{\xi} = \text{grad}_x \lambda + \text{curl}_x \vec{\chi}$ where $\vec{\chi}$
is completely arbitrary and λ must satisfy $\nabla_x^2 \lambda = \delta(\vec{x}-\vec{r}_1)$
$- \delta(\vec{x}-\vec{r}_1')$. The unique solution for λ (which vanishes at
infinity) is

$$\lambda(\vec{x}-\vec{r}_1, \vec{x}-\vec{r}_1') = -\frac{1}{4\pi} \left(\frac{1}{|\vec{x}-\vec{r}_1|} - \frac{1}{|\vec{x}-\vec{r}_1'|} \right)$$

with   (7)

$$\text{grad}_x \lambda = \frac{1}{4\pi} \left(\frac{\vec{x}-\vec{r}_1}{|\vec{x}-\vec{r}_1|^3} - \frac{\vec{x}-\vec{r}_1'}{|\vec{x}-\vec{r}_1'|^3} \right) .$$

We shall call the solution with $\vec{\chi} = 0$ the 'maximal' current, for
the following reason. If $\vec{\xi}$ fell off more slowly than $\text{grad}_x \lambda$
for large x, then its contribution to the bremsstrahlung matrix
element M_2 would become infinite as $k \to 0$. Such behavior is
generally thought to arise only from external emission diagrams

(which only contribute to $\vec{M}_1$). As it is, the x^{-3} falloff of $\text{grad}_x\lambda$ is so slow that $\int d^3x \exp(-i\vec{k}\cdot\vec{x}) \, \text{grad}_x\lambda$, while finite at $k = 0$, is not analytic there. That is, the value of the integral depends upon the direction of $\vec{k}$ as $k \to 0$. Indeed a direct evaluation of this integral yields

$$i \, \frac{\hat{k}}{k} \, (e^{-i\vec{k}\cdot\vec{r}_1{}'} - e^{-i\vec{k}\cdot\vec{r}_1}) \; , \; \hat{k} \equiv \frac{\vec{k}}{k} \qquad \text{(MAXIMAL)}$$

and in the limit $k \to 0$ this becomes $[(\vec{r}_1{}' - \vec{r}_1)\cdot\hat{k}]\hat{k}$. In the course of the derivation of the soft photon theorem for bremsstrahlung[7] it was assumed that the internal emission amplitude is analytic at $k = 0$. For any nonlocal potential, therefore, the choice $\vec{\xi} = \text{grad}_x\lambda$, or anything which falls off less rapidly will violate the theorem, and this will be shown explicitly in Section III, Example 2. The maximal current is shown on Fig. 4.

For any choice of $\vec{\xi}$ which falls off with some power of $1/x$ greater than 3, the limit of $\int d^3x \exp(-i\vec{k}\cdot\vec{x})\vec{\xi}$ as $\vec{k} \to 0$ is independent of the direction of $\vec{k}$ and is simply $\vec{r}_1{}' - \vec{r}_1$, in agreement with the 'minimal' value. <u>All such choices are consistent with the soft photon theorem.</u>

Since the bremsstrahlung matrix element, Eq. (1), involves only that part of $\vec{M}$ which is orthogonal to $\vec{k}$, and since $\vec{\xi} = \text{grad}_x\lambda$ results in a contribution to $\vec{M}$ which is in the $\vec{k}$ direction, <u>the maximal choice for the current does not alter the matrix element.</u> This is obvious in the Coulomb gauge, $\varepsilon^0 = \vec{\varepsilon}\cdot\vec{k} = 0$, but is independent of the choice of gauge. All the bremsstrahlung calculations which have not done anything about the spin-orbit and other nonlocal potentials can therefore be said to have chosen the maximal current.

III. FOUR EXAMPLES WITH NO ELASTIC SCATTERING

We now look at a series of problems which illustrate some of the arbitrariness in calculating bremsstrahlung if the only constraint is current conservation. These examples arose quite naturally in the course of studying different potentials which fit the elastic scattering data in precisely the same way. Case 2 is of the type discussed above in which the charge density is unaltered. In problems 3 and 4, ρ is also allowed to change. To bring the interesting physical questions into sharp focus, we start off with a world in which there is no elastic scattering.

EXAMPLE 1. One potential which certainly gives this answer is $V = 0$, and the corresponding states are solutions of the free particle Schrödinger equation $H_0\Psi = i\partial\Psi/\partial t$. If the normal

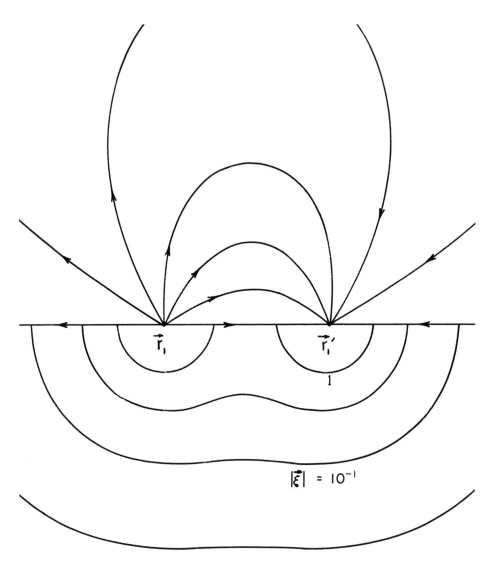

Fig. 4.   The Maximal Current, $\vec{\xi} = \nabla_x \lambda$, Eq. (7). The upper half of
the figure shows the direction of $\vec{\xi}$ along the curves
which emerge from $\vec{r}_1$ at angles of $0°$, $30°$, $60°$, $90°$,
$120°$, $150°$, and $180°$ to the symmetry axis. The lower
half of the figure shows contours of constant $|\vec{\xi}|$. The
separation of the source and sink is taken to be one
unit of length.

electromagnetic current operators $\vec{\rho}_1, \vec{j}_1$ from Eq. (2) are used, then the bremsstrahlung matrix element is a certain δ-function which is incompatible with energy conservation. This is the often made statement that a free particle (or two free particles) cannot radiate. This information is summarized in column 1 of Table I.

We now look for other potentials which produce the same elastic scattering and bound states as $V = 0$.

EXAMPLE 2. Choosing the new potential to be $\overline{V} = UH_0U^{-1}-H_0$ where U is any unitary transformation which approaches the iden- tity at large separations of the two particles, guarantees that $\overline{V}$ is Hermitean and that the phase shifts and binding energies are unaltered. The requirements of translation and Galilean invariance for the new potential are certainly met[8] by writing

$$U = I - 2\sum_k |\alpha_k\rangle\langle\alpha_k| \qquad (8)$$

TABLE I

	1	2	3	4
Potential	0	$UH_0U^{-1}-H_0$	$UH_0U^{-1}-H_0$	0
States	Ψ_0	$U\Psi_0$	$U\Psi_0$	Ψ_0
Charge and Current Densities	$\rho_1, \vec{j}_1$	$\rho_1, \vec{j}_1+\vec{j}_2$	$U\rho_1U^{-1}, U\vec{j}_1U^{-1}$	$U^{-1}\rho_1U, U^{-1}(\vec{j}_1+\vec{j}_2)U$
Brems- strahlung ?	No	Yes	No	Yes

Four Different Examples with No Elastic Scattering

1 and 3 have different potentials but exactly the same brems- strahlung matrix element because the states and the current opera- tor undergo the same unitary transformation. (The same is true for examples 2 and 4.) Examples 1 and 4 have the same potential but different bremsstrahlung matrix elements. (The same is true for 2 and 3.)

where it is understood that U is the identity insofar as the motion of the center of mass is concerned, and the summation is over any number of orthonormal states $|\alpha_k\rangle$ of the internal motion of the system. When the center of mass coordinate is eliminated using Eq. (3), we get

$$\langle \vec{r}|\overline{v}|\vec{r}'\rangle = -2 \sum_k \left(\langle \alpha_k|h_0|\vec{r}'\rangle \alpha_k(\vec{r}) + \langle \vec{r}|h_0|\alpha_k\rangle \alpha_k^*(\vec{r}') \right)$$

$$+ 4 \sum_{k,j} \langle \alpha_k|h_0|\alpha_j\rangle \alpha_k(\vec{r})\alpha_j^*(\vec{r}') \tag{9}$$

where $\alpha_k(r) \equiv \langle \vec{r}|\alpha_k\rangle$, and h_0 is the kinetic energy operator for the relative motion, $\langle \vec{r}|h_0|\alpha_k\rangle = -M\nabla^2\alpha_k(\vec{r})/2m_1 m_2$. $\overline{v}$ will be invariant under time reversal if all the $\alpha_k(r)$ are real functions. This potential is very nonlocal.

The solutions of the new Schrödinger equation $(h_0 + \overline{v})\overline{\phi} = i\partial\overline{\phi}/\partial t$ are given by $\overline{\phi} = U\phi$, so that a state which becomes a plane wave at large separations with momentum $\vec{p}$ for particle one in the center of mass system is just

$$|\overline{\phi}_{\vec{p}}\rangle = U|\vec{p}\rangle = |\vec{p}\rangle - 2 \sum_k \langle \alpha_k|\vec{p}\rangle|\alpha_k\rangle \quad . \tag{10}$$

The second term is normally referred to as the scattered wave, but we take $\alpha_k(r)$ to fall off rapidly at infinity so the phase shifts are all zero.

The half off-shell T-matrix can be written down and is just

$$\langle \vec{p}'|\overline{t}(p)|\vec{p}\rangle = \langle \vec{p}'|\overline{v}|\overline{\phi}_{\vec{p}}\rangle = \frac{M}{m_1 m_2}(p'^2 - p^2) \sum_k \langle \vec{p}'|\alpha_k\rangle\langle \alpha_k|\vec{p}\rangle$$

and $\tag{11}$

$$\langle \vec{p}'|\overline{t}(p')|\vec{p}\rangle = \langle \overline{\phi}_{\vec{p}'}|\overline{v}|\vec{p}\rangle = -\langle \vec{p}'|\overline{t}(p)|\vec{p}\rangle$$

and clearly vanishes on-shell.

To calculate the bremsstrahlung produced by the potential $\overline{v}$, it is necessary to evaluate M_0 and $\vec{M}$ according to Eq. (6). Different choices for $\vec{\xi}$ will result in different values for $\vec{M}_2$. In the limit $k \to 0$ the soft photon theorem[7] says there is no bremsstrahlung since only the on-shell T-matrix and its energy

derivative enter the calculation, and these are identically zero.
We now show that if $\vec{\xi}$ is chosen to be $\mathrm{grad}_x\lambda$, the bremsstrahlung
matrix element does <u>not</u> vanish as $k \to 0$; but if $\vec{\xi}$ falls off more
rapidly the theorem <u>is</u> satisfied. The momenta of the particles are
shown on Fig. 1, and the conservation equations are $p_1 + p_2 = p_1'$
$+ p_2' + k$. The initial momentum of particle one in the initial
center of mass system is $\vec{p}_i = (m_2\vec{p}_1 - m_1\vec{p}_2)/M$, and its final momentum in the final center of mass system is $\vec{p}_f = (m_2\vec{p}_1' - m_1\vec{p}_2')M$.
Using Eq. (10) we can break down the contributions to Eq. (6) ac-
cording to external and internal emission (See Fig. 1). External
emission arises from the ordinary current $\rho_1, \vec{j}_1$ with one plane
wave and one scattered wave.

$$M_0^{ext} = -2e_1 \sum_k \left[\langle \alpha_k | \vec{p}_i \rangle \langle \vec{p}_f + \frac{m_2}{M} \vec{k} | \alpha_k \rangle + \langle \vec{p}_f | \alpha_k \rangle \langle \alpha_k | \vec{p}_i - \frac{m_2}{M} \vec{k} \rangle \right]$$

$$\tag{12}$$

$$\vec{M}_1^{ext} = -\frac{2e_1}{m_1} \sum_k \left[\langle \alpha_k | \vec{p}_i \rangle \langle \vec{p}_f + \frac{m_2}{M} \vec{k} | \alpha_k \rangle (\vec{p}_1' + \tfrac{1}{2}\vec{k}) \right.$$

$$\left. + \langle \vec{p}_f | \alpha_k \rangle \langle \alpha_k | \vec{p}_i - \frac{m_2}{M} \vec{k} \rangle (\vec{p}_1 - \tfrac{1}{2}\vec{k}) \right] \quad .$$

The internal emission receives contributions from the ordinary current with two scattered waves, and also from $\vec{j}_2$ with all four
combinations of initial and final plane and scattered waves.

$$M_0^{int} = 4e_1 \sum_{k,j} \langle \vec{p}_f | \alpha_j \rangle \langle \alpha_k | \vec{p}_i \rangle \int d^3 r \alpha_j^*(\vec{r}) \alpha_k(\vec{r}) e^{-i\frac{m_2}{M}\vec{k}\cdot\vec{r}}$$

$$M_1^{int} = \frac{2e_1}{m_1} \sum_{k,j} \langle \vec{p}_f | \alpha_j \rangle \langle \alpha_k | \vec{p}_i \rangle \left[\frac{m_1}{M}(\vec{p}_i + \vec{p}_f) \, d^3 r \alpha_j^*(\vec{r}) \alpha_k(r) e^{-i\frac{m_2}{M}\vec{k}\cdot r} \right.$$

$$\tag{13}$$

$$\left. - i \int d^3 r e^{-i\frac{m_2}{M}\vec{k}\cdot\vec{r}} \left(\alpha_j^*(\vec{r}) \overset{\leftrightarrow}{\nabla} \alpha_k(r) \right) \right] \quad .$$

Before evaluating $\vec{M}_2^{int}$, note that as $k \to 0$,

$$M_0^{ext} + M_0^{int} = \mathcal{O}(k)$$

$$\vec{M}_1^{ext} + \vec{M}_1^{int} = \vec{N} + \mathcal{O}(k)$$

where   (14)

$$\vec{N} = - \frac{2e_1}{m_1} \sum_k \langle \vec{p}_f | \alpha_k \rangle \langle \alpha_k | \vec{p}_i \rangle (\vec{p}_i + \vec{p}_f)$$

$$- \frac{2ie_1}{m_1} \sum_{k,j} \langle \vec{p}_f | \alpha_j \rangle \langle \alpha_k | \vec{p}_i \rangle \int d^3 r \, \alpha_j^*(\vec{r}) \vec{\nabla} \alpha_k(\vec{r})$$

It was shown above that in the limit $k \to 0$,

$$\vec{F} \equiv \int d^3 x \, e^{-ik \cdot x} \vec{\xi}(\vec{x} - \frac{m_2}{M} \vec{r}, \quad \vec{x} - \frac{m_2}{M} \vec{r}')$$

has two possible values: for the maximal current where $\vec{\xi} = \mathrm{grad}_x \lambda$, $\vec{F} = (m_2 \hat{k}/M)[\hat{k} \cdot (\vec{r}' - \vec{r})]$; and for any $\vec{\xi}$ which falls off more rapidly for large x, $\vec{F} = (m_2/M)(\vec{r}' - \vec{r})$. Putting these two possibilities into Eq. (6), we obtain for the maximal case

$$\vec{M}_2 = \vec{M}_2^{int} = - (\vec{N} \cdot \hat{k})\hat{k} + \mathcal{O}(k)$$

and for the other cases

$$\vec{M}_2 = \vec{M}_2^{int} = - \vec{N} + \mathcal{O}(k)$$

where $\vec{N}$ is defined in Eq. (14). Adding $\vec{M}_1$ and $\vec{M}_2$ it is seen that all choices of $\vec{\xi}$ which fall off faster than x^{-3} produce no bremsstrahlung as $k \to 0$, in agreement with the soft photon theorem (since this potential produces no elastic scattering); but if $\vec{\xi} = \mathrm{grad}_x \lambda$,

$$\vec{M}_1 + \vec{M}_2 = \vec{N} - (\vec{N} \cdot \hat{k})\hat{k} + \mathcal{O}(k) \quad ,$$

and this non-analytic result, although gauge invariant, violates the theorem.

EXAMPLE 3. Up to this point the unitary transformation U was just a convenient device for generating a potential $\overline{V}$ which produces the same elastic scattering as a potential V. Since $\overline{V}$ is nonlocal it was necessary to modify the electromagnetic current, and we did this by adding $\vec{j}_2$ on to the ordinary current. It was pointed out by Foldy[9] that there is another way to generate a conserved current for use with the potential $\overline{V}$, and that is by

applying the unitary transformation U to the current as well as
to the states. This obviously results in exactly the same bremsstrahlung matrix element as the original current gave with the original states (Example 1). It is clear, furthermore, that the current obtained in this way is distinct from that in Example 2, no
matter what choice for $\vec{\xi}$ is made there, since the charge density
is now modified as well as the current density. These new densities are two-body operators.

EXAMPLE 4. For the potential $\overline{V}$ we now have not only the
arbitrariness in the choice of $\vec{\xi}$ (Example 2), but the possibility
of an entirely different current as well (Example 3). This is summarized in Table I. By applying the inverse unitary transformation[10] U^{-1} to the states and the current in Example 2, we come
back to the original potential V = 0, but with a new current and a
non-vanishing bremsstrahlung matrix element! [It is precisely the
same matrix element as in Example 2. M_0 and M_1 are given in
Eqs. (12) and (13).] These free particles are able to radiate because the charge and current distributions in space are altered as
they pass each other.

By exploiting the freedom allowed by current conservation we
have reached the unsatisfactory situation in which it is possible to
have two different potentials which produce the same elastic scattering and the same bremsstrahlung. This is true of Examples 1 and
3 which have no bremsstrahlung, and 2 and 4 which do, with the latter pair having the additional freedom of the choice of $\vec{\xi}$. Examples 2 and 3, on the other hand, have the same potential but different bremsstrahlung, and the same is true for 1 and 4. Example 4
is especially painful because the potential is zero and still there
is bremsstrahlung.

These ambiguities are still present if there is elastic scattering; one starts with a non-zero V in Example 1 and goes through
the same set of transformations.

IV. CONNECTION WITH MESON THEORY

It is clear that some additional physical ingredient is needed
to decide on the proper current. Just as our present limited understanding of the elastic nucleon-nucleon problem has been achieved
by incorporating meson exchange effects explicitly, starting with
the longest range parts of the interaction, we expect that a similar procedure will have to be followed for nucleon-nucleon bremsstrahlung. We shall have to suppose that the electromagnetic interaction of each of the particles which enter into the nucleon-nucleon
interaction (including the nucleons themselves, mesons, N^*'s, etc.)
is known, possibly as given by the minimal rule. And just as one

and two meson exchange diagrams get converted into nucleon-nucleon potentials for use in some equation such as Schrödinger's, we propose that a photon be attached wherever it is permitted in these diagrams and then convert these matrix elements to electromagnetic interactions for use in the same equation. Such a procedure can include all the mesonic information that one boson exchange calculations[11] do, with the additional important property that it is unitary. At small separations of the nucleons it will be necessary to treat the current in some phenomenological way, just as one now uses a hard core or boundary condition or other parameterization of the potential.

In proton-proton scattering, starting at the largest distances comes the exchange of one neutral π-meson. The potential which this produces to order g^2 becomes local at distances which are large compared to the nucleon Compton wavelength. The corresponding diagrams in which a photon comes off one of the proton legs (See Fig. 5) are properly represented by the normal current $\rho_1, \vec{j}_1$ together with a magnetic moment term $\vec{j}_1{}'$.

In neutron-proton scattering, already at the largest distances there is a non-local potential due to the exchange of a single charged π-meson. We now consider the mathematically simpler problem of the exchange of a single charged <u>scalar</u> meson of mass μ and coupling constant g between spinless nucleons. (See Fig. 6.)

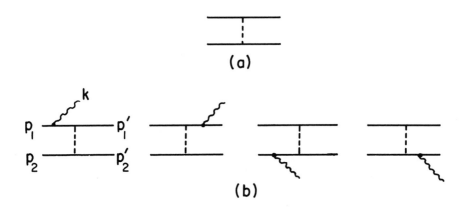

(a)

(b)

Fig. 5. Single neutral meson exchange between two protons.
(a) The diagram which creates the potential. The potential becomes local at large separations. (b) The diagrams which contribute to bremsstrahlung. The ordinary current is sufficient.

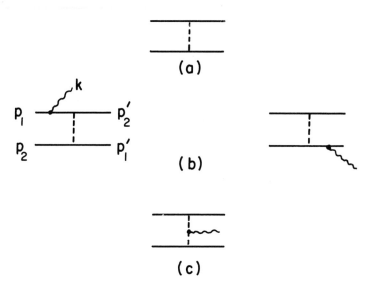

Fig. 6. Single charged meson exchange between a neutron and a
proton. (a) The diagram which creates the potential.
Even at large separations the potential is nonlocal (exchange). (b) and (c) The diagrams which contribute to
bremsstrahlung. The ordinary current in (b), and the
extra current in (c).

The potential which yields this matrix element to order g^2 is

$$\langle \vec{r}|v|\vec{r}\,'\rangle = -\frac{g^2}{4\pi}\,\frac{e^{-\mu r}}{r}\,\delta(\vec{r}+\vec{r}\,')\quad .\tag{15}$$

In the non-relativistic limit, the bremsstrahlung matrix elements
of order eg^2 where the photon is attached to the initial or final
proton (Fig. 6b) are correctly given by M_0 and $\vec{M}_1$, Eq. (6), in
other words by the normal current. We now show that the matrix
element of order eg^2 where the photon is coupled minimally to the
charged meson (Fig. 6c), which in the non-relativistic limit becomes

$$e_1 g^2\,\frac{\vec{p}_1 - \vec{p}_2\,' + \vec{p}_1' - \vec{p}_2}{[\,(\vec{p}_1-\vec{p}_2')^2 + \mu^2][\,(\vec{p}_1' - \vec{p}_2)^2 + \mu^2]}\quad ,\tag{16}$$

is correctly given by leaving the charge density unaltered and choosing $\vec{\xi}$ to be

$$\vec{\xi}(\vec{x} - \tfrac{1}{2}\vec{r}, \vec{x} + \tfrac{1}{2}\vec{r}) = \frac{re^{\mu r}}{4\pi}\left[\frac{e^{-\mu x_-}}{x_-} \overset{\leftrightarrow}{\nabla}_x \frac{e^{-\mu x_+}}{x_+}\right] \qquad , \qquad (17)$$

where we have used the equality of the mass of the nucleons (which is essential for an exchange potential), and $\vec{r}' = - \vec{r}$; $x_\pm \equiv |\vec{x} \pm \vec{r}/2|$. The first step of the proof consists of the straightforward verification that Eq. (17) does satisfy Eq. (5). The second step involves putting Eqs. (15) and (17) into Eq. (6) with plane waves for the initial and final states, $\phi_i(\vec{r}') = \exp(i\vec{p}_i\cdot\vec{r}')$, $\phi_f(\vec{r}) = \exp(i\vec{p}_f\cdot\vec{r})$, $\vec{p}_i \equiv (\vec{p}_1-\vec{p}_2)/2$, $\vec{p}_f \equiv (\vec{p}_1'-\vec{p}_2')/2$. If we have made the correct choice for $\vec{\xi}$, $\vec{M}_2$ will be equal to (16). The proof is trivial after each Yukawa factor in Eq. (17) is written as a Fourier transform.

We have now found the correct modification of the electromagnetic current when the nonlocal potential is due to the exchange of a charged scalar meson. This choice for $\vec{\xi}$, Eq. (17), is between the minimal and maximal cases, falling off exponentially for large x, and is shown on Figs. 7, 8 and 9. [Observe how increasing the mass of the meson brings the pattern closer to that of the minimal current.] This is a very interesting, but hardly surprising result. Even if the minimal coupling is correct for all the particles in a fundamental theory, by the time the meson coordinates have been eliminated, as in the Schrödinger equation, the minimal prescription is no longer correct.

For neutron-proton bremsstrahlung it will be necessary to repeat the above calculation with a pseudoscalar meson and find the appropriate $\vec{\xi}$. While it will undoubtedly differ from Eq. (17), the exponential falloff for large x will remain.

Continuing in to smaller distances in both the p-p and n-p problems, there are uncorrelated two-π-meson, vector meson and other exchanges. These give rise to a spin-orbit potential and other nonlocalities at smaller distances. In the proton-proton case, one might try to say that the exchange of a single neutral vector meson should not require any $\vec{M}_2$ since there are no new places for a photon to be attached. But it must be remembered that the spin-orbit potential is a relativistic term, being of order $(mr)^{-2}$ compared to the leading term, and consequently considerable care must be exercised when M_1 is compared with the diagrams in which the photon is attached to a proton. There could very well be a difference of order m^{-2}, and $\vec{M}_2$ will have to compensate for this.

In the p-p potential model calculations which Dr. Halbert referred to in his fine summary,[12] no extra terms were added for

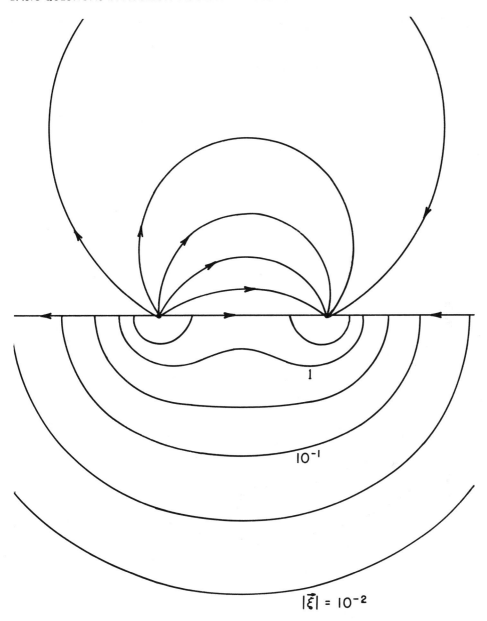

Fig. 7. The current due to the exchange of a charged scalar meson
of mass μ = 0, Eq. (17). See the caption to Fig. 4 for
a description of the two halves of the figure.

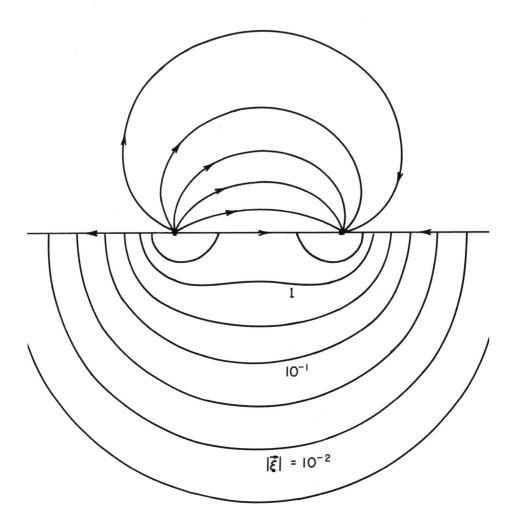

Fig. 8. The current due to the exchange of a charged scalar meson
        of mass μ = 1, Eq. (17). See the caption to Fig. 4 for
        a description of the two halves of the figure.

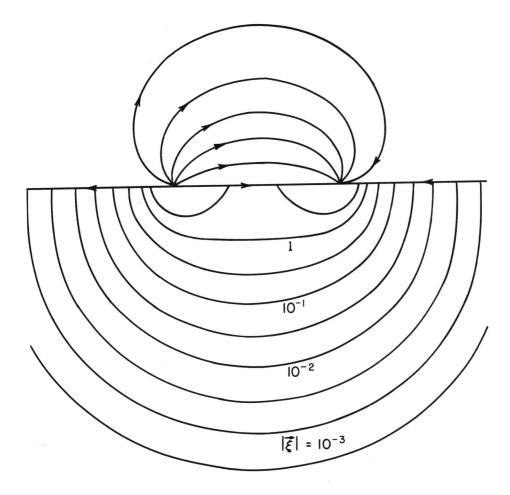

Fig. 9. The current due to the exchange of a charged scalar meson of mass μ = 2, Eq. (17). See the caption to Fig. 4 for a description of the two halves of the figure.

the spin-orbit potential. This is equivalent to choosing the maximal current, as we have shown. Whether or not this is justified
theoretically will depend upon a careful examination of the diagrams
which produce the spin-orbit potential. Short of this, one can take
a phenomenological spin-orbit potential (and quadratic spin-orbit),
try a variety of different currents and see how much difference
they make for the bremsstrahlung matrix element. This procedure is
not exhaustive, but can serve as a guide to the accuracy with which
phenomenological potentials can calculate bremsstrahlung. We plan
to do this.

V. SUMMARY

There are a variety of ways in which the Schrödinger equation
with a nonlocal potential can be made gauge invariant. In one set
the charge and current densities are both altered. In the other
set only the current density is modified following the procedure of
Osborn and Foldy.[4] There are still a continuum of possibilities
ranging from the minimal one in which each gradient operator is
replaced by $\nabla_r - ie \vec{A}(\vec{r})$, to the maximal one in which the bremsstrahlung matrix element is not altered at all! The maximal choice
violates the soft photon theorem[7] because the internal emission
amplitude is not analytic at $k = 0$, but gauge invariance is
satisfied.

By studying a simple Feynman diagram in which a charged scalar
meson is exchanged between (spinless) nucleons, we find the correct
way to modify the electromagnetic current to take account of this
nonlocal (exchange) potential. For non-relativistic nucleons the
charge density can be left unmodified, and the extra current density is intermediate between the minimal and maximal choices falling
off exponentially for large x. It appears very likely that any
nonlocality due to meson exchanges will have this property.

This type of calculation will have to be repeated for those
diagrams which are known to be important in nucleon-nucleon scattering, including the various single boson exchanges, and two-π-
meson exchange. Only by knowing the source of a given nonlocal
potential can one decide how to treat it in electromagnetic
processes.

ACKNOWLEDGMENTS

Throughout the development of these ideas I have benefited from regular discussions with Dr. M. Rich. Conversations with Professors H. A. Bethe, H. Feshbach, L. L. Foldy, and P. Signell helped to stimulate and clarify this work. Drs. A. M. Bolsterli, H. W. Fearing, W. M. Visscher, and G. G. Zipfel, Jr. contributed useful comments. I also want to thank Dr. M. L. Halbert for sending me an advance copy of his excellent talk. Mrs. M. Menzel kindly did the numerical calculations and made the graphs.

REFERENCES

* Work supported by the United States Atomic Energy Commission.

1. B. Bakamjian and L. H. Thomas, Phys. Rev. 92, 1300 (1953); L. L. Foldy, Phys. Rev. 122, 275 (1961).
2. H. Osborn, Phys. Rev. 176, 1514 (1968); R. A. Krajcik and L. L. Foldy, Phys. Rev. Letts. 24, 545 (1970).
3. If there are one or more gradients present we call that a nonlocal potential.
4. R. K. Osborn and L. L. Foldy, Phys. Rev. 79, 795 (1950).
5. This was done by J. A. Wheeler, Phys. Rev. 50, 643 (1936), for an exchange potential.
6. R. G. Sachs, Phys. Rev. 74, 433 (1948).
7. F. E. Low, Phys. Rev. 110, 974 (1958).
8. H. Ekstein, Phys. Rev. 117, 1590 (1960). This is not the most general unitary transformation; see G. A. Baker, Phys. Rev. 128, 1485 (1962), and F. Coester, S. Cohen, B. Day, and C. M. Vincent, Phys. Rev. C 1, 769 (1970) for others.
9. L. L. Foldy (private communication). I am indebted to Professor Foldy for this observation.
10. This suggestion was made by Dr. G. G. Zipfel, Jr. (private communication).
11. R. Baier, H. Kühnelt, and P. Urban, Nucl. Phys. B11, 675 (1969).
12. M. L. Halbert, "Review of Experiments on Nucleon–Nucleon Bremsstrahlung," invited paper for the Gull Lake Symposium on the Two-Body Force in Nuclei.

OFF-ENERGY-SHELL EFFECTS IN MANY-NUCLEON SYSTEMS

Frank Tabakin

Physics Department

University of Pittsburgh

INTRODUCTION

A multitude of potentials can reproduce the same two-nucleon elastic scattering data. For many of these potentials the two-body wave functions can be radically different and can alter dramatically the predictions of many-body nuclear calculations. The basic problem therefore is to find both theoretical and experimental restrictions on the possible forms of the two-nucleon wave function.

It is not easy to determine the short-range part of the two-nucleon wave function, especially once one has opened a Pandora's box filled with all possible local and nonlocal forms. Hope remains, however, that constraints can be found. Perhaps the three body problem, proton-proton bremsstrahlung, and/or increased knowledge of the meson origins of the nuclear interaction will provide the necessary constraints.

In this talk, I would like to report on recent many-body calculations which demonstrate the great need for improved knowledge of the short-range two-nucleon wave function. Three main points are to be emphasized:

1) OFF-SHELL FREEDOM. There is still great freedom in the choice of nucleon-nucleon interaction. Different off-shell behaviour and different choices of high energy phase shifts are the main origins of that freedom.

2) LARGE SYSTEMATIC CHANGES OCCUR IN MANY-BODY CALCULATIONS. Large changes in the nucleon-nucleon interaction generate large,

but systematic changes in results of low order Brueckner
calculations for finite nuclei and for nuclear matter.

3) ADDITIONAL CONSTRAINTS ARE NEEDED. In addition to the
elastic scattering data, an OPEP tail, and hermiticity, further
constraints on the nuclear interaction are needed to make many-
body calculations reliable.

Several authors have explicitly shown that great freedom
exists in the choice of interaction even if all phase shifts were
accurately known at all energies. It has been shown that sets of
interactions yielding the same phase shifts can be generated by
using the freedom permitted in the off-shell behaviour of the
interaction. In particular, some of the ideas contained in
references 1-9 concerning the off-shell freedom will be reviewed.
The related idea of generating sets of phase shift equivalent
potentials using unitary transformations will also be discussed.
Following this survey, I will present some recent results[22]
obtained by P. Sauer, M. Haftel, and E. Lambert who used families
of phase shift equivalent potentials (generated from the Reid
potential) in parallel Brueckner calculations of the binding
energy of O^{16} and of nuclear matter. Systematic rules relating
the binding of finite nuclei to equivalent nuclear matter are
revealed by these calculations. Verification of the usefulness
of the excitation parameter κ as a measure of correlations is
also found.

In these studies it is assumed that even when the basic
interaction is greatly altered, it is still reasonable to use
lowest order Brueckner theory. That assumption must of course
be tested further by evaluation of the contribution of three and
four-hole diagrams. Recently, Wong and Sawada have incorporated[9]
Dahlblom's and Day's estimates of three and four-hole diagrams in
a nuclear matter calculation with phase shift equivalent
potentials. Large changes in binding were produced by Wong's set
of potentials even when these higher order diagrams were included.
In the calculations that I will discuss, only low order Brueckner
theory is used.

Finally, I should like to present a viewpoint concerning
practical means of controlling and restricting the permissible
interactions by invoking not only theoretical and experimental
constraints but also the demands of simplicity.

OFF-SHELL FREEDOM

The freedom available in the choice of two-nucleon potential
is made clear by considering the full off-shell transition matrix
$(k'|T_\ell(\omega)|k)$. For an uncoupled partial wave in the center of
mass system, T is a function of three variables, i.e. k,k' and

the energy ω. Elastic scattering information (including all cross-section, polarization, double and triple scattering experiments , etc.) simply specifies the on-shell value of T

$$(k|T_\ell(k^2+i0)|k) = -(2k/\pi)\sin n_\ell(k)e^{in_\ell(k)} , \qquad (2.1)$$

where the conventions of BGMS are used.[1] In fact, the phase shifts $n_\ell(k)$ are not completely known, especially above the meson production threshold.

Experiments that depend on close nucleon-nucleon encounters, such as proton-proton bremsstrahlung and photo-disintegration of the deuteron, measure the wave function $|\Psi_k^{(+)}>$. The wave function is related to the off-diagonal or half-shell elements of T, $(k'|T(k^2+i0)|k)$, by

$$<k'|\Psi_k^{(+)}> = <k'|k> + \frac{<k'|T(k^2+i0)|k>}{k^2 - k'^2 + i0} \qquad (2.2)$$

Unfortunately, only the on-shell elements of T are known with reasonable accuracy and knowledge of the on-shell elements does not suffice to determine the dynamics of many-nucleon (N > 2) systems. One needs to know the half- and off-shell elements of T. In binding calculations, for example, the reference approximation G matrix is simply $T(\omega)$ evaluated at negative energies, $\omega \sim - 20$ MeV.

The on-shell constraint (2.1) is not sufficient to fix the needed off-shell elements of T. The customary way of restricting the off-shell behaviour is to select a reasonable hermitian potential and use it to fit the phase shifts. Reid has done this with particular success. An alternate means of introducing some, but fewer restrictions has been proposed by BGMS.[1] Their significant discovery was to isolate the off-shell freedom into a real symmetric function of two variables. Time does not permit me to do full justice to this and subsequent papers; instead, let me simply outline the basic ideas and results.

The basic constraint BGMS[1] placed on the transition matrix is the requirement that the wave functions form a complete, orthonormal set. Equivalently, one assumes the existence of an unspecified hermitian potential from which $T(\omega)$ is constructed using the Low equation

$$T(\omega) = V + V(\omega-H)^{-1}V. \qquad (2.3)$$

or in once-subtracted form

$$T(\omega) = T(\omega') + V[\frac{1}{\omega-H} - \frac{1}{\omega'-H}]V.$$

Instead of dealing with the real standing wave solutions of $R(\omega)=$
$V + VP(\omega-H)^{-1}V$, BGMS introduce another real set of wave functions
defined by

$$|\Psi_k^0\rangle = e^{-i\eta(k)}|\Psi_k^{(+)}\rangle = \cos\eta(k)|\Psi_k^P\rangle . \qquad (2.4)$$

The important point is that the real set $|\Psi_k^0\rangle$ obeys the usual
completeness and orthogonality conditions, whereas the standing
wave solutions $|\Psi_k^P\rangle$ include an awkward $\cos\eta(k)$ factor. In terms
of a real matrix defined by

$$(k'|W|k) \equiv (k'|\Psi_k^0) = (k'|S|k) + (k'|A|k), \qquad (2.5)$$

the restriction that the wave functions $|\Psi_k^0\rangle$ form a complete
orthonormal set is simply

$$WW^+ = 1 - |B\rangle\langle B| \qquad (2.6a)$$

$$W^+W = 1. \qquad (2.6b)$$

Here $|B\rangle$ denotes the bound state wave function. The real matrix
W has been split into symmetric (S) and antisymmetric (A)
functions of k and k'. These real quantities can be related to
the real function $\phi(k|k')$, which is the half-shell T matrix times
a phase factor

$$\phi(k|k') = (k'|T(k^2+i0)|k)e^{-i\eta(k)} = \cos\eta(k)(k'|R(k^2)|k)$$

$$= \sigma(k|k') + \alpha(k|k'). \qquad (2.7)$$

Here ϕ is also split into symmetric (σ) and antisymmetric (α)
terms. On-shell σ is fixed by the phase shifts $\sigma(k|k)= -(2k/\pi)$
$\sin\eta(k)$, but one is free to make reasonable choices for $\sigma(k|k')$
off-shell.

   For partial waves in which <u>no bound state</u> occurs, the
relationship between S, A and σ,α is

$$A(k|k') = (k^2-k'^2)^{-1}\sigma(kk'), \qquad (2.8a)$$

$$S(k|k') = \delta(k-k')\cos\eta(k) + (k^2-k'^2)^{-1}\alpha(kk'). \qquad (2.8b)$$

These conditions are simply obtained from (2.2). The steps
involved in constructing T or ϕ are basically as follows (see
figure 1)

   a) pick a reasonable $\sigma(k\ k')$;
   b) use (2.8a) to get A(kk');
   c) find a $S(k|k')$ that satisfies unitarity-(2.6) without
the bound state;
   d) from S find $\alpha(k|k')$, which in turn yields $\phi(k|k')$ and
hence the half-shell T-matrix (2.7).

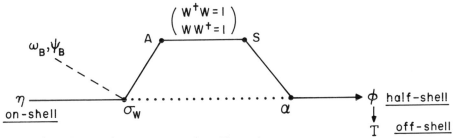

Fig. 1.--Steps in constructing T or ϕ.

Using the BGMS procedure the full T-matrix can be obtained starting from $\sigma(k|k')$-the symmetric part of the half-shell T matrix. Can these procedures be extended to include a bound state and noncentral forces?

The problem of extending the BGMS method to include a bound state has been solved by Haftel.[2] Haftel begins by incorporating the bound state properties in a manner that is reminiscent of Weinberg's quasi-Born method for isolating the bound state contribution to the transition matrix T. However, the novel aspect of Haftel's idea is to include the bound state information in a way that permits direct use of the BGMS procedures. I will not go through his derivation but would only like to indicate that he succeeds in isolating the off-shell freedom into a real symmetric function σ_W, but now knowledge of the binding energy and of the wave function is also required. The restriction of forming a complete orthonormal set has been satisfied. On fig. 1, I have indicated that the assumed wave function ψ_B and binding energy ω_B form the additional input to the BGMS procedure.

As far as I know, no one has carried out these steps for constructing T including a bound state. The main obstacle has been the tensor force; however, Sauer has recently included a tensor force as well as the bound state.[3] It should therefore be possible to generate sets of phase shift equivalent interactions using various choices for $\sigma(k|k')$ off-shell.

For each choice of σ a complete set of states ψ is generated, another choice $\tilde{\sigma}$ produces another set $\tilde{\psi}$. Thus one can relate the various models of the off-shell behaviour by giving a unitary transformation

$$U = \sum_i |\tilde{\psi}_i><\psi_i| \ .$$

The fact that different choices of σ generate unitary transformation permits me to turn to a different idea that has been presented in the literature for the same goals!

TRANSFORMED POTENTIALS

In addition to the transformations defined by different choices of $\sigma(kk')$, unitary transformations have been used to produce sets of phase shift equivalent potentials in an alternate way. From any potential V which fits the elastic scattering data, one can generate an infinite set of phase shift equivalent potentials using a unitary operator

$$U^{+}U = UU^{+} = 1. \tag{3.1}$$

The idea of using U to generate phase shift equivalent potentials originates with Bell, Baker, Ekstein, Coester et al., Lomon, and Signell et al. (see ref. 9-17)

Acting on the Schroedinger equation defined by V and the kinetic energy H_o ($H = H_o + V$)

$$H|\psi> = E|\psi>, \tag{3.2}$$

the unitary transformation produces a new Hamiltonian

$$\tilde{H} = UHU^{+} \tag{3.3}$$

and new states

$$|\tilde{\psi}> = U|\psi>. \tag{3.4}$$

The spectrum is however unchanged

$$\tilde{H}|\tilde{\psi}> = E|\tilde{\psi}>. \tag{3.5}$$

If operators for all other two-body observables are also transformed by the unitary transformation

$$\tilde{F} = UFU^{+}, \tag{3.6}$$

one has a canonical transformation which is simply a change in description without changing the expectation value of observables. However, at this stage, we break the canonical transformation. Special assumptions are now introduced to provide phase shift equivalent Hamiltonians, for which the expectation value of some observables will be altered.

Thus we are not defining full two-body canonical transformations. Instead, some operators will not be transformed. For example, it is assumed that the density operator ρ is not subject to a unitary transformation; consequently, the transformed density can differ from the original density

$$<\tilde{\psi}|\rho|\tilde{\psi}> \neq <\psi|\rho|\psi>$$

Similarly the quadrupole moment will differ from the original value

$$<\tilde{\psi}|Q|\tilde{\psi}> \neq <\psi|Q|\psi>$$

The transformed Hamiltonian H therefore enables us to define a new potential

$$V = \tilde{H} - H_o = UH_o U^+ - H_o + UVU^+ \tag{3.7}$$

which is distinguishable from the original potential with regard to certain selected two-body observables.

Although V and $\tilde{V}$ differ in some ways, they are now forced to have identical asymptotic wave functions. One can always write U in the form

$$(r|U|r') = \delta(r-r') - 2(r|\Lambda|r'). \tag{3.8}$$

If Λ is required to be short-ranged (less than 1.0fm), then Ψ and $\tilde{\Psi}$ will be equal asymptotically and hence produce the same phase shifts. To have U unitary, the short-ranged operator Λ, must satisfy

$$\Lambda^+ + \Lambda = 2\Lambda^+\Lambda. \tag{3.9}$$

The transformed and original angular momentum operators are required to be equal for reasons made clear later

$$\tilde{L} = ULU^+ = L \text{ or } [\vec{L},\Lambda] = 0. \tag{3.10}$$

The translational invariance of V and $\tilde{V}$ are assured simply by using relative coordinates. Finally, the requirement that both H and $\tilde{H}$ be invariant under the time reversal transformation (K_o) is assured by having

$$[K_o,\Lambda] = 0. \tag{3.11}$$

Various possible forms of Λ which satisfy these general restrictions will be discussed later.

Any Λ satisfying the above conditions will generate a new potential $\tilde{V}$ whose phase shifts are identical to those produced by the original V. Furthermore, if F has the proper OPEP tail then so will V, provided Λ is of sufficiently short range.

The wave functions Ψ and $\tilde{\Psi}$ differ only at short distances; therefore, the corresponding transition matrices T and $\tilde{T}$ are equal on-shell, but differ off-shell. Expressions for the change induced in the half-shell T matrix by the unitary transformation have been published by Monahan, Shakin and Thaler[4]. In terms of Λ, the change induced in T is

$$(k'|\tilde{T}(k^2+i0)|k) = (k'|T(k^2+i0)|k)$$

$$+ 2(k'^2-k^2)(k'|\Lambda|\psi_k^{(+)}) \tag{3.12}$$

Here $\psi_k^{(+)}$ is given by (2.2). Multiplication of (3.12) by the common phase shift factor $e^{-i\eta(k)}$ permits us to express the change

induced in the real half-shell function $\phi(k|k')$ defined by BGMS
as

$$\tilde{\phi}(k|k') = \phi(k|k') + 2(k'^2-k^2)(k'|\Lambda|\psi_k^0).\qquad(3.13)$$

The equivalence on-shell is clearly seen in both (3.12) and (3.13).

The significance of (3.13) is that one can start with a
reasonable half-shell T matrix and generate an infinite number of
phase shift equivalent amplitudes using a proper, short-range
function Λ. To define the original "reasonable" $\phi(k|k')$ one can
either use the Reid potential, use the BGMS construction procedure,
or in some way insert the restrictions implied by meson theory,
proton-proton bremsstrahlung experiments, etc. However one
defines ϕ there remains the possibility of transforming to $\tilde{\phi}$ and
generating a new set of phase shift equivalent interactions. Thus
one can use various choices for Λ to test the role of various off-
shell models in many-nucleon systems. The simplest case would be
to consider only small changes about some "reasonable" initial
choice for ϕ and to avoid drastic changes. Part of the problem is
to learn how to control the changes in ϕ produced by the function
Λ. How then shall we chose Λ?

Equation (3.12) is essentially the form proposed by Amado,[5]
who advocated separating the off-shell modifications from the
original T matrix. Picker et al.[7] have also made a separation,
but of a different kind. They prove that T and $\tilde{T}$ are related by

$$(k'|\tilde{T}(k^2+i0)|k) = (k|T(k^2+i0)|k) + (k^2-k'^2)\int_0^\infty dr r^2 j_\ell(k'r)\Delta_\ell(kr),$$

$$\qquad(3.14a)$$

$$\Delta_\ell(kr) = \psi_{\ell k}^{(+)}(r) - e^{i\eta_\ell(k)}[\cos\eta_\ell(k)j_\ell(kr) + \sin\eta_\ell(k)n_\ell(kr].$$

$$\qquad(3.14b)$$

It is significant that they are easily able to restrict their
wave function to be free of anomalous nodes or bumps. Their
approach however does not provide for a complete orthonormal set-
for that condition they propose using the BGMS[1] procedure.
With that step (3.14) becomes a useful alternative to (3.12).

The off-shell freedom has therefore been demonstrated and
placed into the choice of σ and/or the choice of Λ!

A SPECIAL CASE

The goal is to generate a set of phase shift equivalent
potentials and then to test the role of the off-shell elements of
T in many-body calculations. For the initial V, the Reid potential
is adopted.

Various choices of Λ can be made. It appears that a form originally suggested by Ekstein[12] is a reasonable first choice. Therefore, in a nuclear matter study Haftel and I used a short-range projection operator of the form

$$\Lambda_E = g(r)g(r') \tag{4.1}$$

The above Λ produces a unitary transformation since it is a hermitian projection operator and therefore satisfies (3.9)

$$\Lambda^+ = \Lambda = \Lambda^2 \tag{4.2}$$

with $|g>$ normalized as

$$<g_\ell|g_\ell> = 1. \tag{4.3}$$

The condition that Λ commute with the time reversal operator (K_0) is satisfied by (4.1), since g is taken to be real. Although all of the requirements discussed earlier are properly satisfied, we shall see later that the use of a hermitian projection operator leads to undesireable consequences. At this stage, however, the motivation for using (4.1) is its apparent simplicity.

The forms used for the S and D-wave transformation are

$$g_0(r) = C_0 e^{-\alpha_0 r}(1-\beta_0 r) = h_1 - \beta h_2$$
$$g_2(r) = C_2 r e^{-\alpha_2 r}(1-\beta_2 r), \tag{4.4}$$

where the parameters were chosen not only to satisfy (4.3) and to keep Λ short-ranged, but also to induce appreciable changes in the short-range wave functions. Although the original reason for introducing the β parameters in (4.4) was to produce weaker as well as stronger potentials, its main significance is that for certain values of β the bound state wave function is unchanged by the unitary transformation. If β is chosen to give $<g_0|B> = 0$ or

$$\beta = \frac{<h_1|B>}{<h_2|B>} \tag{4.5}$$

Then the Reid bound state $\psi_B(r)$ and, consequently, the deuteron form factor will be unchanged.

TABLE I.   PARAMETERS OF Λ_E

		α_0^{-1} (Fermis)	β_0^{-1} (Fermis)	α_2^{-1} (Fermis)	β_2^{-1} (Fermis)	Q	$(k_F=1.12)^K$
$^1S_0$	UT1	.333	.833	--	--	--	.225
	UT6	.333	1.111	--	--	--	.102
$^3S_1$	UT8	.417	1.25	--	--	.282	.089
	UT10	--	--	.333	1.176	.279	.114
$^3D_1$ +	UT11	--	--	.417	1.39	.282	.181
	REID	--	--	--	--	.282	.097

In Table I, values of β are given which satisfy (4.5) only
approximately. All of the ranges α_0^{-1} are less than .5F. The
values of the transformed quadrupole moment $<\Psi|Q|\Psi>$ are also
shown for UT8, UT10, and UT11 of reference 17. The UT1 and UT6
cases modify the $^1S_0$ partial wave, whereas UT8, UT10, and UT11
change the $^3S_1$ + $^3D_1$ coupled channel. As a consequence of (3.10)
the D-state probability P_D is unchanged by these transformations.
(The effect of altering P_D using other methods has been studied
by Clement, Afnan, and Serduke[20] for nuclear matter and by
Levinger[21] for the three-body problem.)

In figures 2a and 2b the S and D deuteron wave functions in
momentum space are presented. Note the small change induced by
Λ. Only above $k\sim2$ are changes seen on these semi-log plots.
Correspondingly, the electric form factor is only slightly
changed; the values are within the experimental uncertainties of
present electron-deuteron scattering experiments (fig. 2c). For
UT10 and UT11, the form factors are equal to the Reid form factor
on this scale.[17] Also shown in fig. 2c is UT25 which is a
case that can be rejected by the known form factor. Thus the
electric form factor provides some, but incomplete restrictions
on the possible potentials V. The question arises: what do
these wave functions look like in configuration space? In fig.
3a, the deuteron wave function $\Psi_B(r)$ is shown for the Reid and
for the UT8 transformed potentials (UT8 changes the $^3S_1$-wave part
only). For the UT10 and UT11 cases (changes in the $^3D_1$ part only)
the wave functions $\Psi_B(r)$ are given in fig. 3b. (These wave
functions were calculated by Haftel using Λ directly in r-space).

Based on the similarity of the S-wave deuteron wave functions
in momentum space (fig. 2), one might expect the wave functions
in r-space to be quite similar. However, an unexpected node
appears in the ground state wave function (fig. 3). It was clearly

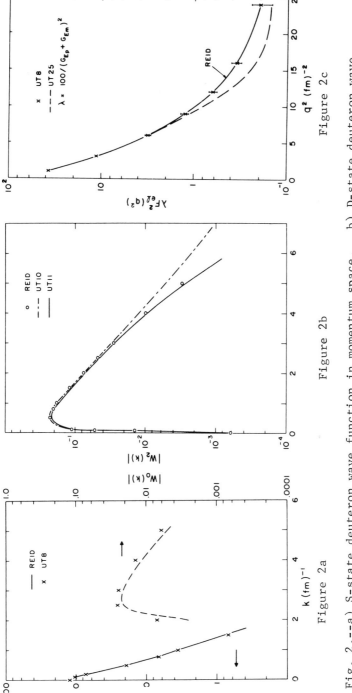

Fig. 2.--a) S-state deuteron wave function in momentum space. b) D-state deuteron wave function in momentum space. c) Electron scattering form factor for the deuteron.

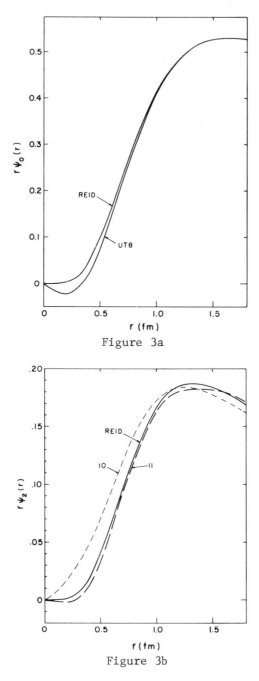

Figure 3a

Figure 3b

Fig. 3.--a) S-state deuteron wave function. b) D-state deuteron
wave function.

difficult to anticipate this result working in momentum space, but it is simply a consequence of the special definition used for Λ.

These nodes and bumps in the short-range wave function are changes that are greater than we wished to produce. How can we modify Λ to produce only mild changes in the wave function? Later a method for restricting Λ and for avoiding such anomalies will be suggested.

Of course, the nodes in Ψ_B can now be avoided by using values of β which leave the bound state exactly in its original form. However, it is still not clear what the transformed continuum wave functions will look like. The continuum states are shown in fig. 4 for the $^1S_0$ (UT1 and UT6) and $^3S_1 + {}^3D_1$ (UT8) cases. Again nodes and bumps appear for the simple reason that the overlap $<g|\Psi_k>$ can be large and positive to give a node, or large and negative to give a bump. The node produced by UT8, and seen in the bound state wave function persists in the continuum. That fact, and the $^1S_0$ nodes indicate that the use of a special β will not solve our problem. The nodes are there - and we have been rather extreme in the choice of Λ!

Despite the extreme variations provided by the special choice made for Λ, one can learn something about the sensitivity of many-body problems to off-shell changes. Also one should really seek firm reasons for rejecting the above transformed wave functions, perhaps based on the notion that extremely wrong wave functions should give extremely wrong answers in many-body systems.

OXYGEN-16

The off-shell transition matrix $\tilde{T}$ differs dramatically from the T matrix produced by the Reid potential. What effect will these off-shell changes have on calculated binding energies and spectra? Several authors have already shown that for nuclear matter large binding energy changes ($\sim$10 MeV/A) can be induced by off-shell changes[9,13-17]. Simple rules relating the change in binding to the change in excitation parameter κ were obtained. Can similar rules be found for finite nuclei? Are the results of finite nuclei and nuclear matter calculations closely related even for greatly altered potentials? Can improved results for binding energies and nuclear radii be obtained by changing the potential? These questions can be answered, in part, using the special case of Λ.

The calculation of the binding energy of $O^{16}_{(22)}$ that I shall present is the work of P. Sauer and E. Lambert[22]. A related paper dealing with the spectrum of O^{18} and F^{18} has been submitted to this conference by H. C. Pradhan, P. U. Sauer, and J. P. Vary. In the O^{16} calculation several approximations were

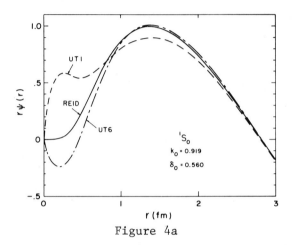

Figure 4a

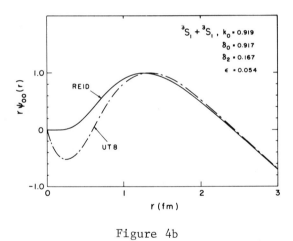

Figure 4b

Fig. 4.--a) $^1S_0$ continuum wave function. b) $^3S_1 + ^3D_1$ continuum wave function.

used and several problems, such as three-body clusters and the
proper choice of intermediate particle spectrum, were ignored.
It is hoped that, for the limited purpose of determining
sensitivity to off-shell changes, these calculations suffice.

The Brueckner self-consistency condition was satisfied for
the occupied (0s and 0p) single nucleon levels. For the
unoccupied levels the single particle potential energy was set
equal to zero. The single nucleon wave functions were not
determined self-consistently, but were taken to be oscillator
wave functions of fixed oscillator energy $\hbar\Omega$ = 14.02 MeV. Thus a
Brueckner, but not a Brueckner-Hartree-Fock, calculation was
performed at a fixed nuclear density.

Sauer[18] solved for the G-matrix by matrix inversion using
the reference G-matrix (i.e. $T(\omega)$) in

$$G(\omega) = T(\omega) + T(\omega)[Q(\omega-QH_oQ)^{-1} - (\omega-H_o)^{-1}]G(\omega). \qquad (5.1)$$

In using QH_oQ in the above propagator, he carefully orthogonalized
the particle and hole states. The Q is an angle-averaged, harmonic
oscillator Pauli operator assumed to be diagonal in relative and
center of mass quantum numbers. Sauer's procedure is ideally
suited for general local and nonlocal potentials because $T(\omega)$
enters directly as input. With this G-matrix, the binding energy
of O^{16} was evaluated using the methods (and codes) developed by
R. J. McCarthy, K.T.R. Davies and M. Baranger[19].

As a result of this calculation, one obtains the Bethe-
Goldstone wave functions for the relative motion of two-nucleons
in O^{16}. These wave functions (fig. 5) again exhibit the
infamous nodes and bumps. For the ^1S_o (n=0) wave function, UT1
again produces the bump it displayed in the free wave functions;
UT6 has maintained its node. As the relative collision energy is
increased (n=1), these properties persist, but the wound
$|\phi-\psi_{BG}|$ decreases for UT1 and increases for UT6. These changes
in wound correspond to a strong dependence of the effective
interaction (G) on the state of relative motion. Similar
results[22] are found for the $^3S_1 + {}^3D_1$ Bethe-Goldstone wave
functions; namely, the same short-range behaviour appears as in
the free wave functions and there is a strong state dependence of
the wound.

The energy per nucleon calculated for O^{16} is shown
on figure 6 as a function of χ_{AV}. Here χ_{AV} and κ_{AV} are
related quantities, which provide the essential smallness
parameter used in the Brueckner expansion. The excitation
parameter κ is defined as the integral over the wound.

$$\kappa = \rho \int |\phi - \psi_{BG}|^2 d\tau \simeq .14, \qquad (5.2)$$

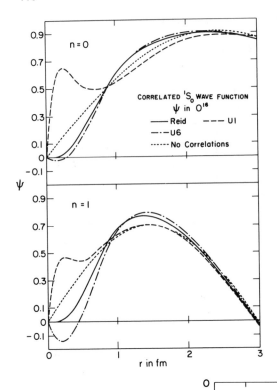

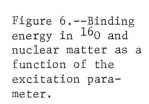

Figure 5.--Bethe-Goldstone wave functions for ^{16}O.

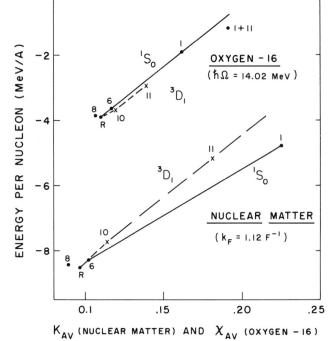

Figure 6.--Binding energy in ^{16}O and nuclear matter as a function of the excitation parameter.

where ρ is the density, ψ_{BG} the relative wave function in the nucleus, and ϕ is the unperturbed wave function. Clearly, nodes and bumps in ψ_{BG} will greatly change κ, mostly producing larger values of κ characteristic of stronger potentials.

The definition of χ_{AV} is based on the idea of occupation probabilities, P_A, which were introduced into Brueckner theory by Brandow, Baranger, Becker, Davies, and McCarthy[19]. The occupation probabilities account for the fact that nucleons in occupied levels are not always there – sometimes they are excited. In defining the single particle potential one accounts for the partial unavailability of a nucleon for interaction and thus the binding energies are changed by including occupation probabilities.

Now the χ_{AV} used in fig. 6 is defined as

$$\chi_{AV} = 1 - <P>_{AV} = \frac{1}{16}[4(1-P_{s1/2}) + 8(1-P_{p3/2}) + 4(1-P_{p1/2})], \quad (5.3)$$

where an average over the "occupied" or hole states is used. Hence χ_{AV} measures the average probability for excitation of nucleons above the Fermi sea. The excitation parameter κ has a similar interpretation; for a given level one relates κ and χ using (5.3) and $P_A \simeq [1 + \kappa_A]^{-1}$. Both χ_{AV} and κ_{AV} measure the strength of the effective two–nucleon interaction in the nuclear medium. Changes in the off–shell T–matrix induce changes in these quantitites. Therefore off–shell effects on nuclear binding are examined by relating binding changes to changes in the excitation parameter.

In fig. 6, the calculated binding for 0^{16} and nuclear matter are presented as functions of χ_{AV} and κ_{AV}, respectively. A systematic dependence of nuclear binding on the quantities χ_{AV} and κ_{AV} is found and some preliminary rules can be extracted from fig. 6:

a) Stronger potentials (i.e. increased χ_{AV}) lead to less binding energy at this density.
b) The slopes of the $^1S_0$ and $^3D_1$ energy changes induced by changes in χ are approximately equal and given by

$$\Delta E_{0-16}(^1S_0) \simeq 38.5 \; \Delta\chi_{AV}(\text{MeV}), \quad (5.4a)$$

$$\Delta E_{0-16}(^3D_1) \simeq 32. \; \Delta\chi_{AV}(\text{MeV}). \quad (5.4b)$$

Nothing can be concluded from the UT8 case because the values of χ_{AV} and E/A are so close to the Reid (R) value.

EQUIVALENT NUCLEAR MATTER

A nuclear matter calculation parallel to the O^{16} case has been performed by Haftel[22] . He also treats general nonlocal potentials by a matrix inversion solution for the G-matrix. Brueckner self-consistency for holes, zero potential energy for particles, and an angle-averaged Pauli operator are also used in the nuclear matter calculation[17]. To further parallel the O^{16} calculation a fermi momentum of $k_F = 1.12F^{-1}$ was chosen by equating the average kinetic energy of O^{16} with the average kinetic energy of nuclear matter. The Bethe-Goldstone wave functions for nuclear matter[22] again closely follow the free behaviour (fig. 4) and at equivalent relative energies they are very close to the ψ_{BG} for oxygen (fig. 5).

The binding energy per nucleon versus the total excitation parameter κ_{AV} for nuclear matter is also shown in fig. 6. Again UT8 tells us nothing. For the $^1S_0$ and $^3D_1$ off-shell changes a systematic behaviour for nuclear matter is found. The preliminary rules are:

$$\Delta E_{NM}(^1S_0) \approx 29. \ \Delta\kappa_{AV}(MeV), \tag{6.1a}$$

$$\Delta E_{NM}(^3D_1) \approx 37. \ \Delta\kappa_{AV}(MeV). \tag{6.1b}$$

Once again stronger potentials produce less binding. The finite and infinite nucleus results are indeed closely related, with one noteworthy difference. Nuclear matter is found to be more sensitive to the $^3D_1$ changes than is O^{16}. (See (5.4), (6.1) and fig. 6.) The suppressed role of D-waves for O^{16} is a consequence of the low weight (down by $\sim 1/3$) assigned to D-waves in transforming to the relative matrix elements as compared to the corresponding nuclear matter weights.

CONCLUSION

Appreciable, systematic changes in calculated binding energies are obtained when the excitation parameter κ_{AV} (or χ_{AV}) is altered. The induced changes in binding are given by (5.4) and (6.1) for O^{16} and nuclear matter. These rules are simply extensions to larger $\Delta\kappa$ of a formula given earlier by Wong[23]. Both κ and χ serve here as simple measures of short-range correlations and of off-shell characteristics of the interaction in the nuclear medium.

The systematic relationship between κ and nuclear binding suggests that knowledge of the correct κ will significantly restrict the off-shell freedom that was shown to be available in describing the two-nucleon interaction. Earlier nuclear matter studies[13-17](and the present results) indicate that off-shell changes can increase the binding, but always at the cost of increasing the equilibrium density. It appears, therefore, that short-range alteration of the potential will not in itself solve the problem of simultaneously obtaining the correct nuclear binding and radii.

If nuclear matter is used as a constraint on the off-shell freedom, then only values of κ_{AV} near that of the Reid potential seem to be permitted. Nevertheless, independent experimental determination of short-range correlations and of κ is needed.

It is satisfying that the O^{16} and nuclear matter results, including the Bethe-Goldstone wave functions, are so closely related (except for the D-wave suppression effect). To further illustrate the parallel nature of the nuclear matter and O^{16} results, the binding energies obtained using various transformed potentials are seen in fig. 7 to be closely correlated. The correlation between changes in nuclear matter and O^{16} binding, which is induced by off-shell alteration of the interaction, is given by

$$\Delta E_{O-16}(^1S_0) \;\approx\; \frac{3}{5}\,\Delta E_{NM}(^1S_0), \qquad\qquad (7.1a)$$

$$\Delta E_{O-16}(^3D_1) \;\approx\; \frac{1}{3}\,\Delta E_{NM}(^3D_1). \qquad\qquad (7.1b)$$

Nuclear matter is seen to be somewhat more sensitive to off-shell effects than is O^{16}. In fig. 7 the arrows indicate the effect of using self-consistent occupation probabilities in the O^{16} calculation. For nuclear matter the effect of occupation probabilities has not yet been studied.

It remains to be seen if the rules (5.4), (6.1) and (7.1) presented here will be maintained when three and four-hole diagrams are included. Wong's[9]recent results indicate that only slight modifications will be needed.

Finally, the special transformation Λ_E (4.1), which is used here to generate phase shift equivalent potentials, produces rather extreme off-shell alterations of the T-matrix. To fully establish the significance of off-shell effects in many-nucleon systems, one should examine simpler, milder changes about some reasonable initial choice of T. Some additional control over the unitary transformation is therefore needed to assure the possibility of generating small changes in the T-matrix. We learn from this study that, for the sake of simplicity, one should use unitary transformations which can be infinitesimally close to the unit operator. We require that $U(\theta) \to 1$ smoothly as $\theta \to 0$, where θ is a free parameter. A hermitian projection

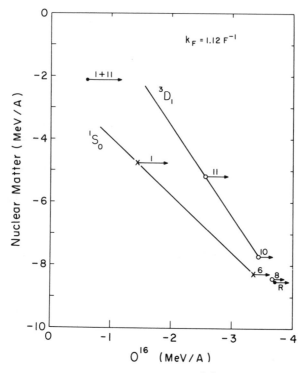

Fig. 7.--Binding energies in ^{16}O and nuclear matter.

operator, $\Lambda_E^+ = \Lambda_E$ (4.2), violates this continuity condition, since it is a discrete transformation, analogous to the improper Lorentz transformations. However, both Baker and a Coester-type transformation $^{(11,13)}$,

$$\Lambda_c = \frac{1-e^{i\theta}}{2}\,|g><g| + \frac{(1-e^{-i\theta})}{2}\,K_0|g><g|K_0,$$

do satisfy the above continuity condition. For large values of θ, the Baker transformation can produce bumps, and the Coester transformation can produce nodes and/or bumps, in the two-body wave function. However, the Baker and Coester transformations have the advantage of providing in θ the control needed to restrict future studies to mild off-shell changes.

ACKNOWLEDGMENT

I am grateful to Drs. P. U. Sauer and M. I. Haftel for their help and for permitting me to discuss their work before publication.

REFERENCES

1. M. Baranger, B. Giraud, S. K. Mukhopadyay, and P. U. Sauer, Nucl. Phys. A138 (1969) 1.
2. M. I. Haftel, Phys. Rev. Letters 25 (1970) 120.
3. P. U. Sauer, Nucl. Phys. A170 (1971) 497.
4. K. L. Kowalski, J. E. Monahan, C. M. Shakin, and R. M. Thaler, Phys. Rev. C3 (1971) 43; J. E. Monahan, C. M. Shakin, and R. M. Thaler, Phys. Rev. C4 (1971) 289.
5. R. D. Amado, Phys. Rev. C2 (1970) 2439.
6. W. Van Dijk and M. Razavy, Nucl. Phys. A159 (1970) 161.
7. H. S. Picker, E. F. Redish, and G. J. Stephenson Jr., Phys. Rev. C4 (1971) 289.
8. D.W.L. Sprung, Hercog Novi Summer School Lectures (1970) and references therein.
9. C. W. Wong and T. Sawada (1971-preprint).
10. J. S. Bell in "The Many-Body Problem" First Bergen International School of Physics - 1961 (W. A. Benjamin Inc., New York)
11. C. A. Baker, Phys. Rev. 128 (1962) 1485.
12. H. Ekstein, Phys. Rev. 117 (1960) 1590.
13. F. Coester, S. Cohen, B. Day and C. M. Vincent, Phys. Rev. C1 (1970) 769.
14. M. Miller, M. Sher, P. Signell, N. Yoder, and D. Marker, Phys. Letters 30B (1969) 157.
15. E. Lomon, Bull. Am. Phys. Soc. 14 (1969) 493.
16. M. Ristig and S. Kistler, Z. Physik 215 (1968) 419.
17. M. I. Haftel and F. Tabakin, Nucl. Phys. A158 (1970) 1 and Phys. Rev. C3 (1971) 921.
18. P. U. Sauer, Nucl. Phys. A150 (1970) 467.
19. K.T.R. Davies and R. J. McCarthy, Phys. Rev. C4 (1971) 81. (Additional references on the occupation probability formalism are given in this paper.)
20. D. M. Clement, F.J.D. Serduke, and I. R. Afnan, Nucl. Phys. A139 (1969) 407.
21. T. Brady, M. Fuda, E. Harms, J. S. Levinger, and R. Stagat, Phys. Rev. 186 (1969) 1069.
22. P. U. Sauer, M. I. Haftel, and E. Lambert (to be published).
23. C. W. Wong, Nucl. Phys. 56 (1964) 213 and Nucl. Phys. 71 (1965) 385.

NEUTRON-PROTON BREMSSTRAHLUNG INCLUDING GAUGE-INVARIANT TERMS ARISING FROM THE NUCLEAR POTENTIAL*

V. R. Brown and J. Franklin

Lawrence Radiation Laboratory, Livermore, California

and Temple University, Philadelphia, Pennsylvania

In a previous calculation of $np\gamma$ using potentials[1] the coupling of the electromagnetic field to the nucleon currents was obtained in the usual way using the principle of minimal electromagnetic coupling in the kinetic-energy part of the hamiltonian only. This prescription is not fully gauge invariant because of the exchange nature and momentum dependence of the two-nucleon potentials. The hamiltonian resulting from this minimal coupling is shown in Eq. (1).

$$H_N + V_{em}^{(1)} = -\frac{\nabla^2}{2\mu} + V_N + a_e(e^{-i\vec{K}\cdot\vec{r}/2} - e^{i\vec{K}\cdot\vec{r}/2})\, i\,\hat{\varepsilon}\cdot\vec{\nabla}$$

$$+ ia_e(e^{-i\vec{K}\cdot\vec{r}/2}\mu_1\vec{\sigma}_1\cdot\vec{K} \times \hat{\varepsilon} + e^{i\vec{K}\cdot\vec{r}/2}\mu_2\vec{\sigma}_2\cdot\vec{K} \times \hat{\varepsilon}), \tag{1}$$

where $a_e = \frac{e}{m}\sqrt{\frac{2\pi}{K}}$, $\mu_n = -1.913/2$ and $\mu_p = 2.793/2$

A momentum-dependent one-boson-exchange potential such as that of Bryan and Scott[2] is defined such that when inserted into the Schrodinger equation it yields in Born approximation the same scattering matrix element as the Feynman diagram for one-boson exchange. A potential so defined is easily written down in momentum space. In a non-relativistic model a method of solution is to Fourier transform everything to configuration space and solve the Schrodinger equation. The correspondence of such a potential with Feynman diagrams suggests a method[3] for introducing minimal coupling.

In the present work the minimal-gauge-invariant interactions are introduced in momentum space in the non-local framework according to the prescription[4]

$$\vec{p} \rightarrow \vec{p} - \frac{e}{c} \vec{A} \ . \tag{2}$$

For simplicity the present discussion is restricted to the long-wavelength dipole approximation[5] for the exchange term. For npγ which is a single-photon process only terms linear in $\vec{A}$ are of interest so that it is sufficient to consider a Taylor expansion in $\vec{A}$ keeping only the first two terms. The first term is just the nuclear potential with no photon emission, and the second term is the additional electromagnetic potential which we call $V_{em}^{(2)}$ to distinguish it from $V_{em}^{(1)}$ of Eq. (1). A transformation of $V_{em}^{(2)}$ to configuration space and a rearrangement of terms yields

$$V_{em}^{(2)} = \sqrt{\frac{2\pi}{K}} \ i \ e \ \hat{\varepsilon} \cdot \{\frac{1}{2} \ (\vec{r}' - \vec{r}) \ V_N(r,r') - \vec{r}' \ V_N^{ex} \ (r,r')\} \ , \tag{3}$$

where $V_N^{ex} \ (r,r')$ is the exchange potential for np scattering still in the non-local representation. The np potential can be written in terms of a direct plus an exchange part as

$$V_N = V_N^d + V_N^{ex} \ T^{ex}, \tag{4}$$

where T^{ex} is the isospin-exchange operator given by

$$T^{ex} = \frac{1}{2} \ (1 + \vec{\tau}_1 \cdot \vec{\tau}_2) \tag{5}$$

The first term of $V_{em}^{(2)}$, which consists of direct plus exchange, exists only for non-local or certain momentum-dependent potentials. The second term of $V_{em}^{(2)}$, which involves just the exchange potential arises because of the interchange of a proton and a neutron through the exchange of a charged meson.

The results of npγ including the various contributions are calculated with the Bryan-Scott potential at 200 MeV for coplanar symmetric angles of 30° to compare with the experiment of Brady[6] et al. The terms arising from the explicit momentum dependence of the Bryan-Scott potential are negligible. A comparison of the results for external radiation scattering alone and those including the exchange-bremsstrahlung term in the present approximation is shown in Fig. 1. The cross section integrated over the photon angular distribution is 35 $\mu b/(sr)^2$ as compared to the experimental result of Brady et al. of 35 $\pm$ 14 $\mu b/(sr)^2$.

REFERENCES

1. V. R. Brown, Physics Letters 32B (1970) 259.

2. R. A. Bryan and B. L. Scott, Phys. Rev. 177 (1969) 1435.

3. V. R. Brown and J. Franklin, Bull. Am. Phys. Soc. 16 (1971) 560. The present work is an extension of Ref. 3.

4. With this prescription the resulting amplitude corresponds to the OBEγ Feynman diagram.

5. Higher multipoles of the electromagnetic interaction can be obtained by expanding in powers of the photon momentum. Preliminary results indicate that the extreme dipole approximation is correct to within 20%.

6. F. P. Brady, J. C. Young, and C. Badrinathan, Phys. Rev. Letters 20 (1968) 750; F. P. Brady and J. C. Young, Phys. Rev. C 2 (1970) 1579.

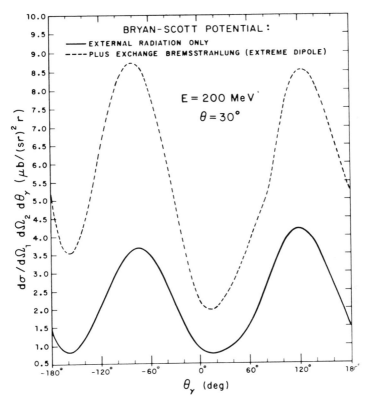

Fig. 1.--The coplanar-symmetric npγ cross section for E = 200 MeV and θ = 30° calculated with the Bryan-Scott potential and comparing the results for external-radiation scattering alone with those including the exchange-bremsstrahlung contribution. The latter is calculated in the long-wavelength dipole approximation.

*Work performed under the auspices of the U. S. Atomic Energy Commission.

OFF-SHELL EFFECTS IN THE ^{18}O and ^{18}F SHELL-MODEL SPECTRA*

H. C. Pradhan, P. U. Sauer and J. P. Vary

Massachusetts Institute of Technology

I.   INTRODUCTION

Adequate knowledge of the two-nucleon interaction, both on-shell and off-shell, and a proper many-body theory to employ that knowledge are the ingredients for a microscopic description of nuclear phenomena. At present, neither of these essentials is reliably established. However, the problem of the effective interaction is beyond the scope of this study. We retreat to the crude and even questionable[1] but common assumption that the effective interaction can be approximated by the sum of the bare reaction matrix G and the three-particle one-hole (3p-1h) core-polarization correction. Within this limited model we want to examine the degree of sensitivity which the low-lying shell-model spectrum exhibits to changes in the off-shell behavior of the free two-nucleon interaction.

A review of the work of Elliott et al.[2] and of Lynch and Kuo[3] suggests a lack of sensitivity. On the other hand, the ^{18}F spectra computed from the non-local Tabakin[4] and the local Hamada-Johnston potential[5] differ in the low-lying 1$^+$ states substantially[6] which casts some doubt on the lack of sensitivity.

Here, using the same technical apparatus everywhere, we compare spectra obtained from potentials exactly phase-equivalent with the Reid soft-core[8] potential. These non-local potentials were obtained by Haftel and Tabakin[9] by short-range unitary transformations (1, 6, 8, 10, and 11).

126

II.   CALCULATIONAL TECHNIQUE

The model space for the two valence nucleons consists of the 1s-0d oscillator orbitals with the 0s-0p orbitals completely filled. Experimental single-particle energies are employed with $(0d_{5/2}, 1s_{1/2}, 0d_{3/2}) = (0.0, 0.87, 5.08)$ MeV respectively. Relative-center of mass (RCM) matrix elements of $G(\omega)$ are calculated according to the method of ref.[11] which is especially suited for non-local potentials. An "angle-averaged" Pauli operator appropriate for shell-model calculations is used. All intermediate states are taken to be purely kinetic. For the oscillator energy we take $\hbar\Omega = 14.02$ MeV. The available energy ω in the bare reaction matrix is -10 MeV. The RCM matrix elements characteristic for the phase-equivalent potentials employed here are listed in Table 1. Untabulated matrix elements for the other partial waves are those of the Reid potential[11]. For the c.m. variables only a dependence on the combination 2N+L of the oscillator quantum numbers is maintained.

In the core-polarization correction of the effective inter- action excitations to the 3p, 2f, and 1h oscillator orbitals are permitted so that the contributions of all intermediate particle- hole states through $6\hbar\Omega$ excitation energy are included. In reaction matrices for core-polarization the starting energy has to be shifted by $1\hbar\Omega(\omega=-25$ MeV$)$ or $2\hbar\Omega(\omega=-40$ MeV$)$ depending on whether a 0p state core particle or a 0s state core particle is excited. This is done for all results of Section 3 and, as seen in Figure 1, it amounts to a sizeable effect when compared with a calculation of common practice which uses the same starting energy in G for all higher- order diagrams. In the present study this is significant for two reasons. First, for those partial waves $(^3S_1-^3D_1$ and $^1S_0)$ in which off-shell variations are generated the reaction matrix is strongly

T=0	ℓ	ℓ'	n	n'	N	L	Reid	8	10	11	Tabakin
$^3S_1$	0	0	0	0	2	0	-9.52	-9.52	-9.52	-9.47	-10.43
			1	1	1	0	-6.73	-4.46	-6.70	-6.58	- 8.15
			2	2	0	0	-3.32	2.47	-3.27	-3.07	- 6.20
$^3(S\text{-}D)_1$	0	2	1	0	1	0	-2.91	-2.92	-3.24	-4.46	- 2.26
			2	1	0	0	-3.31	-3.43	-4.00	-5.76	- 2.34
$^3D_1$	2	2	0	0	1	0	1.31	1.30	3.63	9.77	3.95
			1	1	0	0	1.46	1.42	5.91	12.91	3.69
T=1								1	6		
$^1S_0$	0	0	0	0	2	0	-6.76	-4.92	-6.52		- 7.05
			1	1	1	0	-4.55	-4.47	-3.25		- 4.62
			2	2	0	0	-2.05	-2.82	0.84		- 2.52

TABLE 1.--RCM G matrix elements in MeV. Matrix elements of the Reid potential are compared as well as those of the Tabakin potential, $\hbar\Omega= 14.02$ MeV and $\omega= -10$ MeV.

ω-dependent. The unitary transformations can dramatically change[10] the wound

$$<\chi(\omega), \, n\ell | \chi(\omega), n\ell> = - <n\ell \left| \frac{\partial G(\omega)}{\partial \omega} \right| n\ell>$$

in an oscillator state $|n\ell>$ of relative motion. We therefore anticipate alterations in this ω-dependence. Second, the core-polarization correction of the ^{18}O ground state is especially sensitive to the $^3S_1-^3S_1$ matrix elements which usually exhibit the strongest ω-dependence. This is verified by analyzing the results in Table 2 with respect to those in Table 1.

Reid	-0.82
1	-0.71
6	-0.87
8	-0.35
10	-0.78
11	-0.73
1+11	-0.62
6+8	-0.48

TABLE 2.--The shift in calculated ground-state energy of ^{18}O due to the inclusion of core-polarization in the effective interaction.

III.   RESULTS AND DISCUSSION

We shall present the results on three levels. First, we discuss the changes of important RCM reaction matrix elements due to unitary transformations. Second, we describe the resulting ^{18}O and ^{18}F spectra and third, we discuss the results and compare them with parallel calculations of ^{16}O .

The variations (Table 1) in the RCM matrix elements are much stronger than believed possible in view of the calculations of Lynch and Kuo[3] and of Elliott et al.[2] A strong state dependence of the changes is to be noted; e.g. transformation 1 removes attraction from the diagonal n=0 and n=1 $^1S_0$ matrix elements and adds some to n=2. A somewhat contrary trend occurs with transformation 6. In the $^3S_1 - ^3D_1$ partial wave, transformation 8(10 and 11) only changes the $\ell_1=0$ ($\ell_1=2$) components of the deuteron and of the two-nucleon scattering wave functions, $<r\ell_1 | \Psi^+(k)\ell_2>$, where k is the momentum of relative motion. Consequently, the $^3S_1-^3S_1$ ($^3D_1-^3D_1$) partial wave components receive the largest shifts.

Figures 2 and 3 display a selection of the calculated T=1 spectra of ^{18}O and the T=0 spectra of ^{18}F respectively. For comparison we present the experimental spectra and the spectra resulting from the Tabakin potential.

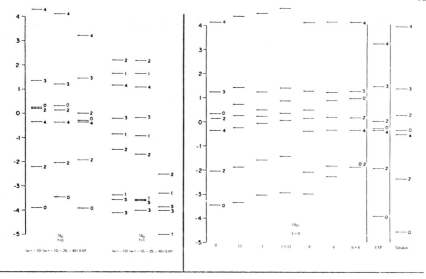

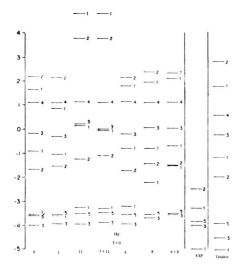

Fig. 1.---T=1 spectra of ^{18}O and T=0 spectra of ^{18}F calculated with
        the Reid potential using different prescriptions for ω
        as discussed in the text.

Fig. 2.--T=1 spectra for ^{18}O. Results for the Reid potential
        (force 0) and various non-local potentials (labelled as
        in Ref. 9) phase-equivalent with the Reid potential are
        presented. Also shown are the experimental spectrum and
        that of the Tabakin potential.

Fig. 3.--T=0 spectra for ^{18}F. See caption to Fig. 2.

Transformations 1 and 6 (8 and 11) affect both the bare and the core-polarization contribution to the effective interaction for ^{18}O (^{18}F). The resulting spectra appear with significant shifts, usually beyond those obtained with those forces (8 and 11 for ^{18}O, 1 and 6 for ^{18}F) that affect core-polarization alone. When the transformations are combined to yield the forces labelled 1 + 11 and 6 + 8, the spectra are further shifted from the Reid spectra. We especially note, that the combined force 6 + 8 makes the ground state and the first excited state of ^{18}O degenerate. For the Reid potential these levels are 1.40 MeV apart. In ^{18}F force 6 + 8 makes the first 1^+ state and the first 2^+ state almost degenerate. They were separated by 2.06 MeV for the Reid potential.

Since we use a limited model[1] for the effective interaction, agreement with experiment would mean little and should not be expected. The off-shell changes studied here push the low-lying levels up in energy, i.e. normally further away from their experimental positions, in both ^{18}O and ^{18}F. Increased repulsion is also a consequence of these transformed potentials in nuclear matter[9] and ^{16}O[10]. However the detailed trends have little else in common. The reason is: the changes in the reaction matrix are strongly state-dependent, and the ^{16}O ground-state and the shell-model states are sensitive to different parts of the reaction matrix. In addition, even the different shell-model states tend to exploit the components of the effective interaction with varying weights. The violent changes in the $^3D_1$-$^3D_1$ matrix elements for forces 10 and 11 are almost unfelt in the low-lying spectra, whereas the comparatively smaller changes in the S waves of forces 6 and 8 have a devastating effect on the same levels. We conclude strong sensitivity of the low-lying ^{18}O and ^{18}F spectra with respect to the off-shell behavior of the nucleon-nucleon interaction in the relative S waves. This conclusion contrasts the implications of the works by Lynch and Kuo[3] and Elliott et al.[2]

REFERENCES

1. M. W. Kirson, preprint and references therein.

2. J. P. Elliott, A. D. Jackson, H. A. Mavromatis, E. A. Sanderson and B. Singh, Nucl. Phys. A121 (1968) 241.

3. R. P. Lynch and T. T. S. Kuo, Nucl. Phys. A95 (1967) 561.

4. D. M. Clement and E. U. Baranger, Nucl. Phys. A108 (1968) 27.

5. T. T. S. Kuo, Nucl. Phys. A103 (1967) 71.

6. E. U. Baranger, Proc. Int. School of Physics Enrico Fermi (1967) Course 40, p. 643.

7.   P. Signell, Phys. Rev. <u>C2</u> (1970) 1171.

8.   R. V. Reid, Ann. of Phys. <u>50</u> (1968) 411.

9.   M. I. Haftel and F. Tabakin, Phys. Rev. <u>C3</u> (1971) 921.

10.   M. I. Haftel, E. Lambert, F. Tabakin and P. U. Sauer, to be published.

11.   P. U. Sauer, Nucl. Phys. <u>A150</u> (1970) 467.

12.   P. U. Sauer, Nucl. Phys. <u>A170</u> (1971) 497.

*This work supported by funds provided by the Atomic Energy Commission under contract AT(30-1)-2098.

TRITON BINDING ENERGY PREDICTIONS FOR PHASE-SHIFT EQUIVALENT

POTENTIALS

Michael I. Haftel
Naval Research Laboratory, Washington, D. C.   20390

We compare theoretical predictions of the triton binding energy
for a set of exactly phase-shift equivalent potentials. The potentials
are generated through a rank one unitary transformation of the S-
wave, spin independent Malfliet-Tjon V potential. We find that
potentials that give drastically different (up to 27 MeV) nuclear
matter binding energies, but nearly identical deuteron wave functions,
give only small (1 MeV) variations in triton binding energy. For
the potentials studied, large variations (up to 4.55 MeV) in triton
binding energy occur only when there are significant differences
in the deuteron wave functions. We find that potentials with
enhanced deuteron charge form factors give increased binding in
the triton.

AN EFFECTIVE REACTION MATRIX DERIVED FROM NUCLEAR MATTER CALCULATIONS

H. A. Bethe

Cornell University, Ithaca, New York

I'm going to talk about the nuclear interaction as derived from nuclear matter theory. I have written here some of the concepts which I shall use

$$G = v - v \frac{Q}{e} G \qquad (1)$$

$$G \phi = v\psi \qquad (2)$$

The first equation determines the so-called reaction matrix, G, which was introduced by Brueckner[1] and which is the basis of the theory. It is calculated from the potential v which is supposed to be given. Q is the Pauli operator and e is the energy difference between the excited and normal states. The sum of the diagonal elements of the G-matrix gives the potential energy of a given piece of nuclear matter or of a finite nucleus.

The quantity ϕ is the undisturbed wavefunction, e.g. the plane wave $\exp(i\vec{k}\cdot\vec{r})$. Or, if you have a potential v which depends on the angular momenta, you resolve it and so ϕ is the spherical Bessel function, the undisturbed radial function for free particles. ϕ is the true wavefunction only at large distances. The actual wave-function ψ goes asympotically into the undisturbed wave function without a phase shift and therefore it is convenient to introduce the defect function,

$$\zeta = \phi - \psi, \qquad (3)$$

the difference between the undisturbed and the actual wave function.

Finally we introduce κ which is known as the defect integral and
which is the integral of the square of the defect wave function,

$$\kappa = \int \zeta^2 \, dt \tag{4}$$

κ is a small parameter; it may be considered as the parameter in
which the expansion proceeds; for the Reid potential[2] at normal
nuclear density it is 14%.

In Fig. 1 is shown perhaps the most interesting of the defect
functions, namely that for the ^3S state. The actual wave function
is the undisturbed wave function minus the defect funcion, i.e.
$\psi = \phi - \zeta$. The undisturbed wave function for an s-state is simply
sin kr, where k has been chosen to be an average momentum in the
Fermi sea for the case of normal nuclear matter density. At small
distances the defect function essentially cancels the undisturbed
wave function because there is strong repulsion. Then later on there
are oscillations with decreasing amplitude. The dashed curve
represents the reference spectrum approximation which you see is
very good at small distances and very bad at large distances. At
large distances it exaggerates the deviation from the unperturbed
wave functions.

Figure 2 gives the main result of the theory. It is the
energy of nuclear matter calculated with the Reid[2] potential in the
first approximation in which you only take into account the correlation between pairs of nucleons. It is assumed from the beginning
that the only forces are forces between pairs, no three-body forces
are put in at this point. In addition it is assumed that correlations

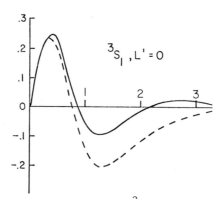

Fig. 1.--The defect function for the $^3S_1$ state. The abscissa is the
radius in F.

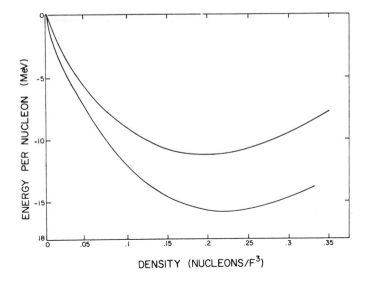

Fig. 2.--The energy per nucleon of nuclear matter.

exist only between pairs. With this approximation you get the
upper curve as a function of the density. The density scale goes
from 0 to 0.35 nucleons per F^3. The minimum of this curve is at
a density of about 0.2 which is about right for nuclear matter.
From observations 0.17 might be even better. A density of 0.2
corresponds to a Fermi momentum of a little over 1.4 F^{-1}. The
minimum of the curve is at an energy of about -11 MeV per particle.
You know of course that this is the difference between a large
kinetic energy of about 24 MeV per particle and a still larger
potential energy of -35 MeV per particle. The lower curve is obtained
by simply multiplying the potential energy by a constant,1.12, with
the intention of making the energy at minimum equal to -16 MeV which
is the observed value for nuclear matter.

A number of corrections are shown in Table I. The first line
is what I have just discussed, the result obtained from the two-body
correlation, -11 MeV. Then there are 3-body correlations, when
three nucleons are close together, giving -1.4 MeV. An additional
point is the hole-hole interaction. I will not go into the details,
but at one time some people thought that this should be treated on
a par with the basic pair interaction. It is in fact only 1% of the

TABLE I.--Contributions to the energy of nuclear matter.

Two-Body Correlations	-11.25
Three-Body Correlations	- 1.4
Hole-Hole Interaction	- 0.35
Four-Body Correlations	- 1.1
Three-Body Forces	- 1.1
Minimal Relativity	- 0.5
	-15.7

basic interaction between pairs, and therefore it is far more reasonable, as is done in the Brueckner-Goldstone expansion, to treat the hole-hole interaction as a perturbation and only the fundamental interaction between existing nucleons as the first approximation. There is a contribution from the four-body inter- action of about -1 MeV, a correction for minimal relativity by Brown[3] and collaborators of -0.5 MeV, and finally the contribution of 3-body meson forces. Brown and Green[4] calculated that a couple of years ago. In the meantime McKellar and Rajaraman[5] have made an improved calculation, the main advantage of which is that the high-momentum cutoff which has to be used for any such 3-body force is provided naturally by the two-body wave function going to zero at small distances. This was predicted by Brown sometime ago and put into a very nice analytic form by McKellar and Rajaraman. This has decreased the 3-body contribution from over -2 to about -1 MeV, so the total is now -15.7 MeV, very close to the observed -16 MeV. Even better is that with the McKellar-Rajaraman calculation the three-body forces no longer depend strongly on density. They increase only very slowly with density, if at all, so this correc- tion will probably not shift the energy minimum as a function of density by very much.

The same can now be said about the 3-body correlations with 2-body forces because it was found by Day[6] and emphasized recently by Wong[7] that the 3-body correlation correction is itself corrected in the next order by a term of relative order κ with a large coefficient, and the bigger κ becomes, the less contribution do you get from the 3-body correlation. This is very welcome because otherwise these corrections might shift the minimum of the energy to a much higher density, in conflict with the observations.

The contribution of ^1S and ^3S states to the nuclear matter potential energy is depicted in Fig. 3. The most conspicuous point is that the ^1S potential energy decreases monotonically with density, whereas the ^3S saturates. This is because the tensor force acts only in second order. Thus at high density the Pauli principle prevents any strong action of the tensor force and makes the ^3S contribution saturate. This is the most important contribution to

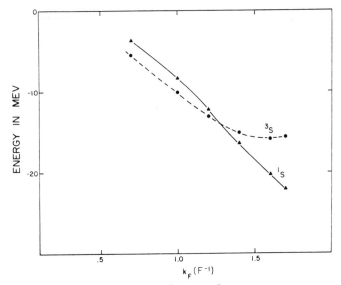

Figure 3.--Contributions of the ^{3}S and ^{1}S states to the nuclear matter potential energy.

saturation of nuclear forces, and without the tensor force you would not get an energy minimum at the observed density.

Figure 4 shows the contribution of S-, P- and D-states. You see that the P-states are repulsive and the D-states are attractive. It is therefore not a bad approximation to consider the S-states alone.

So, these are the present results of nuclear matter theory for infinite nuclear matter with the Coulomb force turned off, and equal numbers of neutrons and protons.

Now how do we apply this to finite nuclei? There's one observation from nuclear matter calculations which is relevant to this. Namely, when you calculate the ratio:

$$g = v\psi/\phi, \tag{5}$$

you find that this ratio is nearly independent of the momentum at any given density. This is not so hard to understand because at large distances ψ is not very different from the unperturbed ϕ, so

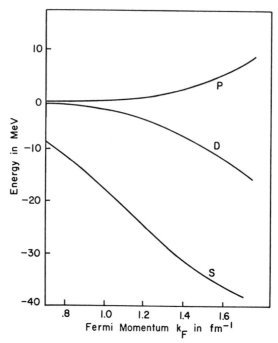

Fig. 4.--Contributions of the S, P and D state to the nuclear matter potential energy.

you just get v, independent of k. And at small distances the dominating thing is the repulsive core which is so strong compared to any kinetic energy that you get essentially the same result independent of k. That's the reason why g(r) is almost independent of k. It does depend on the density, and it still depends somewhat on k. To eliminate this dependence we average over all the occupied states

$$g(\rho,r) = \frac{\int \phi(kr)v(r)\psi(k,r)d^3k}{\int \phi^2(kr)d^3k} \, . \tag{6}$$

This was suggested by Siemens[8] and has the great advantage that it removes any possible trouble from zeros in the undisturbed wave-function ϕ, because the denominator of (6) is of course positive

definite. In this manner we get an effective potential
$g(\rho,r)$ which when multiplied by the unperturbed wave function gives
you the correct effect of the potential in the case of nuclear matter.
In other words when you integrate this g over the position, you get
back the diagonal elements of the G-matrix which give the correct
energy in the case of nuclear matter. So this $g(r)$, used together
with unperturbed wave functions, has the nature of an effective
potential in nuclear matter which can be used in Born approximation.
That is, it gives correct energy when integrated over position
without any corrections. This procedure was tested by Negele[9] who
found that not only the diagonal matrix elements of the G-matrix
but also the nondiagonal matrix elments, including the tensor matrix
elements, are given fairly accurately by this procedure in nuclear
matter.

Now we use this effective potential in finite nuclei. The
potential has been averaged already over k but it still depends,
as the underlying Reid potential does, on the angular momenta LSJ
of the interacting particles, on the density and on the distance
between the particles. Now what do we do with this? First of
all we average over J for the triplet states. This was done
primarily with the view to simplification of the calculations for
finite nuclei. It may be, however, that in this averaging process
there is contained some unwitting cleverness, since we get good
results and other people who don't average over J don't get good
results. We lose however in this process the spin-orbit interaction
which of course is contained precisely in the J-dependence, and
therefore we have to put it in at the end. I'll come back to that
later.

So we average over J and then we still have a dependence on L
and S. Now, the potential for all even L is pretty similar, and
likewise for all odd L. In fact for odd L, $L\neq1$, you may just about
use the one-pion-exchange-potential (OPEP). For even L we then
average over L=0 and L=2 with appropriate weights. Figure 5 gives
an exploratory calculation by Negele[9] which shows that the percentage of S-state, P-state and D-state is not very different in
nuclear matter (dashed curve) and for harmonic oscillator wavefunctions (solid curve). R is the distance from the center of
the nucleus, and r the distance between the two nucleons. Figure 5
justifies averaging over the even states to construct the potential.
By this averaging we get a g which is appropriate to even parity
states, and a g appropriate to odd parity states but still depending on S. When you know the parity and know S you also know T so
you can say that it depends on the isospin rather than on the
ordinary spin.

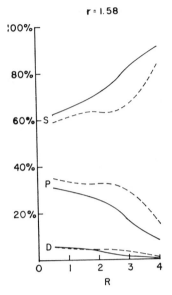

Fig. 5.--Weighting of S, P and D states of the two-body interaction
in ^{40}Ca.

Once you have an even and an odd-parity interaction it is
natural and convenient, especially for a Hartree-Fock calculation,
to write them in the old fashioned way of the 1930's, namely as the
sum of exchange and ordinary interactions

$$g_{even} = g_D + g_X \tag{7a}$$

$$g_{odd} = g_D - g_X \, . \tag{7b}$$

In this way, we define the direct and the exchange potential.

The density dependence doesn't need to be known very accurately
and therefore it is expressed in a simple manner, namely

$$g = g_0(r) + g_1(r)\rho^\alpha \tag{8}$$

where the constant α is taken to be 1/3 for the attractive, and
1 for the repulsive part of the potential. To use this, we must
know the Fermi momentum in a finite nucleus. At this point we
make the local density approximation (LDA). That is, we
say that the interaction at any point in the nucleus is determined by the density at that point. Now, of course, interaction
is at some distance, the forces have some range, and therefore
we have to decide which point to use. In Negele's published paper[9]

he took the density at the mid-point between the two points at which the two nucleons sit. But it is more convenient for computation to take the average of the densities or the average of k_F at the two points,

$$\rho(\vec{r}_1, \vec{r}_2) = 1/2\rho(\vec{r}_1) + 1/2\rho(\vec{r}_2). \tag{9}$$

So, having a force for like particles and for unlike particles, you take some standard radial dependence (the g_0, g_1 of equation (8)) for the term independent of density and another for the term dependent on density, and then find coefficients which make it right for the exchange term and the direct term.

In Table II is shown the <u>unlike</u> interaction which is approximately twice the interaction between like nucleons. I have listed here the volume integral of the various parts of the potential. The short range potential which comes from the repulsive part of the interaction has a volume integral of 26 MeV-F^3. The volume integral of the direct attractive interaction is only a little bit bigger than the repulsive part. The big attractive interaction is in the exchange term which is the difference between the even and the odd parity state terms, and is -50 MeV-F^3.

Figure 6 gives the main features of the direct interaction, non-exchange, and you see this has two parts. The short range part which is repulsive is the residue of the repulsive core of the Reid interaction. The three curves correspond to interactions between like particles (long dashes), unlike particles (short dashes) and the average (solid). The average potential never goes over 600 MeV. So by taking the effective interaction we have essentially reduced the Reid repulsive core, which is quite strong, to something of manageable size. At larger distances the direct interaction attraction goes down to about -25 MeV. The exchange

TABLE II.--Volume integrals† of various parts of the potential for unlike particles.

$k_F(F^{-1})$	Short (MeV-F^3)	Direct (MeV-F^3)	Exchange (MeV-F^3)
1.0	+26	-40	-50
1.4		-33	

†Volume integral of v is defined as $\int vr^2 dr$.

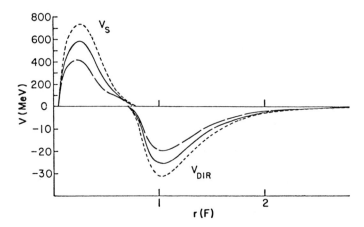

Fig. 6.--The direct interaction.   From Ref. 9.

interaction is shown in Fig. 7.   It is much stronger as you see:
for the average curve it gives an attraction of about 60 MeV.   The
nuclear forces being as they are, if you replace the interaction
by something simpler, then the most important thing is an exchange
interaction.   You won't get realistic answers if you take only
a non-exchange interaction.

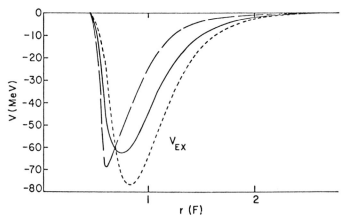

Fig. 7.--The exchange interaction.   From Ref. 9.

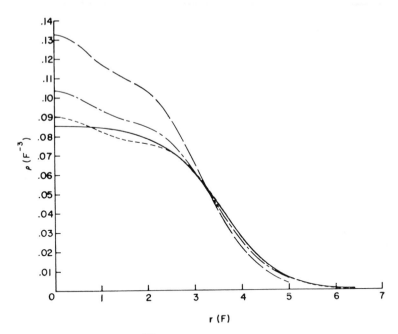

Fig. 8.--^{40}Ca charge density distributions.

Negele[9] then inserted this force into the Hartree-Fock theory
for various closed shell nuclei. The result shown in Fig. 8 is for
^{40}Ca. The ordinate is the density in protons/F^3, and the abscissa
is the distance from the center. The long dashes correspond to a
force which is independent of density. You see this gives a very
high density in the center, much higher than the experiment (shown
as the solid line) would indicate, and correspondingly the radius
of the nucleus is too small. The density dependent force gives you
the dash-dot curve which is a great deal better and which is calculated without any adjustment. This is just deduced from nuclear
matter theory in the manner I have indicated, and it gives about the
correct density. Then Negele adjusted his force in two ways, which
I will describe below. After these adjustments he gets the dashed
curve which is very close to the observed curve as deduced from the
Stanford electron-scattering experiments.[10] You see that the agreement is exact as far as the radius of the calcium nucleus goes.
This radius, however, has been adjusted in the manner I will still
talk about, so this agreement is not an achievement. What is an
achievement is that the slope is also exactly right. The electron
scattering experiments give essentially two numbers, namely the
radius at about half density and the slope of the density versus r

curve at that point. This slope is exactly given by Negele's calculations. The calculation of course gives a maximum at the center,
and also wiggles which correspond to the shell structure. The
central maximum is due to the 0s and 1s wave functions. The shoulder
at about 2.5F is due to the states 0d and 1s, and the minimum at
about 1.5 F corresponds to the location of the 0p protons. There
are relatively few of these, and therefore we have a relative
minimum here.

The adjustment I mentioned earlier, was made in the repulsive
part of the direct interaction. The philosophy behind adjusting
just that part is that we think that the long range part is fairly
well known because we know the nuclear forces quite well at medium
and large distances, while we don't know them well at short distances.
Secondly, the ratio ψ/ϕ is nearly one at large distances, so we have
essentially the uncorrected nuclear force and so we don't want to
make any corrections at large r. But we do think that the short
range behavior will be considerably influenced by the higher order
interactions: by the 3-body correlations, 3-body forces and so on.
Adjustments were made for two purposes. First, to get the correct
binding energy for nuclear matter; this is well justified because
in nuclear matter we know that only with corrections do we get about
the right binding energy. Second, to get the radius of the calcium
nucleus to agree with experiment. This is essentially an adjustment
in the density dependence of the effective nuclear force, and is
not justified by anything in nuclear matter calculations. It is
the one purely arbitrary adjustment which was made. It was of
course necessary to get good agreement with the radius. What was
done was simply to multiply the short range repulsive interaction
by a constant chosen to be 0.67 for $k_F=1.0F^{-1}$ and 0.84 for $k_F=1.4F^{-1}$.
Reduction of the repulsion means we get more binding, as we should
for nuclear matter, but the other adjustment means that we have
shifted the minimum of Siemens'[8] density curve (Fig. 2) to lower
density in order to accommodate the observed density of calcium.

Figure 9 presents the results of a calculation by Negele[9] of
the electron scattering from Ca^{40} at 750 MeV[10]. The solid curve
is a fit to the data by the Stanford group, while the dashed curve
follows from Negele's calculations. At small angles Negele's theory
isn't as good as the empirical fit; at large angles on the other
hand the theory is better. I should mention that this empirical
fit assumed a perfectly smooth electron distribution. Later the
Stanford people put some wiggles into the density distribution and
then, of course they can do as well as they like. But Fig. 9 proves
that Negele's wiggles are at the right place and of about the right
magnitude to wash out the third maximum in the appropriate manner.

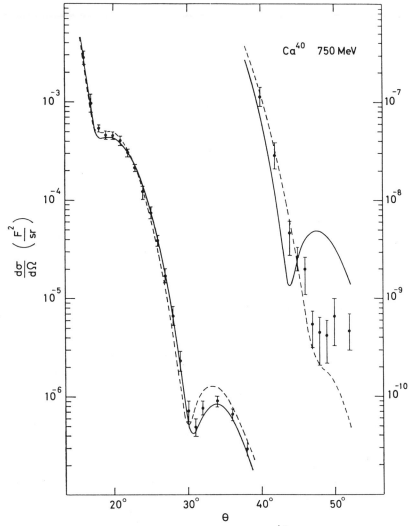

Fig. 9.--Electron scattering from ^{40}Ca at 750 MeV.

In Fig. 10 are shown the calculated and observed distributions for some closed shell nuclei. The observed distribution, the solid line, agrees very well with the calculated distribution out in the falling part of the curve. Which means that the theory gives the correct radius for lead, once you adjust the radius correctly for calcium. So you need to make only <u>one</u> adjustment in density and then all nuclei come out correctly. This is not true in the so-called Hartree-Fock-without-Brueckner calculations. That is, in calculations which try to use some soft core potential directly

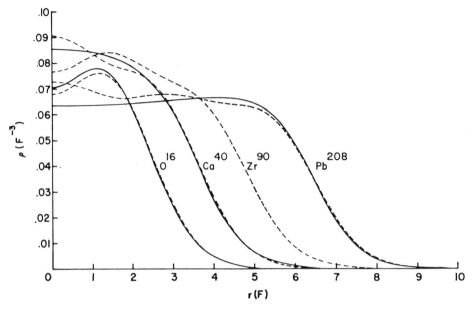

Fig. 10.--Charge-density distributions for some closed shell
nuclei.

without going through nuclear matter theory. These calculations
always predict lead nuclei which are too small if they have the
right size for the oxygen nucleus. This I think is to be expected
if you do not use the density dependent force which nuclear matter
theory gives you.

There are now experiments on zirconium which don't agree
well with the closed shell calculation. However, the forty protons
in zirconium are not in a good closed shell. Therefore, you should
not be surprised that some of the protons are promoted into the
next shell, that is from $1p_{1/2}$ to $0g_{9/2}$. If you promote a
pair of protons in this manner then the agreement is very good
indeed.[11]

Figure 11 compares the neutron distribution and the proton
distribution in the same closed shell nuclei. Figure 12 is the
same thing on an enlarged scale; it gives the difference between
the neutron and proton distributions in the case of ^{208}Pb. The
neutron density has been renormalized by multiplying it by 82/126,
i.e. we plot

$$\Delta\rho = (Z/N)\rho_n(r) - \rho_p(r). \tag{10}$$

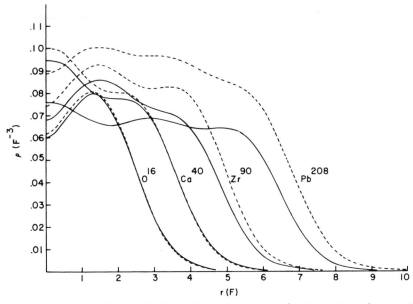

Fig. 11.--Neutron (dashed lines) and proton (solid lines) density distributions for some closed shell nuclei.

You see that in the inside of the nucleus you have some wiggles of $\Delta\rho$ corresponding to the shell structure. But on the whole you have fewer neutrons than protons. Outside you have more neutrons than protons, because the neutrons have a longer tail and this is confirmed by some experiments on K^- capture. At about the nuclear radius where the density falls off most strongly, the $\Delta\rho$ curve also goes to zero.

Table III gives the energies from Negele's calculations[9] in the case of calcium. Let us first look at the total binding energy. If you don't adjust the strength of the forces, the calculated binding energy is about 3 MeV. The adjustment of the forces to give the

TABLE III.--Single-particle energies and binding energies and binding energies for Ca^{40}, in MeV. Single-particle energies represent barycenters for spin-orbit doublets, and binding energies contain the c.m. correction. DDHF indicates density-dependent Hartree-Fock.

	0s	0p	0d	1s	B.E.
HF, bare	-58.2	-37.3	-16.5	-14.4	3.0
DDHF, bare	-43.6	-25.3	- 9.7	- 8.0	3.2
DDHF, adjusted	-46.9	-30.2	-15.2	-11.9	7.5
Expt	-53±11	-37±6	-16.2	-15.7	8.55

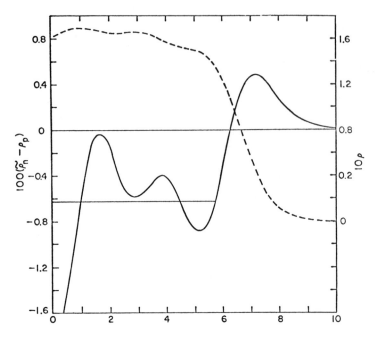

Fig.--The difference between the proton and neutron density distributions for ^{208}Pb (solid line). The dashed line shows the neutron density distribution. The abscissa is the radius in F and the ordinates are nucleons/F^3.

correct binding energy for nuclear matter makes it right for calcium within 0.5 MeV, and also gives correctly the energies of the various particle states. That is, the Hartree-Fock energies are essentially equal to the observed removal energies of these particles. Inserting the density dependence means that a least for the higher shells, that is, for those least bound, there should not be a rearrangement correction because this correction is contained in the density dependence. For the innermost shells there may be such a correction, the Brueckner-Goldman[12] correction, but this has not yet been calculated. The agreement for the 1s state can be much improved by making a correction for "starting energy."

Table IV shows the total binding energies for various nuclei. The agreement for lead is perfect which is perhaps not too surprising because lead is much like nuclear matter. The agreement for the lighter nuclei is not so perfect; it's within about 0.5 MeV.

TABLE IV.--Total Binding Energies per Particle (in MeV), and Nuclear
          Radii (in fm), According to the Negele Theory.

Nucleus	O^{16}	Ca^{40}	Ca^{48}	Zr^{90}	Pb^{208}
Theoretical Binding Energy	7.59	7.99	7.96	8.33	7.83
Experimental Binding Energy	7.98	8.55	8.67	8.71	7.87
Proton rms radius Theoretical	2.71	3.41	3.45	4.18	5.44
Proton rms radius Experimental	2.64	3.43	3.42	–	5.44
Neutron rms radius Theoretical	2.69	3.37	3.60	4.30	5.67

Don Sprung[13] tells me that he has an alternative density-dependent
interaction which does equally well for all nuclei, and which might
be even better here. Also interesting is the root mean square radius
of the nucleus for protons and for neutrons. For calcium, of course,
the protons are a little farther out than the neutrons. For lead
the neutrons are farther out than the protons.

Figure 13 gives a fake potential. Take the Hartree-Fock wave
functions of the neutrons in ^{208}Pb, and insert them into a one-
body Schrödinger equation. Then from the wave functions you can
deduce an effective one-body potential which would give the same
wave functions. This is plotted for various states, the Os being
the lowest and the outermost neutrons the highest in the diagram.
The figure shows the Brueckner effective-mass-syndrome, namely that
the less-bound nucleons have a smaller attractive potential. The
main effect here comes from the exchange force, not from the other
energy dependences. Another interesting point is that the potential
remains quite large, about 5 MeV, at a radius of 9F, while the radius
of the nucleus is only about 6F.

These have been some of the results of Hartree-Fock calculations.
The theory of course has a number of drawbacks. One is that it is
necessary to fudge the fundamental interaction. But a more
important one is the spin-orbit term. This was calculated by Negele[9]
in the following way. He took the Hartree-Fock wave functions for
^{40}Ca, and then calculated the expectation value of the full effective
potential g including the J dependence. Thereby he obtained the
difference in energy between an extra $f_{7/2}$ and $f_{5/2}$ nucleon, giving
the spin-orbit splitting. It comes out correctly as does that for

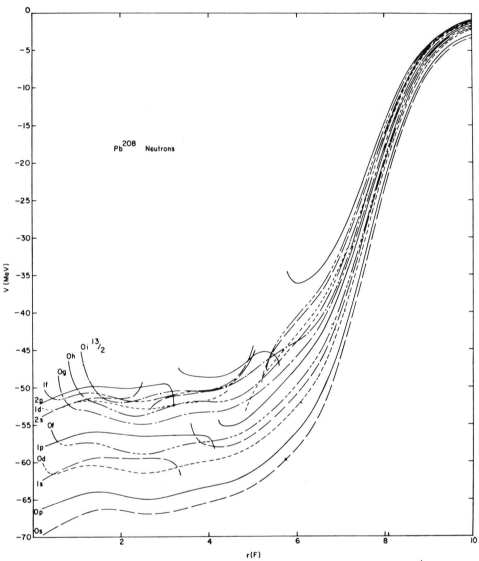

Figure 13.--Equivalent local single-particle neutron potentials for
^{208}Pb.

$d_{5/2}$ and $d_{3/2}$. Then he uses the old Blin-Stoye assumption for the
form of the spin-orbit interaction:

$$\frac{C}{r} \frac{d\rho}{dr} \; \vec{\ell} \cdot \vec{s}. \tag{11}$$

The coefficient C is derived by Negele from calcium, and is then
used for everything. This procedure gives agreement with experiment.

However it doesn't agree with theory. When you look at the spin-orbit interaction in detail you find that a given complete but spin-unsaturated shell like $f_{7/2}$ attracts an additional $f_{5/2}$ nucleon more than an additional $f_{7/2}$ nucleon, i.e. it gives an inverted spin-orbit splitting. That's bad. So if you use the correct theory then you are apt to get much too little spin-orbit splitting for a nucleus like ^{208}Pb-plus-one-nucleon. This I think is one of the major unsolved problems of nuclear physics: to get the correct spin-orbit splitting from theory. We have not been able to do this and neither has anybody else, as far as I know.

I have described just the Negele method of the density dependent force. There is at least one good competitor which is the oscillator basis calculation which has been used by many people starting with Baranger et al.[14], and including McCarthy and Davies[15], Becker et al.[16] and others. A recent paper by Davies and McCarthy[17] on this oscillator basis calculation gives some very good results and some quite bad ones. The good result is that the nuclear radius has the correct dependence on the mass number A, viz. r_o does not depend on A, which is what experiment shows. The Davies-McCarthy calculation gives this result, the same of course as does Negele's, but not the same as many other calculations.

Second, Davies-McCarthy use the occupation number of occupied states which is a very important quantity. The occupation number has an effect similar to Negele's density dependent force in that it gives high total binding energy without an unduly strong binding of individual particle states. If you don't take the occupation number into account then you are apt to get much too high binding energies for individual orbitals. For the $1p_{1/2}$ state of Oxygen, for example, you get 25 MeV or 30 MeV instead of about 15 MeV at the time you get the right total binding energy. Use of the occupation number is certainly well justified in theory, and is a very powerful point which has been especially emphasized also by Brandow[18] and Becker[19].

The bad result is that all nuclear radii are too small by about 15%. I don't think anybody clearly understands the reason for this. I surmise that they may have to correct their force in the way Negele has done, namely bring the saturation to a lower density than the original force wants it to have. This is difficult to achieve by a realistic force; most of the realistic forces which have been proposed go the other way. For instance, the one-boson-exchange potentials have saturation at much too high density, so they are much worse than Reid[2] and much worse than any of the regularly used phenomenological potentials. But you apparently must apply this correction in order to get the right radius of the nucleus.

The binding energies in Davies-McCarthy are not too bad but are generally somewhat low. This could easily be corrected by doing the

fudging that Negele has done, namely reducing the repulsion at
small distance arbitrarily. This can easily be inserted into the
theory, but the density dependence can not easily be inserted into
a theory which is not local but is on an oscillator basis.

There is another type of calculation with a purely phenomenological
approach, started by Skyrme[20] and used extremely successfully by
Brink and Vautherin.[21] I want to mention it because it has been
very successful and is very simple. In this approach the nucleon-
nucleon interaction is put in as an expansion in k which is broken
off after the k^2 term. It includes an exchange and a direct force
implicitly in its formulation. Density dependence is included by
having a term proportional to the cube of the density and otherwise
a delta-function. This method is the main competitor with Negele's
calculation regarding agreement with experiment. Namely, it gives
the right densities, the right radii, the right density distributions, the right binding energy for all nuclei, and the right
energies for individual particle levels. It may not be quite right,
but this point I am not sure about, concerning the symmetry term,
i.e. the difference between like-particle and unlike-particle
interactions.

Finally I want to mention a paper by Wong and Sawada.[22]
Wong and Sawada have calculated just nuclear matter, and have
used one-boson-exchange potentials as well as some phenomenological
ones. The result is that the one-boson-exchange potentials always
give too much binding and too high a density as is shown in Table V.

TABLE V.--Parameters of Nuclear Matter for Several Two-Body Inter-
            actions.

Potential	HJ	RS	UG1	BS	UG3
Binding Energy	10.0	14.0	19	21	25
k_F	1.4	1.50	1.66	1.71	1.85
K	20	130	190	290	290
κ	.207	.137	.087	.091	.069
P_D	7.0	6.5	5.5	5.4	5.0

The first line in Table V is the binding energy. I don't quite
agree with Wong and Sawada in the corrections to the binding energies.
According to them, the Reid soft core $(RS)^2$ gives 14.0 MeV; I
think it gives 15.7 MeV. But the comparison between potentials is
reasonable: the Hamada-Johnston hard core (HJ)[23] gives 4 MeV less
than the Reid soft-core. The one-boson exchange potentials all give

too much binding; especially Ueda-Green number 3 (UG3)[24] which gives as much as 25 MeV binding and of course is quite out of range. Bryan-Scott (BS)[25] is somewhat better. k_F, essentially the density, behaves similarly. The incompressibility parameter (K) for the Hamada-Johnston potential is totally crazy, it's only 20 MeV, which says that the minimum is essentially not visible. In the other cases it goes up as the binding goes up. κ, the percentage of the time that a nucleon is not in its normal state, goes in the usual way, becoming less when the binding energy is greater. The last line gives P_D the percentage D state in the deuteron; it is between 6 and 7 percent. Very likely that's the amount of tensor force which you need in order to get binding at about the right density. The Bressel-Kerman-Rouben[26] results are very similar to Reid's soft core.

REFERENCES

1. K. A. Brueckner, Phys. Rev. 97,1353(1955); K. A. Brueckner, C. A. Levinson, and H. M. Mahmoud, Phys. Rev. 95,217(1954); K. A. Brueckner and C. A. Levinson, Phys. Rev. 97,1344(1955).

2. R. V. Reid, Ann. of Phys. 50,411(1968).

3. G. E. Brown, A. D. Jackson, and T. T. S. Kuo, Nucl. Phys. A133,481(1969).

4. G. E. Brown and A. M. Green, Nuclear Phys. A137,1(1969).

5. B. H. J. McKellar and R. Rajaraman, Phys. Rev. C3,1877(1971).

6. B. D. Day, Phys. Rev. 187,1269(1969).

7. Wong and Sawada, UCLA preprint.

8. P. J. Siemens, Nucl. Phys. A141,225(1970).

9. J. W. Negele, Phys. Rev. C1,1260(1970).

10. J. B. Bellicard, P. Bounin, R. F. Frosch, R. Hofstadter, J. S. McCarthy, F. J. Uhrhane, M. R. Yearian, B. C. Clark, R. Herman, and D. G. Ravenhall, Phys. Rev. Letters 19,527(1967); R. F. Frosch, R. Hofstadter, J. S. McCarthy, G. K. Nöldeke, K. J. van Oostrum, and M. R. Yearian, Phys. Rev. 174,1380(1968).

11. J. W. Negele, Phys. Rev. Letters 27,1291(1971)

12. K. A. Brueckner and D. T. Goldman, Phys. Rev. 117,207(1960).

13. D. W. L. Sprung, private communication.

14. K. T. R. Davies, M. Baranger, R. M. Tarbutton, and T. T. S. Kuo, Phys. Rev. 177,1519(1969).

15. R. J. McCarthy and K. T. R. Davies, Phys. Rev. C1,1640(1970).

16. R. L. Becker, A. D. MacKellar, and B. M. Morris, Phys. Rev. 174, 1264(1968).

17. K. T. R. Davies and R. J. McCarthy, Phys. Rev. C4,81(1971).

18. B. H. Brandow, Rev. Mod. Phys. 39,771(1967).

19. R. L. Becker, Phys. Rev. Lett. 24,400(1970).

20. T. H. R. Skyrme, Phil. Mag. 1,1043(1956); Nuclear Phys. 9,615 (1959).

21. D. Vautherin and D. M. Brink, Phys. Letters 32B,149(1970).

22. Wong and Sawada, UCLA preprint.

23. T. Hamada and I. D. Johnston, Nucl. Phys. 34,382(1962).

24. T. Ueda and A. E. S. Green, Phys. Rev. 174,1304(1968).

25. R. Bryan and B. L. Scott, Phys. Rev. 177,1435(1969).

26. C. N. Bressel, A. K. Kerman and B. Rouben, Nucl. Phys. A124, 624(1969).

THE EFFECTIVE TWO-BODY INTERACTION IN FINITE NUCLEI AND ITS CALCULATION

Bruce R. Barrett

Department of Physics

University of Arizona, Tucson, Arizona

INTRODUCTION

The calculation of the effective interaction and of other effective operators in finite nuclei is an important link in the chain connecting microscopic phenomena, such as the nucleon-nucleon scattering phase shifts, with nuclear phenomena, such as cross sections and excitation energies. But, like all links in a chain, it depends upon the other links for its support. As we will see, the calculation of the effective interaction, which is part of the link having to do with the many-particle shell model (MPSM), is strongly dependent upon links having to do with the theory of the nucleon-nucleon interaction, the treatment of strong short-range correlations and the derivation of the single-particle shell model (SPSM).

This paper is a review of work which has been done and is currently being performed on the theory of effective operators and their calculation. Because of the author's particular association with one aspect of this problem, most of the work referred to will have to do with the effective interaction and its calculation. Space limitations make it impossible to be complete with regard to the tremendous amount of work which has been carried out in this field. Only the major points will be covered, and the reader is referred to the extensive literature on this subject for specific details and a complete list of references.

Effective operators are needed in nuclear physics because of the overwhelming number of degrees of freedom involved in a many-particle (MP) nuclear-structure calculation. In order to make such calculations tractable one must truncate to a limited and manageable number of degrees of freedom. This is a procedure which has been

used for many years in phenomenological MPSM calculations. It is
reasonable to ask whether or not this truncation procedure can be
justified theoretically and, if so, can the effective operators
needed in these truncated spaces be calculated in terms of a microscopic theory involving no adjustable parameters or phenomenological
input.

There are several approaches to this problem, which fall mainly
into two categories, those which emphasize modification of the MP
wavefunction and those which emphasize modification of the MP operator. This review will concentrate on one of the latter approaches,
namely many-body perturbation theory, which has the advantage of being well understood, intuitively appealing and extensively applied
in previous calculations.

BASIC THEORY

The many-body perturbation theory for the nuclear SM has been
formulated by a number of workers using both time-independent methods (mainly Bloch and Horowitz[1] and Brandow[2]) and time-dependent
methods (such as those of Morita[3] and Johnson and Baranger[4]). Although certain aspects of the time-dependent methods are intuitively
appealing, we will discuss only the time-independent method of Bloch
and Horowitz and Brandow, since it is older, more widely known and
understood, and more extensively applied in the literature.

The basic theory for the determination of the effective inter-
action in a truncated model space is as follows. In the infinite
Hilbert space one has an N-body Hamiltonian H, which satisfies the
Schrödinger equation

$$H^N \psi^N = E^N \psi^N, \tag{1}$$

where the superscript N explicitly indicates that the quantities H,
ψ and E are for N particles. From now on this N will be dropped,
but it is to be implicitly understood to be present in the following equations. As usual, H is the kinetic plus potential energy
(T+V) to which we can add and subtract a one-body potential U. We
can then group T and U to form H_0, the unperturbed Hamiltonian, to
which we know the solution, i.e.

$$H_0 \phi_i = \varepsilon_i \phi_i, \tag{2}$$

and (V–U) forms the residual interaction $\overline{V}$, which is assumed to be
small, so that it can be treated in perturbation theory. Since the
solutions of H_0 form a complete set, ψ can be expanded in terms of
its eigensolutions, ϕ_i, i.e.

$$\psi = \sum_i^\infty a_i \phi_i. \tag{3}$$

We cannot solve the Schrödinger equation for ψ and E, since we have too many degrees of freedom. Thus, we want to truncate Ψ into a limited dimensional space which is small enough to make calculations tractable. Let us call this small space our model space and denote it by $\underline{d}$, meaning that it is of dimensionality d. The projection of Ψ into $\underline{d}$ is denoted by Ψ_d, and we want to determine the effective Hamiltonian in $\underline{d}$ such that

$$H_{eff}\Psi_d = E\,\Psi_d, \tag{4}$$

that is, we want to obtain the same excitation energies E in the model space that we would have obtained for the energies of the lowest d states in the full problem. By definition

$$\Psi_d = \sum_{i\in d} a_i \phi_i, \tag{5}$$

and the effective Hamiltonian is of the form

$$H_{eff} = H_o + \mathcal{V}, \tag{6}$$

where $\mathcal{V}$ is the effective interaction to be determined. We now define a wave or model operation Ω, such that

$$\Psi = \Omega\,\Psi_d. \tag{7}$$

We note that Ω is an expansion operator, since it expands Ψ_d into Ψ, which is in the infinite Hilbert space. It then follows that we want to define $\mathcal{V}$ by

$$\mathcal{V} = \overline{V}\Omega, \tag{8}$$

since we want

$$\overline{V}\Psi = \overline{V}\Omega\,\Psi_d = \mathcal{V}\,\Psi_d. \tag{9}$$

To derive Ω and $\mathcal{V}$, we first note that Eqs. (1)-(3) imply that

$$(H-E)\Psi = \sum_i a_i(\mathcal{E}_i - E)\phi_i + \sum_i a_i \overline{V}\phi_i. \tag{10}$$

If we then multiply Eq. (10) on the left by ϕ_j^*, integrate over all space, and use the orthonormality property of the ϕ_i, we find that

$$a_j(\mathcal{E}_j - E) = -\langle \phi_j|\overline{V}|\Psi\rangle = -\langle\phi_j|\mathcal{V}|\Psi_d\rangle, \tag{11}$$

where we have used Eq. (9).

We now want to express Ψ in terms of Ψ_d so that we can determine Ω. To do this we note that

$$(E-H_o)\Psi = (E-H_o)\Psi_d + (E-H_o)\sum_{i\notin d} a_i \phi_i$$

$$= (E-H_o) + \sum_{i\notin d}\langle\phi_i|\mathcal{V}|\Psi_d\rangle\,\phi_i, \tag{12}$$

where we have employed Eqs. (2)-(3), (5) and (11). Dividing through by $(E-H_o)$, we find that

$$\Psi = \left\{ 1 + \sum_{i \notin d} \frac{|\phi_i\rangle\langle\phi_i|\mathcal{V}}{E-H_0} \right\} \Psi_d \equiv \Omega \Psi_d, \qquad (13)$$

so that

$$\Omega = 1 + \frac{Q}{E-H_0} \mathcal{V} \qquad (14)$$

and

$$\mathcal{V}(E) = \overline{V} + \overline{V} \frac{Q}{E-H_0} \mathcal{V}(E), \qquad (15)$$

where

$$Q = \sum_{i \notin d} |\phi_i\rangle\langle\phi_i| \qquad (16)$$

is the projection operator out of the model space. In Eq. (15) it has been explicitly noted that $\mathcal{V}$ is a function of the exact excitation energy for which we are solving.

Thus, the Schrödinger equation which we want to solve for E in the space d is

$$[H_0 + \mathcal{V}(E)-E] \Psi_d = 0. \qquad (17)$$

If the potentials used in determining $\mathcal{V}$ have an infinite hard core, then we must renormalize the interaction to take account of the strong short-range correlations produced by the hard core. In other words we must use Brueckner theory[5] to go from the potential $\overline{V}$ to the two-particle ladder series or G matrix, i.e.

$$G(E) = \overline{V} + \overline{V} \frac{Q_{2p}}{E-H_0} G(E). \qquad (18)$$

Thus, we can replace $\overline{V}$ by G in the definition of $\mathcal{V}$ simply by expanding $\mathcal{V}$ in a power series in $\overline{V}$ and then regrouping the series in terms of the definition of G. Then

$$\mathcal{V}(E) = G + G \frac{Q'}{E-H_0} \mathcal{V}(E), \qquad (19)$$

where $Q' = Q - Q_{2p}$.

Although the perturbation expansion for $\mathcal{V}$ can be worked out without using Goldstone diagrams, it is easier to visualize what is happening physically in terms of these diagrams. Since we want to divide our model space into two parts, a passive core and a small number of active valence orbitals, we will define the vacuum with respect to the closed core. Then an upgoing line represents a particle outside the core and a downgoing line represents a hole in the core, as shown in Fig. 1. A V interaction is represented by a dashed line; a G interaction by a wavy line and a U interaction by an X.

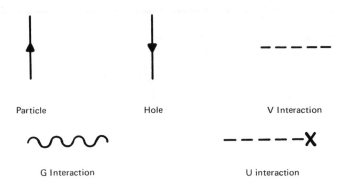

Fig. 1. Notation for the drawing of Goldstone diagrams.

We are now ready to apply the Bloch-Horowitz-Brandow procedure for reducing the problem to that of a few valence particles outside of a closed core. This procedure consists of three steps. First the wavefunction is reduced to that of a set of occupied orbitals in a closed and, therefore, passive core with the remaining small number of particles distributed over a few valence orbitals. For example, one can think of ^{18}O as being two neutrons distributed over the three sd shell orbitals outside an ^{16}O doubly closed-shell core. Having divided the model space into two parts, we note that the excitation energy E can be split into four parts, an unperturbed valence energy E_{0V}, an unperturbed core energy E_{0C}, a valence interaction energy ΔE_V and a core interaction energy ΔE_C. Second the core energy, $E_{0C} + \Delta E_C$, is removed from the problem so that we only have to calculate the excitation energy of a small number of valence particles. And finally the valence interaction energy is eliminated from the energy denominator in $\mathcal{V}$, so that $\mathcal{V}$ is energy independent and can be expressed in terms of known, unperturbed single-particle (SP) valence energies.

The elimination of the core energy and the valence interaction energy proceeds as follows. There are three kinds of diagrams in the expansion for $\mathcal{V}$, as shown in Fig. 2, which is for the case of three valence particles. Diagram (a) is a valence diagram which contains interactions all of which are connected to external valence lines. It should be noted that a valence diagram may be only partially linked, as diagram (a) is. Diagram (b) is a core diagram which contains totally passive valence lines. The third kind of diagram is the mixed diagram which contains both core and valence interaction processes. Diagrams (c) and (d) are examples of mixed diagrams. We now expand the core interaction energy ΔE_C out of all energy denominators in $\mathcal{V}$ and the valence interaction energy ΔE_V out of certain appropriate energy denominators. When this is done, one finds that the diagrams with explicit ΔE_C and ΔE_V insertions identically cancel the mixed terms, as shown by both Bloch and Horowitz[1] and Brandow.[2] When the mixed terms have been eliminated, the model interaction is then of the form of a valence interaction plus a core interaction, the latter being simply the sum of all core diagrams. One can then trivialy show that all core quantities disappear from the problem, so that we have a Schrödinger equation and an effective interaction only in terms of valence quantities. Thus, we have uncoupled the ground state problem and the excitation spectrum problem, which can then be solved separately.

The main problems remaining are that $\mathcal{V}$ still depends on the valence interaction energy ΔE_V and that the diagram expansion for $\mathcal{V}$ contains partially linked diagrams. Brandow[2] showed that by factoring the remaining ΔE_V out of the energy denominators in $\mathcal{V}$ that the diagrams with these insertions cancel all of the partially linked diagrams, so that one has a completely linked diagram expansion for $\mathcal{V}$, which contains only unperturbed SP energies in the energy denominators. It is important to have a completely linked expansion, since it guarantees that $\mathcal{V}$ does not depend upon how the model space is separated into core and valence states. The price which one pays for doing this is the introduction of a new kind of diagram, called a folded diagram by Morita and Brandow. In this formalism folded diagrams correspond to "left over" ΔE_V insertions which were not needed to cancel the mixed terms. Folded diagrams naturally appear in the time-dependent formalism of Morita,[3] Johnson and Baranger[4] and others.[6] It is clear in these time-dependent methods that folded diagrams represent physical processes which must be subtracted from the theoretical instantaneous effective interaction to correct for the known fact that there is a time delay between nuclear interactions.

Thus, we now have a Schrödinger equation which depends only on valence quantities,

$$[H_{ov} + \mathcal{V}(E_{ov}) - E_v] \Psi_{dv} = 0, \qquad (20)$$

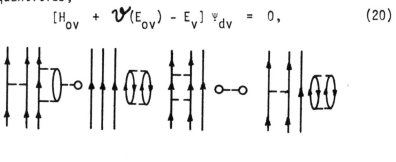

a   b   c   d

Fig. 2. Examples of the three kinds of diagrams contributing to the effective interaction for the case of three valence particles. Diagram (a) is a valence diagram, (b) is core diagram, and (c) and (d) are mixed diagrams.

and an effective interaction which is energy independent,

$$\mathscr{V}(E_{ov}) = G + G \frac{Q'}{E_{ov}-H_{ov}} \mathscr{V}(E_{ov}).$$ (21)

However, $\mathscr{V}$ is no longer Hermitian, since its diagram expansion contains folded diagrams, which are not Hermitian by the nature of the folding process. The process in Fig. 3(a) is not allowed by Q', since it returns to an intermediate state of two valence particles. One obtains a folded diagram from this process by folding diagram 3(a) along the two interaction lines which bracket the forbidden intermediate state. The folded diagram obtained is shown in Fig. 3(b). The downgoing line is circled to indicate that it is a valence particle and not a hole, and one picks up an extra minus sign, which says that this process is subtracted from the interaction. The complete diagram rules for folded diagrams and their phase factors are given by Brandow.[2]

The folding process is non-Hermitian since the folding is always done with regard to the uppermost interaction. This non-Hermiticity is simply a statement of the well-known fact that the projection of orthonormal wavefunctions into a smaller space leads, in general, to non-orthonormal wavefunctions in the smaller space. One can make $\mathscr{V}$ Hermitian by contructing an averaged $\mathscr{V}$, but in this case one loses the simple connection with projected wavefunctions which one would like to maintain.

It should be emphasized that a similar formalism can be worked out to obtain the effective operator in the space $\underline{d}$ for any physically observable operator and that a diagrammatic perturbation expansion can be obtained for these operators. For example, one can obtain an effective charge operator or an effective magnetic-moment

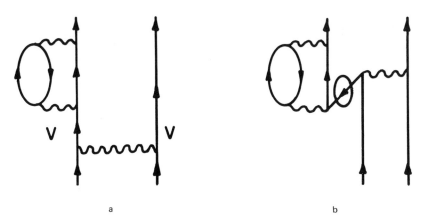

a   b

Fig. 3. An example how one folds a forbidden Goldstone diagram [3(a)] to obtain a folded diagram [3(b)].

operator, as well as an effective interaction, as has been shown
by a number of physicists.[7,8]

PROBLEMS IN CALCULATING $\mathcal{V}$

At this point we have a theory which justifies the truncation
of the wavefunction into a small model space of a few active valence
orbitals and which gives an expression for the effective interaction
in this space which can be expressed in a diagrammatic perturbation
expansion. The question is now whether or not we can perform mean-
ingful calculations to obtain $\mathcal{V}$. There are many problems connected with performing such a calculation, and these problems are
connected with the complete microscopic theory of nuclear phenomena,
as mentioned earlier.

First, the calculation assumes that we know the nature of the
nucleon-nucleon interaction, which we can express in terms of a
potential. Such a theory does not exist at the present time, and
we must rely on phenomenological potentials fit to the nucleon-nucleon scattering phase shifts and to the bound state properties of
the deuteron. There are many such phenomenological potentials, some
with hard cores, others with soft cores; some local, some nonlocal;
some energy dependent, some energy independent. Although these potentials give similar results on-the-energy-shell, they give, in
general, quite different results off-the-energy-shell, and these
off-the-energy-shell effects can be important in nuclear structure
calculations. Thus, the choice of potential can affect the results
obtained and the techniques used for performing the calculation.

Second, one assumes that the nuclear Hartree-Fock (HF) problem
has been completely solved, so that the correct SP energies, $\mathcal{E}_i$, and
wavefunctions, ϕ_i, are known and can be used in the calculation of
$\mathcal{V}$. In practice one does not use HF wavefunctions in performing
SM calculations but assumes that they are well approximated by harmonic-oscillator (HO) wavefunctions, which greatly simplify the
calculations. The spacings between major shells are generally
assumed to be equal to the HO spacing, so that energy denominators
used in the perturbation expansion are multiples of this spacing.
Recent calculations indicate that these assumptions may not be very
good and can introduce sizeable errors into the perturbation theory
calculations.[9,10] Clearly these assumptions should be studied in
more detail.

Connected with the HF calculation is the problem of self-energy insertions and their cancellation by U insertions in diagrams.
If one has performed the HF calculation, then by definition a certain set of U insertions has been identically cancelled by an appropriate set of V self-energy insertions. However, the use of HO
wavefunctions, instead of HF wavefunctions, does not guarantee this

cancellation between U and V self-energy insertions. It is assumed
that HO wavefunctions are close enough to the self-consistent HF
wavefunctions so that the cancellation between U and V self-energy
insertions is large. A recent calculation by Ellis and Mavromatis[10]
indicates that it is not a good approximation to assume that HO
wavefunctions are sufficiently close to self-consistency that self-
energy insertions can be ignored.

Fourth, if one uses a phenomenological potential which produces
strong short-range correlations, then this potential must be renormalized using Brueckner theory[5] to produce a G matrix which has finite matrix elements between the unperturbed wavefunctions. There
are many problems connected with the calculation of the G matrix,
such as the treatment of the projection or Pauli operator Q_{2p} and
the choice of intermediate-state wavefunctions and energies. The
initial or occupied states are approximated by HO bound state wavefunctions, and their energies are taken to be the self-consistent
HF energies. The intermediate or unoccupied states are not well
determined and can be approximated by plane wavefunctions or by HO
wavefunctions, with free-particle kinetic energies used for the former and HO SP energies used for the latter. Neither approximation
is perfect, but there are fewer calculational problems when HO intermediate states are used.[11] Plane wavefunctions and free-particle
kinetic energies lead to many problems in calculating G, such as
non-orthogonality between the initial and intermediate states; approximate treatment of Q_{2p}, which is not diagonal in this basis;
and a large gap between the occupied and unoccupied state spectra
which tends to underestimate the contribution to G of low-lying in-
termiediate states. On the other hand the consistent use of HO
wavefunctions eliminates the non-orthogonality problem, and recent
calculations of the G matrix using matrix inversion by Barrett,
Hewitt and McCarthy[12] (BHM) have shown that the matrix elements of
G can be computed exactly, if a HO basis is used in both the initial
and intermediates states. Using HO wavefunctions they are able to
calculate the matrix elements of G in the two-particle basis, in
which Q_{2p} is diagonal, and hence, they can treat Q_{2p} exactly. BHM
also have the ability in their calculations to shift the entire intermediate-state spectrum with respect to the initial-state spectrum and have been able to study the effects of different intermediate-state spectra on the results obtained for G. Their calculations show that the size of the matrix elements for G can vary
widely for different spacings between the initial- and intermediate-
state spectra, indicating that much more attention should be paid
to what one uses for the intermediate-state spectrum in calculating
G.

Finally, it is not clear that simple perturbation theory is the
best method for calculating $\mathcal{V}$, because of convergence problems
which will be discussed shortly. One could also think of calculating $\mathcal{V}$ by matrix inversion of the basic equation, by expanding the

basis so as to simplify the expansion for $\boldsymbol{\mathcal{V}}$,[13] by iteration methods,[14] and by rearrangement methods.[15,16]

CALCULATIONS

Although the effective interaction theory we have just developed is completely general and holds for any number of valence particles, calculations have been performed for only two valence particles. Calculations for three or more particles would be of interest and worth doing, since they have the possibility of telling us something about three or more body forces in nuclei.

The first step in calculating $\boldsymbol{\mathcal{V}}$ is the determination of G, since most phenomenological potentials have an infinite hard core and/or produce strong short-range correlations. So one has all the problems connected with the calculation of G. The first effort to calculate G for a finite nucleus from a phenomenological potential with no adjustable parameters was made by Dawson, Talmi and Walecka[17] who solved the Bethe-Goldstone equation[5] for G for ^{18}O. They reproduced the ordering of the low-lying spectra for ^{18}O but the 0^+ ground state was roughly 1 MeV too high in energy.

The first major attempt to calculate G and higher-order corrections to it was performed by Kuo and Brown.[18] Although they carried out an improved G matrix calculation, they used plane wave intermediate states which means that they could treat the Pauli operator Q_{2p} only approximately. Recently there has also been found an arithmetic error in their treatment of the tensor term.[19] In any case, their calculations were the first to point out the importance of other terms in the perturbation expansion for $\boldsymbol{\mathcal{V}}$ besides G. Figure 4 shows the three, second-order-in-G terms which they included in their calculations of $\boldsymbol{\mathcal{V}}$. The first diagram is the well-known core-polarization term whose importance was first indicated by Bertsch[20] and confirmed by the calculations of Kuo and

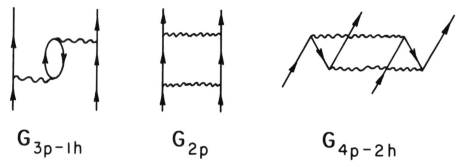

$$G_{3p-1h} \qquad G_{2p} \qquad G_{4p-2h}$$

Fig. 4. The three second-order-in-G diagrams contributing to $\boldsymbol{\mathcal{V}}$ calculated by Kuo and Brown, refs. 18 and 21.

Brown. Core-polarization is an important process since it represents a long-range correlation which is produced by short-range interactions. In this process one of the valence nucleons excites a
particle out of the closed core producing a particle-hole (p-h)
pair. The excited particle then interacts with the other valence
particle and decays back into the core, filling the hole state.
Thus, we note that although we have eliminated all static features
of the core by factoring out the core energy, $\mathcal{V}$ does contain all
dynamical core effects.

Kuo and Brown calculated this term in second-order perturbation theory using their previously computed G matrix elements and
energy denominators equal to $-2\hbar\omega$, representing intermediate-par-
ticle excitations of two major shells. It was assumed that excitations to higher excited levels were negligible. It should be
noted that there is an inconsistency in their calculations (and
in most perturbation theory calculations) at this point, since the
intermediate-state energy spectrum used in the calculation of G
and $\mathcal{V}$ should be the same. Nevertheless, their calculations showed
that the core-polarization term is a significant contribution to
$\mathcal{V}$, and, in particular, that it lowers the 0^+ ground state of ^{18}O,
bringing it into better agreement with experiment.

Later Kuo[21] further improved his G matrix calculations and
suggested the inclusion of two more second-order diagrams to improve the agreement of the calculated spectra with experiment.
These extra diagrams are the second two terms shown in Fig. 4.
The G_{2p} term is simply a second-order ladder diagram with an intermediate excitation of $2\hbar\omega$. In principal, this term should already be included in the calculation of G. However, it was argued
that the use of plane wave intermediate states with an angle-averaged Q_{2p} operator underestimated the contribution of the low-lying
two-particle excitations, which should, therefore, be included as
a perturbation correction. There is obviously some double counting
involved when this term is included; however, the amount of this
double counting is still uncertain, although a calculation by
Kirson[22] indicates that this double counting is 100% and that these
two-particle ladders should be omitted.

The final second-order term has a 4 particle—2 hole (4p-2h)
intermediate state, which represents a deformed state of the nucleus. Such deformations would be expected to be important in $\mathcal{V}$,
and the calculations by Kuo found it to be helpful in depressing
the 0^+, T=1 and 1^+, T=0 ground states in ^{18}O and ^{18}F respectively,
although its effects were not as significant as those for the core
polarization.

If the second-order-in-G terms make such a significant contribution to $\mathcal{V}$, one might reasonably ask whether or not even
higher-order terms in G are important. In particular, one would

expect that the p-h excitation of the core should be summed to all orders, that is, as a string of upward going p-h pairs, known as the
TDA, or as a series of both forward and backward going p-h pairs,
known as the RPA. Calculations by Zamick and Osnes and Warke[23] of
the TDA and RPA series did show that a strong collective enhancement was obtained when these series were summed and that in particular the ground states were greatly depressed in energy, far below
the experimental values. Clearly some compensating effect was needed
to offset this strong negative enhancement.

At the same time that these RPA calculations were being performed, Barrett and Kirson[15] undertook a calculation of all third-
order-in-G terms contributing to $\mathcal{V}$ for A = 18 nuclei. They used
the Kuo G matrices and made the following assumptions. They included (1) no diagrams which contained two-particle ladders, since
these ladders should have been included in the calculation of G,
(2) no diagrams with self-energy insertions, i.e., they assumed HF
self-consistency, and (3) all diagrams were calculated only for intermediate excitations of $2\hbar\omega$. Except for the two-particle ladder
assumption, these assumptions were the same as those used by Kuo
and Brown.[18] The results of their calculation showed that the total
third-order contribution is as big as or bigger than the second-
order contribution and of opposite sign. Their results indicate
that there is no apparent convergence in the perturbation expansion
for $\mathcal{V}$ in powers of G.

However, their calculations did provide some interesting in-
sights regarding possible important terms contributing to $\mathcal{V}$. The
third-order deformed 4p-2h term was found to be an order of magnitude smaller than the second-order term, indicating that there is
apparently no collective enhancement produced by this process. This
conclusion is consistent with a recent calculation by Goode,[24] who
summed this effect to all orders in G and found no enhancement.

Secondly, the large terms in third order fell into three groups.
First, there were those which contributed to number-conserving sets.
These are sets of diagrams which are connected by the conservation

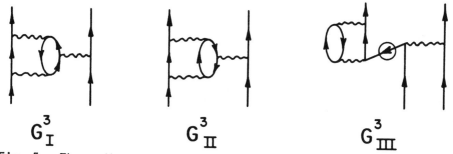

Fig. 5. Three diagrams constituting one of the two number-conserving sets in third order in G.

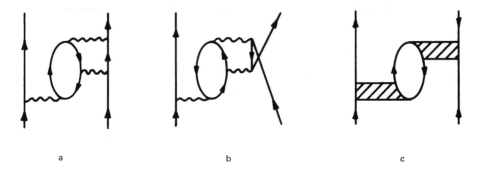

a   b   c

Fig. 6. Diagrams (a) and (b) are third order in G processes which represent a renormalization of the vertex connecting a valence particle to the particle-hole propagator. Diagram (c) is the generalization of this vertex renormalization to all orders in G, which is represented by a black box.

of the number of particles, as pointed out by Brandow,[25] who predicted that the terms in these sets should tend to strongly cancel. An example of such a number-conserving set is shown in Fig. 5. The calculations of Barrett and Kirson[15,26] showed that although individual terms in these sets might be large, the sum of the terms did strongly cancel. It should be noted that the set shown contains a folded diagram and that this folded diagram is important in producing the strong cancellation.

Second, there were the terms contributing to the TDA and RPA series which have already been seen to be important when summed to all orders.

Finally, there was a new type of process, as shown in Fig. 6. This process shown in (a) and (b), was found to be large and positive and was mainly responsible for the large third-order result. Physically, this process represents a modification or renormalization of the vertices which connect the valence particles to the p-h propagator. This vertex renormalization can be generalized to all orders in G. This generalization is represented schematically by the black boxes in diagram 6(c).

It was suggested by Kirson and Zamick[27] that there are two major types of renormalization in calculating $\mathcal{V}$, namely (1) propagation renormalization, i.e. renormalization of the p-h excitation, such as in the TDA or RPA, and (2) vertex renormalization, and that these renormalizations should be calculated to all orders in G, since they are collective effects of opposite sign but of roughly the same magnitude.

Such a calculation was undertaken by Kirson,[28] who not only

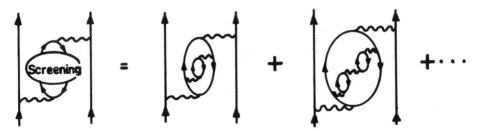

Fig. 7.   Diagrammatic definition of the screening series.

summed the RPA series and the vertex renormalization series but also
a third series indicated in Fig. 7.   Kirson called this process
screening, since the internally excited p-h pairs tended to weaken
or screen the strength and coherence of the outer p-h pair.

Kirson summed the RPA series, the screening series and the
vertex renormalization series to all orders in G and found that the
net result of these three infinite series was approximately zero.
Figure 8 shows the detailed results of Kirson's calculation for
the ^{18}O spectrum.   The TDA and RPA series greatly depress the spectrum.   The inclusion of screening damps the RPA back to the TDA, a
result which was independently obtained by Osnes, Kuo and Warke.[29]
Finally, the inclusion of the vertex or black box renormalization
produces a spectrum which looks like that for the bare G.

At first this would appear to be a depressing result, since we
know that the bare G does not reproduce experimental spectra.   However, there are two points which offer hope.   One is the G matrix
calculations of BHM[12] which indicate that the G matrix can be made
much more attractive by varying the intermediate spectrum, especially by decreasing the gap between the occupied and unoccupied states,
as shown in Fig. 9.

The other is a calculation by Goode[24] in which he solved the
problem of a fully $2\hbar\omega$ excited ^{16}O core and then sandwiched these
excitations between two neutron states to obtain the completely
renormalized propagator for ^{18}O.   The results of his calculation
were that the completely renormalized propagator, which contains
more processes than Kirson's calculation of this effect, is essentially equal to the lowest-order core-polarization term.

Since Goode does not include the vertex-renormalization effect,
it is interesting to conjecture that $\mathcal{V}$ might still be calculated
in a fairly simple manner, as indicated in Fig. 10.   Namely, one
should perform a much improved calculation of G in which the intermediate-state spectrum is correctly treated.   In this case G should
be more attractive.   Then the calculations of Goode and Kirson
would indicate that the major correction term to G is the single

p-h pair connected to the valence particles by <u>completely</u> renormalized vertices. However, this is only a conjecture on the part of
the author, and there are no detailed calculations at this time to
support it. Also this conjecture is based on the assumption that
all processes found to be small in third order remain small in
higher orders and, in particular, that they do not give rise to any
collective enhancement when summed to all orders in G. In this
same regard it is assumed that the number-conserving sets in higher
order strongly cancel and do not produce a significant effect.

Also, as was pointed out earlier, there are many other problems
which must be considered in much more detail, such as the choice of
potential, the choice of SP wavefunctions and energies and the

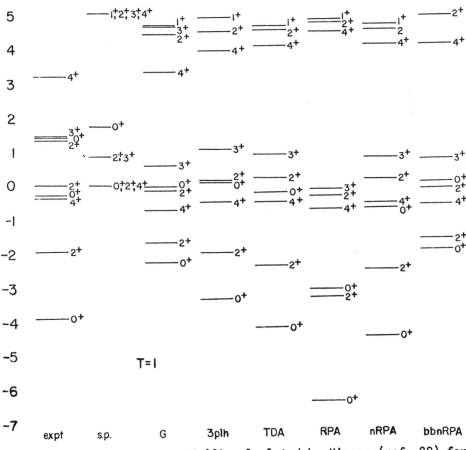

Fig. 8. The T=1 spectrum of ^{18}O calculated by Kirson (ref. 28) for
different combinations of the processes contributing to $\mathscr{V}$. The
symbol nRPA denotes RPA plus screening. bbnRPA indicates the results for the black box (vertex) renormalization, screening and
RPA processes summed to all orders in G.

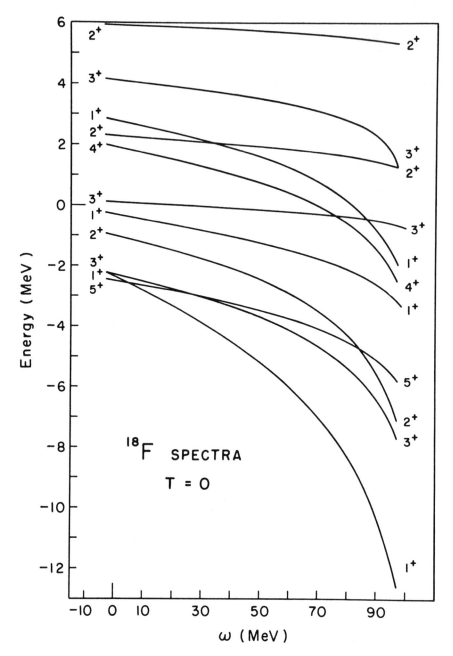

Fig. 9. Energy spectrum of ^{18}F calculated with the BHM G matrix elements (ref. 12) versus the starting energy, ω, in G which is related to the gap between the occupied and unoccupied SP states.

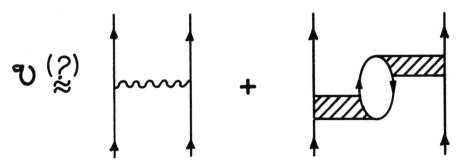

Fig. 10. A conjecture for a fairly simple diagrammatic expansion for $\mathcal{v}$ based on the results of refs. 24 and 28.

treatment of self-energy insertions, before one can honestly say that an accurate calculation of $\mathcal{v}$ has been performed. Unfortunately, present research seems to indicate that careful consideration of these problems will lead to results which are more repulsive in magnitude and, thus, will push the calculated spectra farther away from the experimental results.[9,10]

It should be noted that the calculations by Goode[24] and Kirson[28] do not prove that the perturbation series for $\mathcal{v}$ converges. They only show that when certain subsets of diagrams are summed to all orders in G one obtains a finite result. In fact, Schucan and Weidenmüller[16] have recently investigated the perturbation expansion for $\mathcal{v}$ and algebraically studied its convergence properties. They find that the expansion is likely to diverge or at least to converge poorly if the spectrum of states in the model space d overlaps with the spectrum of states outside d, that is, if the d states calculated in the model space do not in actuality correspond to the lowest d states in the full Hilbert space. This criterion for divergence was also pointed out earlier by Brandow.[25] Unfortunately, this condition occurs in almost all physical cases of interest, indicating that a priori we should have no reason to believe that the perturbation expansion for $\mathcal{v}$ converges.

However, Schucan and Weidenmüller show that it is possible to rearrange the series algebraically, so that convergence is improved. They also suggest that convergence can be improved by omitting states or effects from the calculation which would tend to mix the states in the model space with the states excluded from the model space (such as the 4p-2h intermediate states in ^{18}O). Consequently, they feel that $\mathcal{v}$ cannot be calculated without introducing phenomenological and/or experimental information into the calculation to guide one away from the physical effects which are causing the series to diverge. This point of view has also been advanced by Harvey and Khanna.[8] However, it is not obvious that this procedure greatly improves convergence. For example, even if the 4p-2h intermediate states are omitted in the calculation of $\mathcal{V}$ for ^{18}O, the perturbation expansion does not show any marked improvement in convergence in going from first to second to third order in G. It is possible that a Kirson-like calculation might produce better results if the 4p-2h states are omitted, since the vertex renormalization would be weakened more than the propagator renormalization.

Finally, I would like to mention that extensive calculations have also been carried out to compute the perturbation expansions for the effective charge and magnetic-moment operators and that these expansions show equally poor and, in many cases, worse convergence than the expansion for the effective interaction. However, there is no reason to believe that these different expansions should have similar convergence properties, as Palumbo[30] has demonstrated.

Although the problem of calculating $\mathcal{V}$ has turned out to be far more complicated than it appeared to be earlier, we have learned a great deal of physics from the work that has been done, such as (1) improved methods for treating the hard core and strong short-range correlations, (2) the existence and importance of vertex as well as propagator renormalization and (3) the importance of the intermediate-state spectrum in determining the correct value of the G matrix. With the tremendous amount of research which is being performed at the present time regarding this problem and with continued hard work, we can look forward to learning a great deal more about the nature of the effective interaction in finite nuclei.

REFERENCES

1. C. Bloch and J. Horowitz, Nucl. Phys. 8, 91 (1958).
2. B. H. Brandow, Rev. Mod. Phys. 39, 771 (1967).
3. T. Morita, Progr. Theor. Phys. 29, 351 (1963).
4. M. B. Johnson and M. Baranger, Ann. Phys. (N.Y.) 62, 172 (1971).
5. K. A. Brueckner, C. A. Levinson, and H. M. Mahmoud, Phys. Rev. 95, 217 (1954).
   H. A. Bethe and J. Goldstone, Proc. Roy. Soc. A238, 551 (1957).
   K. A. Brueckner and J. L. Gammel, Phys. Rev. 109, 1023 (1958) and references therein.
6. G. Oberlechner, F. Owono-N'-Guema and J. Richert, Nuovo Cimento 68B, 23 (1970).
   T. T. S. Kuo, S. Y. Lee, and K. F. Ratcliff, preprint and to be published.
7. P. Federman and L. Zamick, Phys. Rev. 177, 1534 (1969).
   A. E. L. Dieperink and P. J. Brussaard, Nucl. Phys. A129, 33 (1969).
   S. Siegel and L. Zamick, Nucl. Phys. A145, 89 (1970).
   P. Goode and S. Siegel, Phys. Lett. 31B, 418 (1970).
   P. J. Ellis and S. Siegel, Phys. Lett. 34B, 177 (1971).
   A. E. L. Dieperink, Nucl. Phys. A160, 171 (1971).

8. M. Harvey and F. C. Khanna, Nucl. Phys. A152, 588 (1970);
   A155, 337 (1970).
   F. C. Khanna, H. C. Lee and M. Harvey, Nucl. Phys. A164, 612
   (1971).
9. S. Kahana, H. C. Lee and C. K. Scott, Phys. Rev. 180, 956 (1969).
   H. C. Pradhan, private communication and M.I.T. thesis, 1971.
10. P. J. Ellis and H. A. Mavromatis, Oxford University preprint,
    1971.
11. An excellent review article on G-matrix calculations is M.
    Baranger, in Proc. Int. School of Physics Enrico Fermi, Course
    40, 1967, M. Jean, ed. (Academic Press, N. Y., 1969), p. 511.
12. B. R. Barrett, R. G. L. Hewitt, and R. J. McCarthy, Phys. Rev.
    C3, 1137 (1971).
13. N. LoIudice, D. J. Rowe, S. S. M. Wong, private communication.
14. F. Coester, in Lectures in Theoretical Physics, Vol. 11B,
    K. T. Mahanthappa and W. E. Brittin, eds. (Gordon and Breach,
    N. Y., 1969), p. 157.
15. B. R. Barrett and M. W. Kirson, Nucl. Phys. A148, 145 (1970).
16. T. H. Schucan and H. A. Weidenmüller, preprint and to be
    published in Ann. Phys.
17. J. F. Dawson, I. Talmi and J. D. Walecka, Ann Phys. (N.Y.)
    18, 339 (1962).
18. T. T. S. Kuo and G. E. Brown, Nucl. Phys. 85, 40 (1966).
19. B. H. Wildenthal, E. C. Halbert, J. B. McGrory and T. T. S.
    Kuo, Phys. Lett. 32B, 339 (1970).
20. G. F. Bertsch, Nucl. Phys. 74, 234 (1965).
21. T. T. S. Kuo, Nucl. Phys. A103, 71 (1967).
22. M. W. Kirson, Phys. Lett. 32B, 33 (1970).
23. L. Zamick, Phys. Rev. Lett. 23, 1406 (1969).
    E. Osnes and C. S. Warke, Phys. Lett. 30B, 306 (1969).
24. P. Goode, preprint and to be published.
25. B. H. Brandow, in Lectures in Theoretical Physics, Vol. 11B,
    K. T. Mahanthappa and W. E. Brittin, eds. (Gordon and Breach,
    N. Y., 1969), p. 55.
26. B. R. Barrett and M. W. Kirson, Phys. Lett. 30B, 8 (1969).
27. M. W. Kirson and L. Zamick, Ann. Phys. (N.Y.) 60, 188 (1970).
28. M. W. Kirson, Ann. Phys. (N.Y.) to be published (de Shalit
    memorial volume).
29. E. Osnes, T. T. S. Kuo and C. S. Warke, Nucl. Phys. A168,
    190 (1971).
30. F. Palumbo, Phys. Rev. C4, 327 (1971).

COMMENTS

G. E. Brown: I would like to comment on several points raised
by Dr. Barrett.

(1)   Although Kuo and I operated somewhat intuitively in our
first paper, the matters criticized by Barrett were cleaned up,
and the numerical values of the matrix elements did not change
sufficiently to warrant publishing new tables.   Specifically, in
later work we went over to a model space encompassing the last two
closed shells and the first two unfilled shells.   Then the intermediate states for the G-matrix calculation lie only outside these
two shells, and it is quite adequate to take plane waves for these.

Not only we, but also Bethe and Sprung and collaborators have
calculated the potential energy of the intermediate states; this
potential energy is very close to zero for intermediate states outside of the model space.   (The potential energy must be calculated
properly off the energy shell.)   The potential energy in intermediate states does not look anything like that given by a harmonic
oscillator potential.   It is really much better taken to be plain
zero.

(2)   I claim that Barrett and Kirson are working in a model
space of the size mentioned above, although they, like the bourgeois
gentilhomme, may not know it.   I outlined this in my Rev. Mod. Phys.
article of January this year.   Although Kirson found that our
G-matrix with plane-wave intermediate states gave about the correct
binding energies, it is ridiculous to claim that it will accurately
include, in the detail needed, intermediate states of excitation
energy $1\hbar\omega$ or $2\hbar\omega$;   these must be put into the model space.   When
this is done, there are terms in third order which Barrett and
Kirson have left out, and which reduce the third-order correction.

(3)   It is silly to believe that the 4 particle-2 hole states
will be brought from their unperturbed energy of about 35 MeV down
to their perturbed energy of about 1 to 2 MeV by perturbation
theory, even if summed to all orders.   The deformed 4 particle-2
hole states should be constructed explicitly and included in the
secular matrix.

(4)   The problem of calculating the effective interaction is
clearly not a purely mathematical one.   The expansion made is
nominally in powers of the density, but one knows that if the
density is decreased sufficiently, nuclear matter would break up
into a gas of alpha particles.   Therefore, the question of convergence cannot be proved mathematically, and one must use physical
arguments.

Kuo and I included the "bubble" term because it was the perturbation-theory expression of the interaction of nucleons through the exchange of vibrations, especially the 2^+ vibration. It is quite inconceivable to me that it makes sense to wipe this out by infinite summations of selected subsets of diagrams. One knows of many cases, in many-body theory where one obtains wrong answers by making such summations of selected subsets.

One has to be guided in what one does by physical arguments.

B. R. Barrett: I would like to make a few comments on Gerry Brown's remarks.

(1) Personally I feel that Gerry Brown and I basically agree on the question of the model space and are simply saying the same thing in two different ways. By advocating a better treatment of the intermediate-state spectrum in the calculation of G and the effective interaction, I am actually calling for a splitting of the space outside the model space into two parts; a high-energy region appropriate for hard-core scattering and a low-energy region appropriate for excitations near the Fermi surface. Although the intermediate-state potential in the high-energy region is best approximated by zero, the intermediate-state potential in the low-energy region is certainly not zero. I feel that it is the intermediate-state potential in the low-energy region which should be treated more accurately in the calculation of G and the effective interaction and that such a treatment would lead to a more attractive reaction matrix.

(2) As mentioned earlier, the work of Schucan and Weidenmüller and also of Brandow indicates that the perturbation expansion for the effective interaction will diverge or at least converge poorly if the states outside the model space d overlap with the states in d. Since this situation occurs in almost all cases of physical interest, we have no reason to believe that the expansion should converge. The suggestion by Schucan and Weidenmuller that one should use phenomenological and/or experimental information to guide one away from processes which are causing the expansion to diverge is totally consistent with Brown's third and fourth comment. Their suggestion makes perfectly good sense to me, since we are physicists and not mathematicians and, as such, our job is to put physics into the abstract mathematics in order to understand the results obtained in experiments.

SOLUTION OF THE BETHE-GOLDSTONE EQUATION IN FINITE NUCLEI FROM N-N PHASE SHIFT DATA*

R. J. W. Hodgson

University of Ottawa, Ottawa, Canada

Recently, considerable interest has centered around nuclear structure calculations which work directly from the observed free N-N phase shift data, rather than from an intermediate potential model [1, 2, 3, 4]. One of the advantages of this new approach is that it offers a method enabling one to perform some structure calculations without introducing an arbitrary off-energy-shell behavior. This, together with some recent developments in techniques for performing finite nuclei calculations [5, 6], results in a fresh new approach to the problem which is fast and flexible.

We present here an outline of the method used to calculate the reaction matrix in finite nuclei which employs harmonic-oscillator matrix elements which have been determined from the free N-N phase shift data [1, 2]. The technique employed follows from a suggestion of S. T. Butler et al. [5] and the work of Barrett, Hewitt and McCarthy [6].

It allows for the exact treatment of the Pauli operator, and for a simple and rapid way of studying the effect of a variation of the intermediate-state energy.

The starting point for our calculation is the dispersion relation for the partial-wave scattering amplitude A_ℓ (E):

$$A_\ell(E) = b_\ell(E) + \frac{1}{\pi} \int_0^\infty \frac{\text{Im } A_\ell(E')}{E' - E - i\varepsilon} \, dE'$$

$$+ \frac{1}{\pi} \int_{-\infty}^{-\mu^2} \frac{\text{Im}[A_\ell(E') - b_\ell(E')]}{E' - E} \, dE' + \frac{\Gamma \delta_{\ell 0}}{E - E_B} . \tag{1}$$

Assuming that $A_\ell(E)$ is known for $E>0$ from the experimental results, and making the approximation

$$\text{Im } b_\ell(E) = \text{Im } A_\ell(E), \quad E<-\mu^2 \tag{2}$$

enables one to express the Born approximation term $b_\ell(E)$ as a function of the phase shift $\eta_\ell(E)$. Here μ is the pion rest mass.

$$b_\ell(E) \simeq \frac{\cos \eta_\ell(E) \sin \eta_\ell(E)}{\sqrt{E}} - \frac{P}{\pi} \int_0^\infty \frac{\sin^2 \eta_\ell(E') dE'}{\sqrt{E'} \, (E' - E)} \tag{3}$$

Thus, from a knowledge of the phase shifts η_ℓ, the Born term can be computed.

Two problems require some further investigation before these results can be used. Experimentally, the phase shifts have only been measured for $E<400$ MeV, so that in order to evaluate the integral in (3), some extrapolation must be employed.

We have found that the harmonic-oscillator matrix elements are not strongly sensitive to the behavior of η_ℓ at high energies [7], and for the present work, have chosen to set $\eta = 0$ for $E>500$ MeV. In addition, special consideration must be given to the treatment of the bound state [8].

Once $b_\ell(E)$ has been computed, it is a fairly straight forward procedure to determine the harmonic-oscillator matrix element

$$\langle n\ell|V|n'\ell'\rangle = \int_0^\infty R_{n\ell}(r)V(r) \, R_{n'\ell'}(r) \, r^2 dr \tag{4}$$

from a series of recurrence relations [7].

The final step is to employ the matrix elements in (4) to solve the Bethe-Goldstone wave equation.

$$\psi_\alpha^{BG}(\omega) = \Phi_\alpha + \sum_\mu^\infty \frac{Q_\mu \Phi_\mu <\Phi_\mu|v|\psi_\alpha^{BG}(\omega)}{\omega - \varepsilon_\mu} \tag{5}$$

where the Φ_α are eigenfunctions of the two-body harmonic-oscillator Hamiltonian H_o,

$$H_o \Phi_\alpha = \varepsilon_\alpha \Phi_\alpha, \tag{6}$$

and Q_μ is the eigenvalue of the Pauli projection operator.

Following the suggestion of Butler et al. [5], the BG wave-function is expanded in terms of the eigenfunctions of the complete two-body Hamiltonian $H_o + V$, so that short range correlations are already incorporated. That is, we consider

$$\psi_\alpha^{BG} = \sum_i^\infty a_{i\alpha}(\omega)\ \psi_i \tag{7}$$

where

$$(H_o + v)\ \psi_i = E_i \psi_i. \tag{8}$$

Then, as has been shown by Barret et al. [6], the reaction matrix

$$G_{\beta\alpha}(\omega) = <\Phi_\beta|v|\psi_\alpha^{BG}(\omega)> \tag{9}$$

can be expressed as

$$G_{\beta\alpha}(\omega) = G_{\beta\alpha}^R(\omega) - \sum_\mu^\infty G_{\beta\alpha}^R(\omega)\ \frac{1-Q_\mu}{\omega-\varepsilon_\mu}\ G_{\mu\alpha}(\omega) \tag{10}$$

where

$$G_{\beta\alpha}^R(\omega) = (\varepsilon_\alpha - \omega)[\delta_{\alpha\beta} - (\varepsilon_\beta - \omega)\sum_i \frac{b_{i\alpha}\ b_{i\beta}}{E_i - \omega}]. \tag{11}$$

Here $b_{i\alpha}$ is an overlap function defined by

$$b_{i\alpha} = <\psi_i|\Phi_\alpha> \tag{12}$$

Equations (10) and (11) form the basis of our calculation. The values of $b_{i\alpha}$ and E_i are first computed by diagonalizing the matrix

$$<\Phi_\alpha|H_o + V|\Phi_{\alpha'}> = \varepsilon_\alpha \delta_{\alpha\alpha'}$$

$$+ \sum C_\alpha\ [n\ell S\bar{J}NL;JT]\ C_{\alpha'}[n'\ell'S\bar{J}NL;JT]$$

$$x <n\ell|V|\ n'\ell'> \tag{13}$$

where $C_\alpha[n\ell S\bar{J}NL;JT]$ are recoupling coefficients [6].

In addition to using the harmonic-oscillator matrix elements derived from the dispersion relation approach outlined above, we have also used those obtained by the Sussex group [1] who employ a cut-off oscillator potential.

The results have been applied to a calculation of the spectra of the A=18 nuclei, using the shell model with an inert ^{16}O core. Fig. 1 shows the spectra for ^{18}O, and Fig. 2 that for ^{18}F.

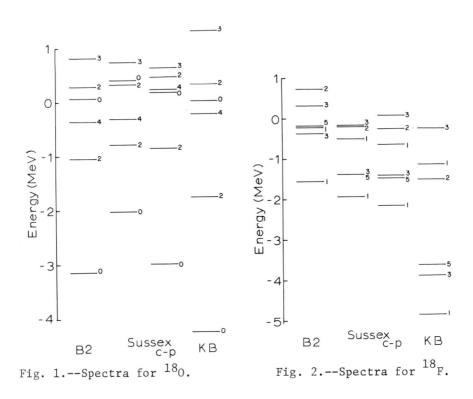

Fig. 1.--Spectra for ^{18}O.   Fig. 2.--Spectra for ^{18}F.

The calculations were done with ω=98.0 MeV, which is approximately self-consistent [6]. The results obtained using the dispersion approximation (3) are labelled B2, and all spectra are compared with the results of Kuo and Brown [9]. The latter include the effects of the core polarization term G_{3p1h}. We have applied this same correction to the Sussex spectra, and the results are included in both diagrams (c-p).

The results for ^{18}O are reasonably good, and we intend to examine the effect on the B2 spectra of including G_{3p1h} shortly. However, the lowest levels of ^{18}F are underbound by about 3 MeV. This appears to be due to the effective interaction in the ^{3}S - ^{3}D states, which is quite different than that obtained from realistic potential models [7]. This seems to be caused by a sensitive dependence of the Born term b(E) on the form of the phase shift near 500 MeV, which is not known experimentally. More work must be done in this area.

REFERENCES

1. J. P. Elliott, A. D. Jackson, H. A. Mavromatis, E. A. Sanderson, and B. Singh, Nucl. Phys. A121 (1968) 241.

2. M. Razavy and R. J. W. Hodgson, Nucl. Phys. A149 (1970) 65.

3. E. Ley Koo, M. de Llano, D. V. Grillot, and H. McManus, Nucl. Phys. A133 (1969) 610.

4. P. Ripa and E. Maqueda, Nucl. Phys. A166 (1971) 534.

5. S. T. Butler, R. G. L. Hewitt, B. H. J. McKellar, I. R. Nicholls, and J. S. Truelove, Phys. Rev. 186 (1969) 963.

6. B. R. Barrett, R. G. L. Hewitt, and R. J. McCarthy, Phys. Rev. C3 (1971) 1137.

7. R. J. W. Hodgson, Can. J. Phys. 49 (1971) 1401.

8. R. J. W. Hodgson; to be published.

9. T. T. S. Kuo and G. E. Brown, Nucl. Phys. 85 (1966) 40.

Work supported by the National Research Council.

THE TWO-BODY INTERACTION IN NUCLEAR SHELL MODEL CALCULATIONS[*]

J. B. McGrory

Oak Ridge National Laboratory

INTRODUCTION

I would like to start by emphasizing that the work I will discuss here is the result of a lengthy and close collaboration between Edith Halbert at Oak Ridge, Hobson Wildenthal at Michigan State, and myself. In recent years, we have been making an intensive study of the properties of low-lying bound states of light mass nuclei in terms of the conventional nuclear shell model. In particular, we have studied the structure of nuclei in the 0p-shell, the 0d-1s shell, and the 0f-1p shell. The discussion here can be broken down into five sections. I will first describe what I mean by a conventional shell-model calculation, and indicate how the effective two-body interaction is introduced in the calculation. I will also discuss here the types of two-body interactions we have used in these calculations, which are typical of interactions used in most shell-model calculations. Second, I will illustrate the type of agreement with experimental data we have obtained for nuclei in the s-d shell for energy levels, strengths in single-nucleon-transfer reactions, and for electric-quadrupole and magnetic-dipole moments and transition rates. Third, I will discuss the sensitivity of these calculated results to the differences between the various interactions we have used. The material in these two sections represent a quick review of a recently published[1] detailed discussion of shell-model calculations of nuclei with A = 18-22. Unless otherwise noted, all references to experimental data will be found in that article. Fourth, I will discuss the renormalization of the effective two-body interaction which occurs when the model space is truncated to include only part of a complete oscillator shell. Fifth, since it is well known that the shell model works because it is possible to "hide" a considerable

amount of configuration mixing which occurs in actual nuclear wave
functions through renormalization of the two-body interaction, I
will try to illustrate what role the individual orbits play in an
s-d-shell-model calculation.

I.   Discussion of the Shell Model and Two-Body
Interactions in such Calculations

In all these calculations we will assume the nuclear wave
functions can be expanded in terms of a finite set of the single-
particle orbits of a harmonic oscillator. We assume that the lowest orbits up to a given level are completely filled with nucleons
to form an inert core. We assume that all single-particle orbits
above a given level are explicitly unoccupied. For example, in the
s-d shell calculations which will be discussed here, we assume that
the 0s and 0p shell orbits, 16 in all, are completely filled with
16 particles to form an inert ^{16}O core. We further assume that all
orbits from the 0f-1p shell and higher are unoccupied. Thus, the
active orbits of the model are the $d_{5/2}$, $s_{1/2}$ and $d_{3/2}$ orbits. In
this model, for a nucleus with mass A, A-16 particles occupy the
$d_{5/2}$, $s_{1/2}$, and $d_{3/2}$ orbits. Thus, we form basis states of the
form

$$\Psi_{J,T} = \Psi(d_{5/2}^{n_1}, s_{1/2}^{n_2}, d_{3/2}^{n_3}, \alpha,J,T)$$

(1)

where
$$n_1 + n_2 + n_3 = A-16.$$

Within the space defined by these basis states, we diagonalize an
effective Hamiltonian which is assumed to be the sum of a one-body
plus a two-body interaction.

$$H_{eff} = \sum_{i=1}^{A-16} H_1(i) + \sum_{i<j}^{A-16} H_{12}(i,j).$$

(2)

In principle, the one-body term introduces the effects due to the
differences of the interaction of particles in the various s-d
shell orbits with the 16 core particles. It is generally assumed
to be a diagonal operator, and the eigenvalues of the one-body
operator are taken from the observed spectrum of the nucleus which
is one particle beyond the closed core in the assumed model space
of the calculation. The two-body part of the effective interaction
is obviously not the bare interaction, and must be appropriately
renormalized to account for effects due to configurations explicitly
omitted from the model space, as has been discussed in previous
talks at this conference. When this effective Hamiltonian is
diagonalized, the eigenvalues are identified with energies of low-
lying states. The eigenfunctions are associated with the nuclear
wave functions, and as such are used to calculate various observables.

The physics of the shell-model calculation involves the selection of the appropriate model space, and the determination of the correct effective interaction for that model space. The mechanics of the calculation involves the construction of the basis states, (1), in the chosen model space, the calculation of the matrix elements of the effective operator, (2), and the diagonalization of the resulting matrix. It is well known that for an effective interaction which is the sum of a diagonal one-body operator plus a two-body operator, the following is true

$$\langle \psi_{JT} | H_1 + H_{12} | \psi'_{JT} \rangle = \sum_{i=1} n_i \, \epsilon(j_i) \, \delta_{\psi_{JT}, \psi'_{JT}}$$

$$+ \sum_{\substack{j_1 j_2 j_3 j_4 \\ \Delta_J \Delta_T}} A(j_1 j_2 j_3 j_4 \Delta_J \Delta_T, \psi_{JT} \psi'_{JT}) \, \langle j_1 j_2 \Delta_J \Delta_T | H_{12} | j_3 j_4 \Delta_J \Delta_T \rangle \tag{3}$$

where n_i is number of particles in the ith active orbit, $\epsilon(j_i)$ is single-particle energy of ith orbit, and $|j_1 j_2 \Delta_J \Delta_T\rangle$ is a renormalized, antisymmetrized state of two particles in the indicated orbits coupled to a total angular momentum Δ_J and total isospin Δ_T. Thus, for the two-body part of the residual interaction, the matrix element of an n-particle state can be written as a linear combination of two-body matrix elements. This decomposition proceeds by straightforward fractional-parentage and Racah-algebra techniques, and there are several computer programs available to carry out this mechanical procedure.[2] We see that we have all the information we need about the residual interaction when we know all the one- and two-body matrix elements for the orbits in the active space. In the complete s-d shell, this involves 66 numbers; 63 two-body matrix elements and 3 single-particle energies.

There have been a variety of methods used historically to fix the two-body part of the effective interaction. In most of the earliest calculations, some sort of central force with Yukawa or Gaussian shape was used, and the strength and exchange mixtures of the interaction varied to fit experimental data. A different approach is to consider the set of two-body matrix elements as free parameters which are determined by requiring that they give a good fit to the energies of a selected set of observed levels. An extreme variation of this approach was used by McCullen, Bayman, and Zamick, and Ginocchio and French[3] who treated nuclei outside ^{40}Ca in terms of pure $f_{7/2}^n$ configurations. In this case there were just 8 matrix elements, and these were taken directly from the observed spectra of ^{42}Sc and ^{42}Ca, the two-particle nuclei in their model. The method of parameterization of the interaction is certainly the most straightforward method for obtaining a renormalized interaction. It has certain more or less obvious difficulties.

One is to obtain a set of data which is sufficient to determine the parameters accurately. Somewhat related to this is the problem that all the parameters may not be linearly independent. The third general method for obtaining the effective residual interaction is to calculate it from the free nucleon-nucleon interaction, as has been discussed in some detail here already. Aesthetically, this is obviously the most satisfactory way to obtain the interaction. In principle, this route leads to a parameter-free calculation of the eigenvalues and wave functions of the shell model.

We have used all three of these general approaches to obtain effective interactions in the A=18-22 calculations. We have used the "realistic" interactions of Kuo and Brown,[4] and of Kuo[5] with single-particle energies taken from the spectrum of ^{17}O. We will refer to these calculations as KB+^{17}O and K+^{17}O, respectively. These are interactions for complete s-d shell calculations, and they are calculated starting with the Hamada-Johnston nucleon-nucleon interaction. The two calculations differ in the way in which the unrenormalized G-matrix elements are calculated, and in the renormalization terms which are included.

We have used several variations of the method of parameterization of the interaction matrix elements. In two calculations we used the Kuo and Kuo-Brown interaction but we treated the single-particle energies as free parameters. The two interactions we thus obtained will be referred to as K+SPE and KB+SPE. The rationale for this is that the single-particle energy reflects the interaction of the active particles with the core particles, and that as particles are added in the active space, the structure of the core particles is altered. This might lead to a situation where the single-particle energies in the A=17 and A=22 systems are significantly different. We have searched on the single-particle energies which fit levels in all nuclei with A = 18-22 simultaneously, so that we have only averaged out this effect. The major effect that we have found by this search is an improvement in the calculated ground-state binding energies when we use the Kuo interaction. It is quite possible that the change in the single-particle energies effectively corrects a deficiency in the two-body interaction, rather than reflecting a real change in the single-particle energies.

The other variation of the parameterization of the interaction is that we used the Kuo interaction as a starting point, and then treated as parameters most of the two-body matrix elements which involve only $d_{5/2}$ or $s_{1/2}$ orbits, plus the three single-particle energies. The rationale here is that the $d_{5/2}$ and $s_{1/2}$ orbits are the lowest orbits in energy. They are within 1 MeV of each other, and about 4 MeV below the $d_{3/2}$ orbit. Thus, we expect the low-lying states to be dominated by these orbits, so the calculation should be most sensitive to the matrix elements involving these orbits. The interaction obtained in this way will be referred to as K+12FP.

Thus, we have used five interactions which all start with a so-called realistic interaction. These have central and non-central components. We have also used two phenomenological forces. The first of these is based on the surface delta interaction (SDI) originally suggested by Arvieu and Moszkowski.[6] The surface delta interaction is essentially a delta force with the further restriction that it acts only if the interacting nucleons are at the nuclear surface. There is the further assumption that all single-particle orbits have the same size and phase at the nuclear radius. These assumptions are equivalent to using an ordinary δ-force and setting all radial integrals equal to one. It is a local central force and it is not a translationally invariant force. The force acts only in spatially symmetric states. From this it follows that the spin exchange can be specified by giving the force different strengths in T=0 and T=1 states. Thus V_{SDI} is a five-parameter interaction, if we parameterize the three single-particle energies. In studies of A=33-39 nuclei, Glaudemans, Brussaard, and Wildenthal[7] found that the ground-state binding energies and the relative spacings of states with different isospin were not well accounted for by the surface delta interaction, but the relative structure of states with the same isospin was reproduced by the interaction. They pointed out that the inclusion of an isospin-dependent monopole interaction alleviated this problem. They called the interaction which is a surface-delta interaction plus an isospin-dependent monopole interaction the modified surface delta interaction,

$$V_{MSDI} = (A_0 \, V_{SDI} + B_0) \delta_{T,0}$$

$$+ (A_1 \, V_{SDI} + B_1) \delta_{T,1}.$$

It can be shown that the isospin monopole terms do not affect the structure of the eigenfunctions. They add a constant to the binding energy of all nuclei with a given A,T. Thus they affect only the ground-state binding energies and the relative position of states with different T.

The seventh interaction which we have used is a central, non-local force parameterized in terms of radial integrals of the form

$$I_{n\ell} = \int R_{n\ell}^2(r) V(r) r^2 dr$$

where $R_{n\ell}(r)$ is the radial part of the harmonic oscillator wave function, and $V(r)$ is the residual interaction. This method of parameterization has been enthusiastically proposed for some years by Pandya.[8] Any two-body matrix element in j-j coupling can be reduced to a linear combination of radial integrals of the form

$$I_{n\ell,n'\ell'} \equiv \int R_{n\ell}^*(r) R_{n'\ell'}(r) V(r) r^2 dr$$

simply by transforming from j-j coupling to ℓ-s coupling with the usual 9-j recoupling coefficients, and then from particle coordinates to relative and center-of-mass coordinates with the usual Moshinsky transformation brackets. If we ignore any symmetry of the Moshinsky transformation brackets, there are 26 integrals of the form $I_{n\ell,n'\ell'}$ in the s-d shell for any translationally invariant central plus tensor force interaction. If we restrict ourselves to central forces, and do consider symmetries of the brackets, then only the integrals of the form $I_{n\ell,n\ell} \equiv I_{n\ell}$ enter, and there are 14 independent parameters which completely specify the 63 two-body matrix elements in the s-d shell. We have determined a central force parameterized in terms of these integrals, and will refer to this as the RIP (radial integral parameterization) interaction.

Kulkarni and Pandya[9] have pointed out an interesting feature of these radial integrals. Calculations have shown that for typical effective interactions, the integrals decrease in magnitude with increasing ℓ, and that $I_{n\ell}$ with $\ell \geq 2$ can be ignored. If odd-state interactions are also assumed to be weak, then only the s-state integrals (I_{no}) are important. There are three such integrals in the s-d shell, I_{00}, I_{10}, and I_{20}. Moszkowski has shown that I_{00} conserves SU(3) symmetry for a degenerate single-particle space. Kulkarni and Pandya have shown empirically that I_{20} conserves SU(3) in the degenerate two-particle system. Thus SU(3) symmetry is destroyed essentially by the I_{10} term, and the spin-orbit splitting. In the RIP T=0 interaction determined in our calculations[1] I_{10} is essentially zero, so the T=0 interaction conserves SU(3). The T=1 interaction has a relatively large I_{10} term.

There is one general comment to be made about all the searched interactions. We consistently find that the T=1 two-body matrix elements are much more accurately determined than the T=0 interaction. In fact, in the searches to determine the 12FP interaction, seven of eight T=0 two-body matrix elements could not be well determined by the fitting procedure we used, and in the radial integral parameterization, 3 of the 7 T=0 parameters were undetermined. One reason for this is that calculations involving pure neutron states are independent of the T=0 interaction. Second, in many of the many-body matrix elements, the two-body matrix elements are multiplied by a statistical factor, (2T+1), which enhances the T=1 contributions.

II. Comparison of Shell-Model Calculations with Experiment

In this section we compare the results of one shell-model calculation with experimental results for nuclei in the mass region A = 18-22. In these calculations we assume an ^{16}O core, and all Pauli allowed states of active particles in the $d_{5/2}$, $s_{1/2}$ and $d_{3/2}$

orbits are included in the model space. The single-particle energies are taken from the observed spectrum of [17]O, which is one
neutron outside an inert core in this model. We use the "realistic" matrix elements designed by Kuo for the complete s-d shell
calculation (i.e. the K+[17]O interaction). The reason for using
this particular interaction for these illustrative purposes is that
it leads to results which are overall in as good agreement with
experiment as any of the interactions we have used. There have
been more recent and somewhat more accurate calculations of the s-d
shell effective interactions by Kuo,[10] but the two-body matrix
elements are very similar to those used here. The Kuo interaction
used here is renormalized to introduce the perturbative effect of
some three-particle one-hole and four-particle two-hole states
which lie within 2 $\hbar\omega$ excitation energy of the s-d shell configurations. We have calculated the ground-state binding energies,
excitation spectra, single-nucleon transfer strengths, magnetic-
dipole and electric-quadrupole moments and transition rates of
states in the A=18-22 region. The nuclei in this region all display a sequence of levels which start with the ground state which
have properties very similar to those predicted for a rigid rota-
tor. For these ground-state rotational bands, the calculated
shell-model properties are in good agreement with the observed
properties of the 18-22 nuclei. For states outside these ground-
state bands, the experimental data **are** far less complete and accurate, and the agreement between theory and experiment is distinctly
less satisfactory in general. To illustrate this summary, consider
the results for [21]Ne. In Fig. 1 is shown the observed spectrum of

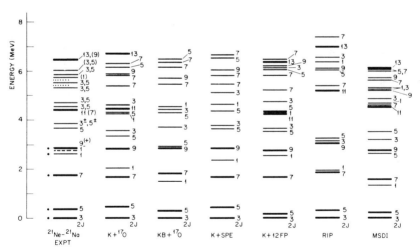

Fig. 1. Observed Spectrum of [21]Ne, and Spectra Calculated with
Different Effective Residual Interactions as Described in Text.
Members of ground state rotational bands are indicated with heavy
lines. See Ref. 1 for sources of experimental data.

^{21}Ne, and the spectrum calculated with the K$+^1$7$_0$ interaction in the
first and second columns, respectively. The ground-state band is
a K=3/2 band, and is indicated with extra thick lines. There is
an obvious strong similarity between the observed and calculated
spectra for the band up to the 13/2 state. Beyond that, there are
extra states observed, and it is difficult to correlate model state
with observed state. The first 1/2$^+$ state may be the first member
of a K=1/2 band, and it appears in the calculated spectrum about
1 MeV below the observed. There is a trend in all the calculated
spectra of A=18-22 nuclei for excited bands to appear at too low
an energy.

A second test of the shell model wave functions is a comparison of the calculated strengths for single nucleon transfer reactions with strengths extracted from observed cross sections by
distorted-wave Born approximation methods (DWBA). The absolute
values of spectroscopic factors extracted from experiment are
accurate to no better than 30%. In general, the calculated S-
factors for strong transitions to members of ground-state bands in
these nuclei are in agreement with S-factors extracted from experiment to within DWBA uncertainties. The difficulties in correlating calculated and observed states outside the ground-state band
preclude any generalization for transfers to such states.

We have calculated the magnetic-dipole observables for these
nuclei with the bare neutron and proton magnetic dipole operators.
Just as the two-body interaction is renormalized by space truncation, the one-body operators should be renormalized. There is an
extensive literature on the subject of effective transition operators, and there are still significant problems in calculating such
renormalizations (see discussion and references on this subject in
Ref. 1). Our justification in using the bare magnetic dipole operator is that a posteriori, it works. Because of various well-known
selection rules, most Ml-transitions are very weak in the lower s-d
shell, so there is very little data with which to compare calculated reduced transition rates for Ml-transitions. In Table I, the
observed and calculated magnetic moments are given. The agreement
is excellent. The most significant discrepancy is for ^{21}Ne, where
the calculated moment is 30% larger than the observed value. The
small value of this moment suggests that the calculated value may
be sensitive to the wave function.

The electric quadrupole observables have been calculated with
an effective E2-operator of the form

$$Q_M^2 = \sum (1+\epsilon_p)(\tfrac{1}{2}+t_z)r^2 Y_M^2 + \epsilon_n(\tfrac{1}{2}-t_z)r^2 Y_M^2$$

(we assume $t_z = +\tfrac{1}{2}$ for protons), and with $\epsilon_p = \epsilon_n = 0.5e$. The
justification here is again purely a pragmatic one. Any general
statement with regard to a comparison of calculated and observed
quadrupole moments and B(E2)-values in these calculations must be

TABLE I. Observed Magnetic Moments and Values Calculated in Shell Model with K+^{17}O Interaction

Nucleus	2J	M(n.m.)		Nucleus	2J	M(n.m.)	
		Expt.	Calc.			Expt.	Calc.
^{17}F	5	4.7	4.8	^{20}F		2.1	1.9
^{17}O	5	-1.9	-1.9	^{21}Ne		-0.7	-0.9
^{18}F	10	2.9	2.9	^{22}Na		1.8	1.8
^{19}F	1	2.6	2.9				

qualified, mainly because of uncertainties in the experimental data. Within these uncertainties the measured quadrupole moments for states in all nuclei with A = 18-39 are satisfactorily accounted for with the model calculations.[11] The general pattern of B(E2)-values observed in transitions within ground-state bands is reproduced by the calculations. It is also true that where there have been discrepancies between the calculated and observed values for ground-state band transitions, when newer more accurate measurements have been made, the improved numbers have always been in much better agreement with the calculations.

Calculations similar to the A=18-22 calculation summarized here have been made for essentially all s-d shell nuclei, and for a number of nuclei in the beginning of the so-called f-p shell. These calculations offer an abundance of evidence that the shell model provides a straightforward way of accounting for a large amount of experimental data on low-lying levels in these light nuclei with one uniform model. These results allow one to say very little about the exact wave functions of individual nuclear states. All the observables we have discussed involve overlaps of initial and final states, and they provide information only about the relative behavior of the wave functions. This, of course, was the point of the "pseudonium" calculations of Lawson and Soper.[12]

III. Sensitivity of the Calculations to the Residual Interaction

We have repeated the calculations of the A=18-22 nuclei with the seven interactions described in Section II. From a comparison of all these calculations we can make the following general

statement. For the properties of the states in ground-state
rotational bands, the calculated results show very little sensitivity to the differences that exist between the types of interactions we have used. There is considerable sensitivity to the
differences in these interactions for the properties of states outside the ground-state bands, but for such states, the experimental
data is not good enough to make such sensitivity useful. No one
of the interactions stands out as clearly superior to the other
interactions.

These general statements are illustrated, for energy level
calculations, by Fig. 1 where the spectra for ^{21}Ne calculated with
these different interactions are shown. We see there that the
ground-state band states are fairly well reproduced in all spectra
in which they are calculated (i.e., calculations for 11/2 and 13/2
states with the KB+^{1}7O and K+SPE were not made). The position of
the K=1/2^{+} excited band is seen to be fairly sensitive to the differences in the interactions.

It is not easy to make a generalization about the spectroscopic factors. There is significant sensitivity to the interaction for these numbers. There is also considerably more uncertainty in the experimentally determined S-factors than in the observed energy levels. If we limit ourselves to strong transitions
for which there is experimental data, and for which the S-factors
in the various models differ by as much as a factor of two, we find
there are fourteen such cases. For these cases, only one S-factor
calculated with the K+^{1}7O interaction disagrees with experiment by
more than 30%, the limit of accuracy we assume for the DWBA theory.
Four of the S-factors calculated with the K+12FP interaction, six
S-factors in the RIP calculation, and seven S-factors in the MSDI
calculation disagree with experiment by more than 30%. This would
suggest that for the calculations of the S-factors, the K+^{1}7O interaction is possibly the best.

The magnetic moments are extremely insensitive to the differences in the interactions we have used. The differences between
the various calculated moments are no larger than the average differences between theory and experiment. The calculated M1's on the
other hand are very sensitive, but there is little experimental
data on these transitions so this sensitivity is not particularly
useful.

The B(E2)-values calculated for transitions within ground-
state bands are very insensitive to variations between the interactions we use. This is illustrated in Table II. For transitions
involving one or two states outside the band there is much more
sensitivity. Comparisons between calculated and measured E2-
values for these transitions do not obviously favor any one of the
interactions.

TABLE II.   B(E2)-values Calculated with Several Different Residual
Two-Body Interactions

Nucleus	J_i	J_f	B(E2)-values (e^2fm^4)			Nucleus	J_i	J_f	B(E2)-values (e^2fm^4)		
			K+^{17}O	RIP	MSDI				K+^{17}O	RIP	MSDI
^{20}Ne	2	0	48	50	46	^{20}F	3	2	27	26	16
	4	2	60	62	54		4	2	11	13	9
	6	4	49	50	41		4	3	21	19	18
	8	6	31	29	22		5	3	17	17	13
							5	4	11	12	12

IV.   Renormalization of Effective Two-Body Interaction for Truncation within One Oscillator Shell

It has been stressed previously that there is a strong inter-
dependence between the effective interaction and the model space.
There are several types of renormalizations that enter a shell-
model calculation. An interesting question is how do these various renormalizations affect the matrix elements of the effective
interaction. One renormalization is due to excitation of particles
to intermediate states at high energy which results primarily from
the short-range repulsion. It is difficult to see what the contribution of these excitations are to the Kuo-Brown numbers, since
the completely unrenormalized matrix elements with a hard core
potential are infinite. Clement and Baranger[13] have deduced an
effective interaction from the Tabakin separable potential, and
have included the effects of the highly excited states in second-
order Born approximation. Their results show that these excitations affect the T=0 matrix elements much more strongly than the
T=1 elements. Many of the T=0 matrix elements are made more
attractive by 15-20%. The second correction to the matrix elements
is for low-lying excitations just outside the model space, e.g.
2 $\hbar\omega$ excitations of one and two particles outside the model space.
For these corrections the largest effect is on the J=0,T=1 states,
which are made significantly more attractive.

The realistic effective interactions are usually designed for
use in the complete space of states in a given oscillator shell.
The renormalizations which have been extensively considered so far
involve relatively high excitations of the model space (e.g. the
lowest excitations are at 2 $\hbar\omega$). These renormalized interactions
are not obviously appropriate for use in a model space which is a
truncation of the full oscillator shell. In most cases it is not
possible to carry out a shell-model calculation in a complete
oscillator space. Thus a question of great interest to the shell

modeler is how to renormalize an interaction to introduce effects
of very low-lying excitations which are omitted when truncation
occurs within a major shell. We have made one empirical study in
this direction already. We have been interested recently in nuclei
at the beginning of the so-called f-p shell. In these, we have
assumed an inert ^{40}Ca core, and in our first studies we include all
four f-p shell orbits, the $f_{7/2}$, $p_{3/2}$, $f_{5/2}$, and $p_{1/2}$ orbits. Kuo
and Brown[14] have calculated an effective interaction for this
region, but for a model space which includes the $g_{9/2}$-orbit. We
have convinced ourselves that we can reasonably neglect this orbit
for studies of the low-lying states. To renormalize for this
omission, we have calculated the low-lying levels of nuclei with
A = 42-44 in a model space which includes all Pauli-allowed states
of from two to four particles in the four f-p orbits, and we have
treated the eight two-body matrix elements,
$\langle f_{7/2}^2 \ J,T | H_{2b-eff} | f_{7/2}^2 \ J,T \rangle$ as free parameters which were varied
to fit a selected set of binding energies and excitation energies.
The main effect of freeing these matrix elements from the original
Kuo-Brown values was to improve the agreement with the observed
ground-state binding energies. With this model, many observed
properties of many low-lying states in these nuclei are reproduced.
As an example, the calculated and observed excitation energies of
^{44}Sc are listed in columns one and three of Table III, and the
calculated and observed strengths in the ^{43}Ca(^{3}He,d)^{44}Sc reaction
are listed in columns four and six of the same table. A study of
these results shows that there is clearly a set of observed states
in ^{44}Sc whose properties are well described by this spherical shell
model.[15] It should be noted that there are observed 18 other
states below 1.5 MeV not accounted for by the model, which appear
to live in very peaceful coexistence with the spherical states.
The calculated ground-state magnetic moment is 2.60 mm as compared
to the measured value 2.56, and with an effective charge of about
0.9, the known quadrupole moments of the nuclei in this region are
accounted for. There are many interesting nuclei in this mass
region with five or more particles outside the ^{40}Ca core, but our
computer capabilities at present preclude the diagonalization of
the complete f-p shell space for more than four particles. It is
possible to diagonalize the space of five and six particles in the
$f_{7/2}$-$p_{3/2}$ space, and there is reason to expect that many of the
observed properties of these nuclei could be accounted for in such
a space. We are thus faced with the problem of an effective interaction for this space.

There are 30 two-body matrix elements in the $f_{7/2}$-$p_{3/2}$ space.
There is not nearly enough experimental data to permit the determination of these thirty matrix elements with any degree of confi-
dence. Therefore we have chosen to use our four-shell shell-model
results as empirical data. A two-shell model which reproduces the
four-shell model results could be expected to give a good idea of
the shell-model results for A = 45 and A = 46. We have chosen from

TABLE III. Calculated and Observed Spectra for ^{44}Sc, and Calculated and Observed Strengths for ^{43}Ca$(^{3}$He,d$)^{44}$Sc Reaction. The columns headed 2-Shell and 4-Shell refer to full and truncated f-p-shell calculations described in the text.

J	Excitation Energy (MeV)			$(2J_f+1)/(2J_i+1)C^2S(^3$He,d$)$		
	4-Shell	2-Shell	Expt.[a]	4-Shell	2-Shell	Expt.
2	0	0	0	0.26	0.31	0.28
6	0.61	0.46	0.27	0.72	0.72	0.73
4	0.43	0.32	0.35	0.51	0.52	0.45
1	0.77	0.63	0.67	0.14	0.13	0.15
3	0.80	0.74	0.76	0.16	0.14	----
7	1.33	1.28	0.97	1.58	1.57	1.62
5	1.23	1.10	1.05	0.41	0.51	0.34
3	1.35	1.30	1.19	0.36	0.33	0.49
5	1.63	1.44	1.53	0.48	0.57	0.76

[a] See Reference 15.

the shell-model spectra obtained in the complete f-p-shell-model space for A=42 to A=44 nuclei 93 excitation energies and six ground-state binding energies. We then tried to reproduce these energies with an $f_{7/2}$-$p_{3/2}$ model by adjusting the thirty two-body matrix elements in this two-shell space to minimize the r.m.s. deviation between the four-shell and two-shell calculated spectra. It was found that seven of the matrix elements could not be determined from this search. These were all T=0 matrix elements, and they all involved at least one particle in the $p_{3/2}$-shell. We fixed these matrix elements at the original Kuo-Brown values, and finally varied 23 two-body matrix elements to fit the 99 pieces of "experimental" data. The final r.m.s. deviation was 90 keV out of a typical excitation energy of 2 MeV. In Table III, the energies calculated in the two-shell model are displayed as are the calculated strengths for the ^{43}Ca$(^{3}$He,d$)^{44}$Sc reaction. We see that the two-shell model does an excellent job of reproducing the four-shell result. In this theoretical "experiment" we know the structures of the four-shell wave function and the two-shell wave function, so we can see directly how much configuration "hiding" exists in the two-shell model. In ^{44}Sc, configurations of the form $(f_{7/2},p_{3/2})^{4}$

constitute about 85% of the total four-shell wave functions, and in ^{44}Ti they constitute about 75%. Preliminary calculations of the structure of ^{45}Ca, ^{45}Sc, and ^{45}Ti as calculated in the $f_{7/2}$-$p_{3/2}$ model are in reasonable agreement with experiment.

The $f_{7/2}$-$p_{3/2}$ effective interaction determined in these calculations is shown in Table IV, as are the matrix elements, in the $f_{7/2}$-$p_{3/2}$ space, of the interaction which produced the four-shell results. For the T=1 interaction, the main effect of renormalizing from the four-shell to the two-shell space is to make the three J=0 matrix elements more attractive by about 500 keV. The relative structure of the rest of the interaction is not greatly altered by the renormalization. For the T=0 interaction, seven of the matrix elements could not be accurately determined in the search. This is similar to the results found in the s-d shell, where the T=0 interaction is not as well determined as the T=1 interaction. In the $f_{7/2}^2$, T=0 interaction there are large renormalizations for the low spin 1^+ and 3^+ states. The other general feature of the renormalized T=0 interaction is that the center-of-gravity of the $f_{7/2}$-$p_{3/2}$ interaction is more attractive by roughly 0.5 MeV as compared to this number in the four-shell interaction.

There are at least two factors which might account for these features of the renormalized interaction. In this region, the predominant orbit is the $f_{7/2}$ orbit. High-spin states will be dominated by pure $f_{7/2}^n$-configurations, e.g. the J,T=7,0 state of the two-particle system is a pure $f_{7/2}^2$ state. The low-spin states are more configuration-mixed with low-j single-particle orbits. When the lower-spin orbits are truncated from the model space, states with large total J are not as affected as states with low J. To compensate for this, the low-J two-body matrix elements have a greater renormalization than the matrix elements with high spin. The renormalization of the J,T=0,1 matrix elements suggests a pairing property for these interactions. As more orbits are added in a pairing model, the ground state becomes more bound. Thus, the effect of truncation when the residual interaction is a pairing interaction is to reduce the splitting of the 0^+ state from the remaining states. Thus, to renormalize a pairing interaction to compensate for truncation, one would merely increase the interaction strength. The increase in the J,T=0,1 matrix elements may reflect such an effect.

V. Contributions of Active Orbits in
Shell-Model Calculations

The basis of the theory of renormalization of the residual two-body interactions is that the contributions of the single-particle orbits outside the active model space can be implicitly included in the shell-model calculation as a perturbation of the two-body interaction. The theory is directed to reproducing energy levels, and

TABLE IV. Matrix Elements, $\langle j_1 j_2 JT | H | j_3 j_4 JT \rangle$, Involving $f_{7/2}$ and $p_{3/2}$ Orbits. K-B are Kuo-Brown numbers, Freed are $7/2^2$ matrix elements fit to experiment in Four-Shell Model. 2-Shell is inter-action obtained by fitting Four-Shell Results in $f_{7/2}$-$p_{3/2}$ space. Matrix elements not listed were undetermined in 2-Shell search, and left at original Kuo-Brown values.

$2J_1$	$2J_2$	$2J_3$	$2J_4$	J T	K-B	Freed	2-Shell
7	7	7	7	0 1	-1.81	-2.22	-2.75
				2 1	-0.78	-1.15	-1.04
				4 1	-0.09	-0.36	-0.36
				6 1	0.23	0.29	0.20
3	3	3	3	0 1	-1.21		-1.72
				2 1	-0.38		-0.64
7	3	7	3	2 1	-0.86		-1.03
				3 1	-0.03		-0.17
				4 1	-0.05		-0.19
				5 1	0.15		0.15
7	7	3	3	0 1	-0.78		-1.22
				2 1	-0.27		-0.58
7	7	7	3	2 1	-0.50		-0.72
				4 1	-0.31		-0.38
7	3	3	3	2 1	-0.32		-0.58
7	7	7	7	1 0	-0.52	-1.45	-2.71
				3 0	-0.21	-1.07	-1.55
				5 0	-0.50	-1.10	-1.04
				7 0	-2.20	-2.42	-2.47
7	3	7	3	2 0	-0.29		-0.60
				4 0	-0.16		-0.70
				5 0	-2.17		-2.77
7	7	7	3	5 0	-0.82		-0.92

there is no a priori reason why the resulting wave functions should be useful for any other observables. But experience has shown that the resulting wave functions are capable of correlating a great deal of the data, at least in light nuclei. We are interested as

to just what is the role of the various orbits in the shell-model
calculations, and what observables are sensitive to what orbits.
We have therefore carried out some theoretical "experiments" aimed
at this question. We considered the nuclei with A = 18-21 in the
s-d shell. We again assumed an inert ^{16}O core, and we made calculations in three different models. The first model was the full
s-d shell model discussed above. In the second model the $d_{3/2}$
orbit was omitted from the calculation, and in the third model, the
$d_{3/2}$ and $s_{1/2}$ orbits were omitted, so that the third model was the
pure $d_{5/2}^n$ calculation. We used the single-particle energies taken
from ^{17}O in all three models. For the interaction we used the
matrix elements derived for the full s-d shell space by Kuo.[5] We
assumed these were the "correct" matrix elements for the full space
calculation. In the $(d_{5/2}\text{-}s_{1/2})^n$ calculation we used the Kuo
matrix elements involving these orbits without any renormalization,
and we did the same for the pure $d_{5/2}^n$ calculation. Thus, in the
two-shell $d_{5/2}\text{-}s_{1/2}$ calculation there are, in principle, no contributions from the $d_{3/2}$ orbit. In the $d_{5/2}^n$ calculation there are
no contributions from the $s_{1/2}$ or $d_{3/2}$ orbits. Thus, a comparison
of the three calculations should suggest what are the contributions
of these orbits to the calculations. We again calculated energy
levels, S-factors, magnetic dipole and electric quadrupole observables, and log ft-values for β-decay.

The comparison of the energy level calculations can be summarized generally as follows. The qualitative structure of the
ground-state rotational bands are present in both the three-shell
and the two-shell spectra, but they are not present in the $d_{5/2}^n$
calculation. As an example the spectra of ^{21}Ne as calculated in
these three models are shown in Fig. 2, as is the experimental
spectrum. We see that the ground-state K=3/2 band is already
clearly present in the two-shell calculation with the correct spin
ordering and relative spacing, but this is not so in the pure $d_{5/2}^n$.
The agreement of the calculated and observed spectra is essentially
as good for the two-shell calculation as for the three-shell model.
This general picture is fairly typical of all the 18-22 spectra
with the exception of two or three states in ^{18}F and ^{19}F. From
these results, it is obvious that the low-lying three-shell spectra
could be reproduced in the two-shell space by a fairly simple renormalization of the interaction.

One interesting feature appears in the structure of the wave
functions in these calculations. This is the fact that the
$(d_{5/2}\text{-}s_{1/2})$ structure in three-shell wave functions is very similar to the structure of the two-shell wave functions. By this I
mean if we take that part of the three-shell wave function that involves only $(d_{5/2},s_{1/2})^n$ configurations, renormalize it, and take
overlaps with the analogous two-shell function, the overlaps are
extremely good. This is seen in Table V, where these overlaps are
listed for some states in the ground-state bands of the A=19, 20,

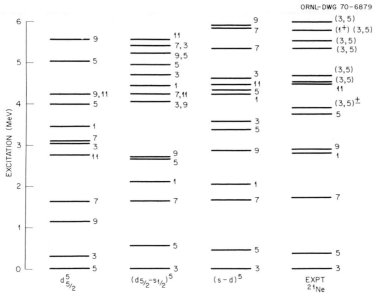

Fig. 2. ^{21}Ne Spectra Calculated in s-d-Shell Models Described in Text, and Observed Spectrum of ^{21}Ne.

and 21 nuclei. We also show there what part of the three-shell wave function is in the $(d_{5/2}-s_{1/2})^n$ space. These numbers show that whenever at least half of the three-shell wave function is in the $(d_{5/2}-s_{1/2})$ space, the overlap of the $(d_{5/2}-s_{1/2})$ part of the three-shell wave function is at least 94%, and more often 99%.

In four of the cases shown, configurations involving the $d_{3/2}$-orbit constitute more than 50% of the wave function. The $3/2^+$ state in ^{19}F has a large $d_{3/2}$-single-particle component. The higher members of the ^{20}Ne band must contain $d_{3/2}$ particles since there are so few ways to form high-spin states in the $(d_{5/2}-s_{1/2})^4$ configuration. The structure of the $7/2^+$ state in ^{19}F is difficult to understand.

The magnetic moments are extremely insensitive to configuration mixing in these s-d shell-model calculations. The differences between values calculated in the three models are not large, and the differences are comparable to the difference between experiment and the value calculated in the full three-shell model. Thus, for this observable, it is obvious that any renormalization of the effective two-body interaction will have little effect on the calculated numbers. The differences between B(M1)'s calculated in the two-shell and three-shell models can be as large as two orders of magnitude. Unfortunately, there is almost no experimental information on B(M1)'s in this region with which to compare to see if this sensitivity is useful.

TABLE V.   Comparison of $(d_{5/2}\text{-}s_{1/2})$ Part of Three-Shell Wave Function with Two-Shell Wave Function.   Fourth column indicates percentage of three-shell wave function which does not involve the $d_{3/2}$ orbit.

A	2T	2J	% in $(d_{5/2}\text{-}s_{1/2})$	Square of Overlap	2-Shell Dimension
19	1	1	69	100	4
		3	40	81	4
		5	72	98	6
		7	30	3	5
		9	78	99	3
	3	1	92	100	1
		3	93	100	2
		5	90	99	3
20	0	0	52	99	6
		4	51	99	10
		8	41	89	9
		12	37	87	5
20	2	4	64	98	10
		6	73	94	12
		8	69	95	8
		10	75	99	7
20	4	0	83	99	2
		4	82	99	4
21	1	3	53	99	18

    The quadrupole moments are somewhat sensitive to configuration mixing as seen in Table VI.   In all these calculations we again use a state independent added effective charge of 0.5e for both protons and neutrons.   We evaluate the oscillator parameter from the expression $\hbar\omega = 41\ A^{-\frac{1}{3}}$.   The moments calculated in the two-shell model and the three-shell model all agree in sign (shape), and the relative numbers for the different quadrupole moments are similar in the two calculations.   The numbers calculated in the pure $d_{5/2}^{\,n}$ model are distinctly different both in shape and magnitude.   The same generalizations apply to the B(E2)-value calculations.   The two-shell results are very similar to the three-shell numbers, and could be made almost identical by a simple increase of the effective charge.   The $d_{5/2}^{\,n}$ numbers are distinctly different in relative

TABLE VI.   Quadrupole Moments Calculated in Three Shell-Models Described
in the Text.   The effective operator used in these calculations
is described in the text.

Nucleus	2J	Calculated Quadrupole Moments (e fm^2)			Expt.
		$(d_{5/2})^n$	$(d_{5/2}\text{-}s_{1/2})^n$	$(d_{5/2}\text{-}s_{1/2}\text{-}d_{3/2})^n$	
^{19}F	5	-8.3	- 8.9	- 9.2	± 11.02 ± 2.0
	9	-5.5	-10.3	-11.8	
^{20}Ne	4	+6.4	-11.7	-14.3	-24 ± 3
	8	+2.7	-11.7	-18.3	
^{20}F	4	+1.8	+ 4.4	+ 7.5	
	6	-0.1	- 3.9	- 3.2	
^{21}Ne	3	-2.3	+ 8.9	+10.3	
	5	+1.4	- 3.6	- 3.4	

value. It seems unlikely that a renormalization of the two-body
interaction could give a one-shell model which would accurately reproduce the three-shell results for the electric quadrupole observables.

The energy level and electric quadrupole operator results here
are consistent with the possibility that to a large extent, many of
low-lying states of these nuclei can be well described in terms of
one state in the SU(3)-representation. The electric quadrupole
operator does not connect different SU(3)-representations so any
small admixtures of SU(3)-states will affect the electric quadrupole
results only in second order. Thus, if SU(3) is a fairly good
representation for both the two-shell and three-shell calculations
one would expect rotation-like spectra in both cases, and the relative B(E2)-values and quadrupole moments would be similar in the
two models. In the complete (s-d)^n space, the ground-state bands
are dominated by one SU(3)-state ($\sim$ 80%). Eliminating the $d_{3/2}$-
orbit (thus implicitly introducing an infinite spin-orbit splitting)
reduces this admixture to about 40-50%, but this apparently is still
a large enough dominance to retain rotation-like features in the
calculation.[16] The leading SU(3) states are strongly admixed in
$\ell=0$ and $\ell=2$ states. Eliminating the $s_{1/2}$ orbit destroys the goodness of SU(3), and the structure of the calculations changes
significantly.

The log ft-values for Gamow-Teller β-decays are extremely

sensitive to $d_{3/2}$ admixtures. The sensitivity for this observable is the same as for the Ml-observables, since the dominant operator in both cases is the operator $\sigma^1\tau^1$. There is a significant amount of experimental data on these decays. The results suggest that the model must include the $d_{3/2}$ orbit in order to account for the observed log ft values. The results are discussed in detail elsewhere.[17]

In this review, I have first illustrated that the shell model can correlate a large amount of information on energy levels, single-particle strengths, and electromagnetic properties. We have shown that for a certain variety of effective residual one-body plus two-body interactions, the calculated results are qualitatively very similar. I then discussed the renormalization of the interaction that takes place when a model space is truncated. In the example used here, in going from four shells to the lowest two shells in the f-p shell the main renormalizations were to increase the attraction of the J=0,T=1 matrix elements, to increase the average interaction of particles in different orbits in T=0 states, and to make the T=0 matrix elements for low J in the $f_{7/2}^2$ configuration more attractive. Finally, since it is now well known that many contributions from orbits outside the model space can be absorbed into the renormalized interaction, I discussed some calculations which attempt to show what the contributions of the various orbits in the s-d shell-model calculations are. The results suggest that most of the properties of the calculation in the full shell can be reproduced in the $d_{5/2}$-$s_{1/2}$ calculation with fairly simple renormalizations of the one-body and two-body operators, but a pure $d_{5/2}^n$ model could not reproduce the full calculations with a simply renormalized operator.

REFERENCES

1. E. C. Halbert, J. B. McGrory, B. H. Wildenthal, and S. P. Pandya, Advances in Nuclear Physics, Vol. 3 (M. Baranger and E. Vogt, eds.) Plenum Press, New York, 1969.

2. J. B. French, E. C. Halbert, J. B. McGrory, and S. S. M. Wong, Advances in Nuclear Physics, Vol. 4 (M. Baranger and E. Vogt, eds.) Plenum Press, New York, 1971; S. Cohen, R. D. Lawson, M. H. Macfarlane, and M. Soga, Methods in Computational Physics (B. Alder, S. Feinbach, and M. Rosenberg, eds.) Academic Press, New York and London (1969).

3. J. D. McCullen, B. F. Bayman, and L. Zamick, Phys. Rev. 134 (1964) B515; J. N. Ginocchio and J. B. French, Phys. Letters 7 (1963) 137.

4. T. T. S. Kuo and G. E. Brown, Nucl. Phys. 85 (1966) 40.

5. T. T. S. Kuo, Nucl. Phys. A103 (1967) 71.

6. R. Arvieu and S. Moszkowski, Phys. Rev. 145 (1966) 830.

7. P. W. M. Glaudemans, P. J. Brussaard, and B. H. Wildenthal, Nucl. Phys. A102 (1967) 593.

8. See S. Cohen, E. C. Halbert, and S. P. Pandya, Nucl. Phys. A114 (1968) 353, and references therein.

9. D. R. Kulkarni and S. P. Pandya, Nuovo Cimento 60B (1969) 199.

10. T. T. S. Kuo, private communication.

11. B. H. Wildenthal, J. B. McGrory, and P. W. M. Glaudemans, Phys. Rev. Letters 26 (1971) 96.

12. R. D. Lawson and J. M. Soper, Proceedings of International Nuclear Physics Conference, Gatlinburg, Tenn., 1966 (R. L. Becker, C. D. Goodman, P. H. Stelson, and A. Zucker, eds.) Academic Press, New York and London (1967).

13. D. M. Clement and E. U. Baranger, Nucl. Phys. A108 (1968) 27.

14. T. T. S. Kuo and G. E. Brown, Nucl. Phys. A114 (1968) 241.

15. J. B. McGrory and E. C. Halbert, to be published in Physics Letters.

16. J. B. McGrory, to be published.

17. J. B. McGrory, Phys. Letters 33B (1970) 327.

*Research sponsored by the U. S. Atomic Energy Commission under contract with The Union Carbide Corporation.

THE EFFECTIVE NUCLEON-NUCLEON INTERACTION DEDUCED FROM NUCLEAR SPECTRA[*]

John P. Schiffer

Argonne National Laboratory, Argonne, Illinois and

University of Chicago, Chicago, Illinois

The subject of this paper is the study of the effective interaction between nucleons as it exhibits itself in the spectra of some of the simplest nuclei. These are nuclei differing from closed shells by only two nucleons. In Fig. 1 we see schematically how the energy level schemes of such nuclei, when combined with the energy levels of adjacent single-particle nuclei, yield the matrix elements of the residual interaction between two nucleons.[1]

The next question is how one knows that the energy levels of such nuclei are indeed representative of "two-nucleon" residual interaction and that they do not contain appreciable core excitations. There are partial answers to these questions on two levels. To the extent that core excitations are important in the "one-nucleon" levels of the closed-shell-plus-one-nucleon nuclei, they are properly included in the effective interaction. The detailed identification of states as belonging to multiplets of a given configuration depends on careful use of reaction data, as will be shown by one typical (in the usual sense) example.

This example is a study of the ^{209}Bi$(d,t)^{208}$Bi reaction.[2] ^{208}Pb is probably the best closed-shell nucleus at our disposal. The nuclei immediately adjacent to it exhibit a simple spectrum, as is shown in Fig. 2; the six orbits that make up the neutron excess appear as ^{207}Pb hole states that had been identified in a number of neutron-pickup experiments on ^{208}Pb. When the neutron pickup on ^{209}Bi was studied, we found a multiplet of states corresponding to each state in ^{207}Pb. The odd proton in ^{209}Bi is in an $h_{9/2}$ orbit. The simplest model would predict that multiplets should appear in ^{208}Bi with cross sections that add up to the pickup cross section to the corresponding hole state in ^{207}Pb and the strengths to individual members of the multiplet should be divided up in prop n

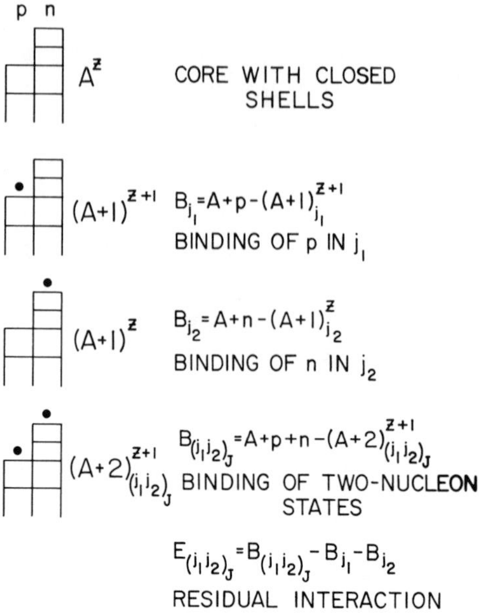

A^z CORE WITH CLOSED
           SHELLS

$(A+1)^{z+1}_{j_1}$ $B_{j_1}=A+p-(A+1)^{z+1}_{j_1}$
           BINDING OF p IN j_1

$(A+1)^{z}_{j_2}$ $B_{j_2}=A+n-(A+1)^{z}_{j_2}$
           BINDING OF n IN j_2

$(A+2)^{z+1}_{(j_1 j_2)_J}$ $B_{(j_1 j_2)_J}=A+p+n-(A+2)^{z+1}_{(j_1 j_2)_J}$
           BINDING OF TWO-NUCLEON
                     STATES

$$E_{(j_1 j_2)_J}=B_{(j_1 j_2)_J}-B_{j_1}-B_{j_2}$$
RESIDUAL INTERACTION

Fig. 1.--Schematic definition of the residual interaction.

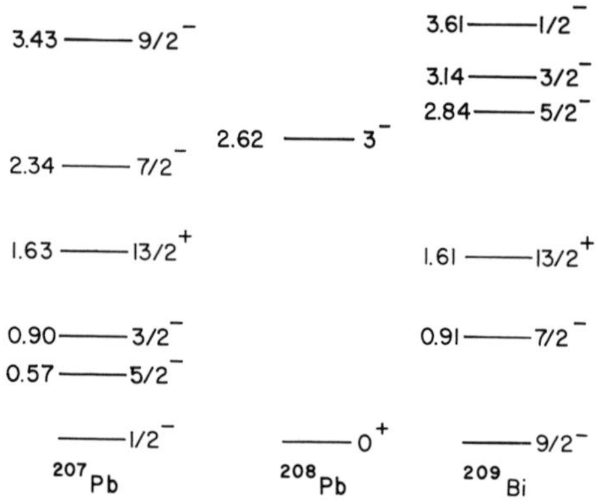

3.43 ——— $9/2^-$ 3.61———$1/2^-$

                                          3.14———$3/2^-$
                   2.62 ——— 3^- 2.84———$5/2^-$
2.34 ——— $7/2^-$

1.63 ——— $13/2^+$ 1.61 ——— $13/2^+$

0.90——— $3/2^-$ 0.91——— $7/2^-$
0.57——— $5/2^-$

       ——$1/2^-$ ——0^+ ——$9/2^-$
   ^{207}Pb ^{208}Pb ^{209}Bi

Fig. 2.--Single-particle states in ^{209}Bi and single-hole state in
^{207}Pb.

portion to 2J+1. The summed cross sections are compared in Table 1
and the relative cross sections are shown in Fig. 3. We see that
the assignment of spins to the members of the multiplet is relatively

TABLE 1.--Ratio of neutron-pickup cross sections.

Orbit	$\dfrac{\sum \sigma(\text{Bi}^{208})^a}{\sigma(\text{Pb}^{207})}$
$p_{1/2}$	0.97
$p_{3/2}$	0.96
$f_{5/2}$	1.07
$f_{7/2}$	1.09
$h_{9/2}$	0.96
$i_{13/2}$	0.95

[a]Summed cross section of a multiplet seen in $^{209}\text{Bi}(d,t)^{208}\text{Bi}$ divided by the corresponding cross section for $^{208}\text{Pb}(d,t)^{207}\text{Pb}$ (from Ref. 2).

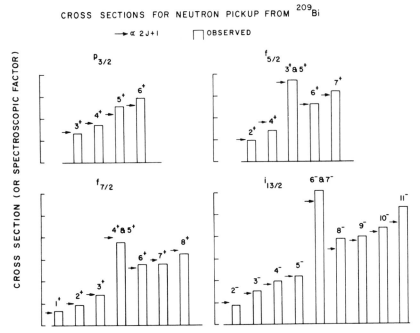

Fig. 3.--Distribution of neutron pickup strengths for reactions on ^{209}Bi. The arrows indicate expected levels of cross sections.

straightforward, and that among them they definitely exhaust the pickup strength of the single state in ^{207}Pb. Thus, whatever the nature of the "hole states" of ^{208}Pb, as represented by ^{207}Pb, the same "hole states" of ^{209}Bi (g.s.) are well represented in the multiplets of ^{208}Bi. To the extent that core polarizations alter the pure single-particle and single-hole states, this is carried over.

Next let us consider a useful way of visually presenting these multiplets. The usual energy-level diagrams are not easy to absorb at a glance. In the spirit of our thinking, we define an angle θ_{12} as shown in Fig. 4, and will plot energy matrix elements against this angle. Such a plot is shown in Fig. 5 for various particle-hole multiplets from both ^{208}Bi and ^{96}Nb. The θ dependence of the energy matrix elements clearly must reflect the range and other properties of the force. A short-range force will have the biggest matrix elements for θ_{min} and θ_{max} (i.e., for J_{max} and J_{min}) because the orbits are most nearly coplanar for these limiting values. For a particle-hole multiplet, the interaction is repulsive and thus a U-shaped plot is expected. The extent of curvature in the U is a measure of the range. A long-range force would give a degenerate multiplet with a constant matrix element. For a particle-particle multiplet, the U becomes inverted since the residual particle-particle interaction is attractive.

One can perform the equivalent of a Legendre decomposition in $\cos\theta_{12}$ by taking the Racah transform of the two-body spectrum,[1] following J. B. French. The result is

$$E_J = \sum_k \Phi \; \alpha_k W(j_1 j_2 j_1 j_2, \; Jk),$$

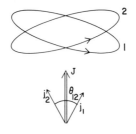

$$\theta_{12} = \cos^{-1} \frac{j_1(j_1+1) + j_2(j_2+1) - J(J+1)}{2\left[j_1 j_2 (j_1+1)(j_2+1)\right]^{1/2}}$$

Fig. 4.--Definition of the angle θ_{12}.

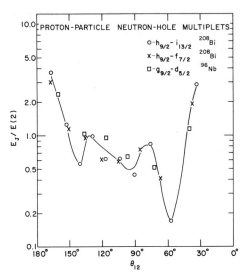

Fig. 5.--Comparison of three particle-hole spectra with nonidentical
    orbits and mixed isospin (n-p). The line is drawn to
    connect the points.

where Φ represents phases and statistical weights as defined in
Ref. 1. The coefficients of odd polynomials can only arise from a
spin-dependent force. A transformation from particle-particle to
particle-hole matrix elements is accomplished simply by changing
the signs of the even-multipole coefficients. This leads to the
Pandya transformation

$$E_J^{p-h} = -\sum_{J'} (2J' + 1)W(j_1 j_2 j_2 j_1, JJ')E_{J'}^{p-p}.$$

In order to compare multiplets from different orbits and nuclei, we
have to correct for the different radial overlaps between orbits.
This is accomplished by dividing every matrix element by the monopole coefficient

$$\alpha_0 = \sum_J (2J + 1)E_J = \overline{E(2)}.$$

This is done in Fig. 5. We may note the remarkable similarity in
the various mutliplets. All of these are (n-p) configurations:
that is to say, the orbits are specified as to which is occupied
by a neutron and which by a proton and the isospin of the pair is
mixed. Thus

$$E_J^{n-p} = \frac{1}{2} (E_J^{T=0} + E_J^{T=1}).$$

Two other sets of data are compared in Fig. 6, where the $(f_{7/2})^2$ particle-hole matrix elements of ^{48}Sc are compared with the $(g_{9/2})^2$ ones of ^{90}Nb. The similarity is again remarkable. Here, since the two orbits are identical, the isospin structure is more complicated. A further check is a comparison between the particle-particle and particle-hole forms of the same multiplet. Such data are known in a few cases, one of which is illustrated in Fig. 7.

All the available (n-p) multiplets, converted into particle-particle matrix elements, are compared in Fig. 8 and the identical-orbit data in Fig. 9. We note that for the large J values, the separation between states that differ from J_{max} by an even or odd integer is a uniform feature of all the data but is more pronounced for the identical orbits. This separation depends on orbital symmetries and is reproduced by even a simple delta-function calculation, as shown in Fig. 10.

What then do we learn from the data displayed in Figs. 8 and 9? First of all we learn that the experimental data selected as representing the matrix elements of the effective interaction do indeed show a remarkably constant behavior throughout the periodic table. Secondly we find that a δ-function force can fit these data very well. This is shown in Fig. 11, where the data are represented

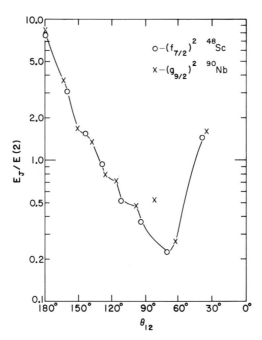

Fig. 6.--Comparison of two identical-orbit particle-hole spectra. The line is drawn to connect the points.

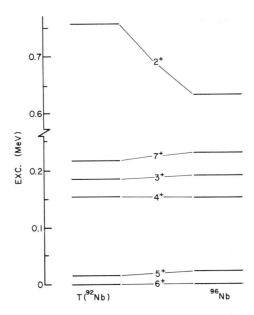

Fig. 7.--Comparison of $g_{9/2}d_{5/2}$ spectra from ^{92}Nb and ^{96}Nb. The former is converted into the particle-particle spectrum by the Pandya transformation, represented by T ( ).

in terms of multipole coefficients. The rate of fall-off of the even coefficients with increasing multipolarity is a measure of the range, and we see that the δ-function force gives excellent agreement with the average values. The odd coefficients were fitted with a spin-exchange force, also of zero range. The identical-orbit data differ somewhat in the quadrupole and the odd-k coefficients.

A year ago my report would have had to stop here because this was the extent of the data. Within the last year, however, additional data have become available. From these the matrix elements for nonidentical orbits can be obtained separately for T=0 and T=1. Figure 12 shows the matrix elements for the $1d_{3/2}-1f_{7/2}$ interaction as deduced in two recent experiments on mass-34 nuclei. We note that the T=0 and T=1 states have remarkably similar patterns of relative spacings, each characteristic of a short-range force, but are separated by a ~2-MeV gap. Thus the isospin splitting is monopole (long-range) in character and sufficiently strong to yield repulsive matrix elements in T=1.

Additional data on the isospin splitting[3] are seen in Figs. 13 and 14. Here we still have an isospin splitting, but the relative spacing of T=1 states is quite different from that of the T=0 ones. The T=0 pattern is still that characteristic of a short-range attractive force. For T=1, however, the matrix elements are much

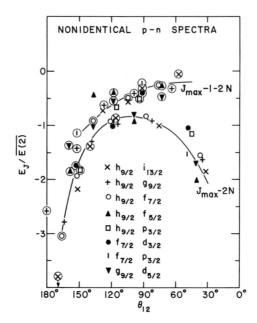

Fig. 8.--Comparison of all known (n-p) matrix elements for non-
    identical orbits, converted into particle-particle energies.
    The circled points differ from J_{max} by an odd integer in
    angular momentum. The lines are drawn to connect the points.

closer together, the plot shows a slight downward curvature characteristic of an attractive force, but their centroid indicates a force
that is slightly repulsive. Thus there is a strong suggestion of
competition between a short-range attractive force and a longer-
range isospin-dependent force that is manifestly repulsive in T=1.
In a light nucleus, such as A=34 (Fig. 12), the long-range force
manifests itself as a monopole term; but in the Pb region the radii
of the orbits are almost twice as large and the repulsion in T=1
has a finite range giving rise to more than just a monopole term.
Therefore much of the curvature within the multiplet is cancelled.
The isospin dependence seems to have a range larger than the A=34
orbits and smaller than those in A=208.

    This is clearly reflected in the isospin-dependent multipole
coefficients (Table 2), defined by

$$E_{J,T} = \sum_{k,\tau} \Phi' \alpha_{k\tau} \; W(\tfrac{1}{2}\tfrac{1}{2}\tfrac{1}{2}\tfrac{1}{2},T\tau)W(j_1 j_2 j_1 j_2, \; Jk).$$

The isovector (α_{k1}) interaction is purely monopole for ^{34}Cl whereas
it has appreciably higher -k components in the Pb region.

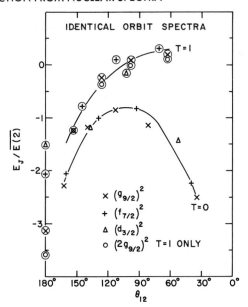

Fig. 9.--Same as Fig. 8 but with identical orbits.

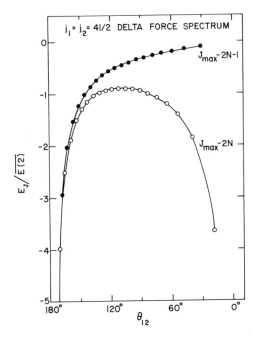

Fig. 10.--A spectrum calculated for a δ-function force. The lines
are drawn to connect points.

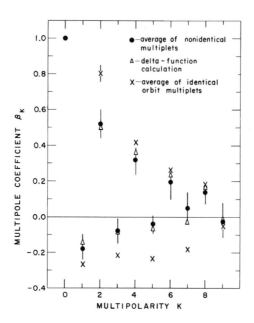

Fig. 11.--Comparison between average multipole coefficients from
the data and those from δ-function calculations.   The
error bars represent the mean fluctuations.

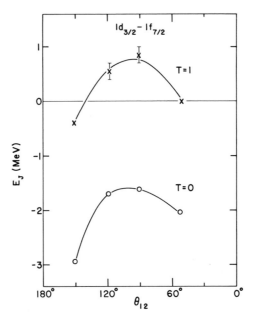

Fig. 12.--The $d_{3/2}f_{7/2}$ matrix elements in A=34.

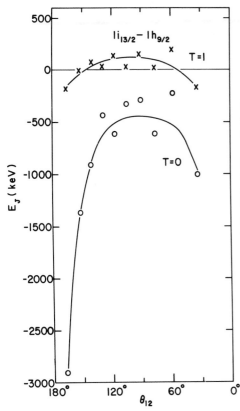

Fig. 13.--The $h_{9/2}i_{13/2}$ matrix elements. For T=1, the points represent the data for ^{210}Po (with Coulomb matrix elements removed). For T=0, the points were obtained from a combination of the Pandya transforms for ^{208}Bi and ^{210}Po. The lines are drawn to emphasize qualitative trends.

TABLE 2.--Isospin multipole coefficients.

	k	$(d_{3/2}f_{7/2})$, A=32		$(h_{9/2}i_{13/2})$, A=208	
		α_{k0} (MeV)	α_{k1} (MeV)	α_{k0} (keV)	α_{k1} (keV)
Even	0	-0.55	1.15	280	350
	2	-0.82	0.00	330	240
	4			210	150
	6			130	100
	8			50	20
Odd	1	0.00	0.10	-50	-70
	3	0.29	0.00	-10	-40
	.			.	.
	.			.	.
	.			.	.

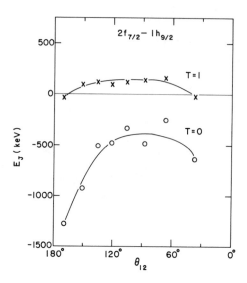

Fig. 14.--Same as Fig. 13 for the $h_{9/2}2f_{7/2}$ matrix elements.

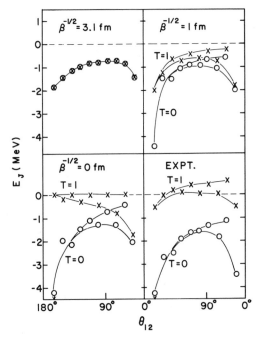

Fig. 15.--The dependence of matrix elements on the range of a
          Wigner force.

Next then I would like to show some attempts to fit these data in terms of an effective potential. This is work I did together with N. Anantaraman.

At first we explored the range dependence of two-body matrix elements, using a Wigner force (no spin or isospin dependence). Such calculations are shown in Fig. 15, where it is interesting to note that the isospin splitting becomes quite marked in the limit of short range. In fact, this isospin dependence is a much more sensitive function of range than is the "curvature" as measured by the quadrupole coefficients, as is shown in Fig. 16. Were it not for a possible explicit isospin dependence in the effective inter-action, this would be a sensitive way of pinning down the range of the force.

Because of such ambiguities, we separated the data into T=1 and T=0 matrix elements and fitted them separately to calculated matrix elements. The T=1 matrix elements then determine the triplet-odd and singlet-even components of the two-body interaction (these would be equal for a Wigner force) and the tensor-odd component.

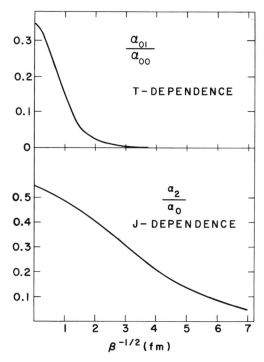

Fig. 16.--The isospin splitting and the quadrupole coefficient as a function of range of a Wigner force.

The T=0 matrix elements are determined by the triplet-even,
singlet-odd, and tensor-even components of the interaction. The
dependence of matrix elements on such components of the force is
shown in Fig. 17 for a 1-fm range. In all that follows, oscillator
wave functions were used and the oscillator constant was kept fixed
so that the bound protons had the same mean-square radius as is
seen in electron scattering. A Gaussian interaction of the form

$$V = V_0 e^{-\beta r^2}$$

was used in computing the matrix elements. The calculations were
carried out with a program supplied by W. W. True.

In attempting to fit the T=1 matrix elements, it was immediately
clear that a single-range potential could not do the job. It was
therefore decided to use two ranges. The shorter range was rather
arbitrarily fixed at $\beta=1$. This was deemed perhaps a little more
"realistic" (in the sense of OPEP) than zero range, and the long
range $\beta=0.1(\beta^{-\frac{1}{2}}=3.2$ fm) was selected as more or less the minimum
range that could have the desired effect. The long-range term was
restricted to a Wigner mixture, but the triplet and singlet components
were allowed to vary separately for the shorter range.

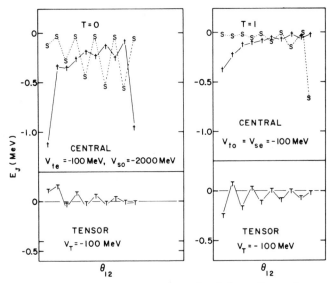

Figure 17.--Matrix elements calculated with a Gaussian potential of
range $\beta^{-\frac{1}{2}}=1$ fm in the $1h_{9/2}$ $1i_{13/2}$ multiplet. Note the
twenty-fold increase in the strength of the singlet-odd
force. The symbols t, s, and T refer to the triplet,
singlet, and tensor matrix elements, respectively.

The results of fitting the $h_{9/2}i_{13/2}$ matrix elements for T=1 is shown in Table 3. It is clear that the tensor force produces no improvement in the data but that the spin dependence in the short-range force does help. For the T=0 matrix elements, the effect of the long-range interaction was similar to that of the short-range one and could not be well distinguished in the data. The singlet-odd interaction was very poorly determined and the tensor force caused only a marginal difference, so it is really only the triplet-even interaction that plays any role.

TABLE 3.--Potential strengths (MeV) for the $h_{9/2}i_{13/2}$ multiplet.

T=1	Triplet odd		Singlet even	TO=SE	Tensor odd	$\sqrt{\frac{x^2}{N}}$
		β=1		β=0.1	β=1	(keV)
	-230		-165	7.2	4	34
	-231		-164	7.2		34
	-145[a]		-145[a]	5.0		62
T=0	Triplet even		Singlet odd		Tensor even	$\sqrt{\frac{x^2}{N}}$
		β=1			β=1	
	-242		-122		-138	183
	-261		- 5.6			195
	-257[a]		-257[a]			198

[a]Constrained to be equal ($V_{triplet} = V_{singlet}$).

The strengths extracted from various individual sets of data are shown in Table 4, where we note that they are indeed similar in magnitude. This leads us to attempt to fit larger groups of data with the same interaction. In order to include (n-p) data we developed the following procedure. We first fitted the parameters of the T=1 interaction using only the pure T=1 data. This seemed necessary because the absolute magnitude of the T=1 matrix element is only about a tenth of the T=0 matrix elements and the weighting in a least-squares fit would tend to destroy any fit to the T=1 matrix elements. After determining and fixing the T=1 interaction strengths, the T=0 and the n-p matrix elements were fitted by varying only the T=0 strengths. The results are shown in Table 5. Since matrix elements tend to be larger for lighter nuclei, a weighting factor of $A^{2/3}$ was included in these global fits.

TABLE 4.--Strengths (MeV) from best individual fits. (No tensor force).

Orbits	A	T=1 β=1		β=0.1	T=0 β=1	$\sqrt{\chi^2/N}$
		TO	SE	TO=SE	TE=SO	(T=1)/(T=0) (keV)
$1h_{9/2}1i_{13/2}$	208	-231	-164	7.2	-257	34/198
$1h_{9/2}2f_{7/2}$	208	-252	-159	6.4	-209	13/166
$(2g_{9/2})^2$	208	-191[a]	-191[a]	3.5		57
$(1g_{9/2})^2$	90	-146[a]	-146[a]	4.6	-139	169/213
$(1f_{7/2})^2$	48	-125[a]	-125[a]	3.5	-162	207/263
$1d_{3/2}1f_{7/2}$	32	-372	-167	11.7	-184	170/419
Weighted fit to all data		-230	-165	6.5	-196	235/363

[a] The identical-orbit cases had to be constrained (TO=SE).

The fit to the $h_{9/2}i_{13/2}$ matrix elements is shown in Fig. 18. One may question the real meaning of these two-body parameters. They do not much resemble the free nucleon-nucleon force. A comparison with the Hamada-Johnston force is shown in Fig. 19. Quite apart from the long-range repulsion, the triplet-odd strength is much weaker in the free nucleon-nucleon force. We must remember that most of the T=1 matrix elements are in fact almost degenerate around zero. The triplet-odd force is required mainly because the 2^-, T=1 matrix element in the $h_{9/2}i_{13/2}$ multiplet and the 1^+ in $f_{7/2}h_{9/2}$ are attractive. The identical-orbit cases could all be fitted quite well without much triplet-odd strength. What we are fitting is the small deviation in the near cancellation between the short-range attractive and the long-range repulsive force. This is clearly a dangerous game, and it is quite possible that a more palatable interaction might be found to fit the observed data. We have explored the tensor-odd force, but it does not seem to be able to substitute for the long-range repulsion in T=1.

An improvement in the fits obtained could be sought to several directions. Perhaps one should use a Yukawa interaction as being more realistic, and perhaps Woods-Saxon wave functions would reproduce radial wave functions--though these effects will probably not matter much. Perhaps one could improve on our way of constraining the range and other parameters of the force.

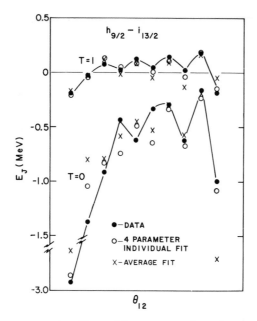

Fig. 18.--The fits to the $1h_{9/2}1i_{13/2}$ matrix elements. They were
    obtained from the best individual fit of Table 3 and the
    average fit of Table 5.

    One would like to see the matrix elements from a two-body
spin-orbit force. As is shown in Table 6 there is some evidence for
differences between the centroids of multiplets involving spin-orbit
partners in the same nuclei. Such a difference could not be explained
by a central interaction, and the required tensor strength would
destroy all the more detailed agreement. It could conceivably be
fitted by a two-body spin-orbit force.

    McGrory has told us how many-particle shell-model fits to data
are sensitive to only certain matrix elements. We have here restricted
ourselves to all we know <u>directly</u> from experiment about diagonal two-
nucleon matrix elements. The success of a surface-delta plus isospin-
dependent monopole interaction for shell-model calculations in the
s-d shell is apparent from what we have said.

    One piece of evidence I would like to mention regards off-
diagonal matrix elements. For states for which [208]Bi experiments[2]
give evidence of admixtures, Fig. 20 compares the observed admixtures
with the corresponding predictions by Kuo.[4] Even remembering that
the plot is biased by the restriction to experimentally observed
admixtures, we see that there is no real correlation--though the
<u>average</u> magnitude of admixtures is about correct.

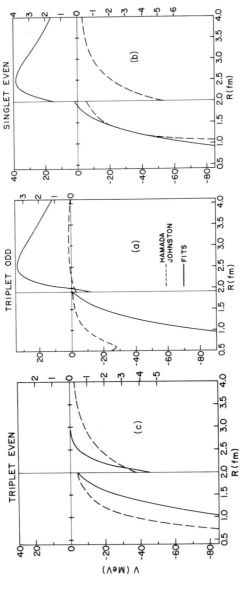

Fig. 19.—Comparison between the average-fit central potentials and the corresponding Hamada–Johnston values.

TABLE 5.--Summary of global fits.

T		All data	Pb only
1	$\begin{array}{c}\text{TO}\\\text{SE}\end{array}(\beta=1)$	−225 MeV −160 MeV	−260 MeV −215 MeV
	$\begin{array}{c}\text{TO}\\\text{SE}\end{array}(\beta=0.1)$	6.4 MeV	7.4 MeV
	Tensor-odd: No evidence, constrained to zero		
0	TE $\beta=1$	−195 MeV	−210 MeV
	SO: No real evidence, constrained to equal TE		
	Tensor-even	−240[a]	260[a]

[a]The inclusion of the tensor-even matrix element caused a slight improvement (from 250 down to 170 keV) in $\sqrt{\chi^2}/N$ within the data in the Pb region (mostly the 0^- state in ^{210}Bi). The triplet-even strength was not appreciably altered with the inclusion of a tensor term.

TABLE 6.--Monopole coefficients α_0 for (n-p) spin-orbit pairs.

	α_0 (keV)		α_0 (keV)		α_0 (keV)
$1h_{9/2}2f_{7/2}$	−216	$1h_{9/2}3p_{3/2}$	−190	$3p_{1/2}2f_{7/2}$	−176
$1h_{9/2}2f_{5/2}$	−140	$1h_{9/2}3p_{1/2}$	−104	$3p_{1/2}2f_{5/2}$	−153

To conclude, we have attempted to show that there is a remarkable similarity in the behavior of the experimental two-nucleon matrix elements throughout the periodic table. The experimental values we have used are appended in Table 7. Our attempts at fitting the matrix elements should not be taken too seriously: they are intended to provoke theorists into doing better calculations. The data suggest that the real effective interaction in nuclear matter is rather well defined by the data and that the large admixtures appearing in shell-model calculations are symptoms of quite uniform effects appearing throughout. One would hope that the apparent simplicity in the data will lead to a correspondingly simple theoretical description.

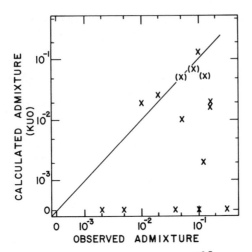

Fig. 20.--Small admixtures to states in ^{208}Bi (from Ref. 2). The
        points in parentheses represent states that were identified
        (very tentatively) on the basis of the correspondence
        between calculated and observed admixtures.

REFERENCES

1. We follow the procedure defined by M. Moinester, J. P. Schiffer,
    and W. P. Alford, Phys. Rev. 179, 984(1969). See also J. P.
    Schiffer, Ann. Phys. (to be published) and Proceedings of the
    Topical Conference on the Structure of $1f_{7/2}$ Nuclei, Padua,
    Italy (to be published).

2. W. P. Alford, J. P. Schiffer, and J. J. Schwartz, Phys. Rev.
    C 3, 860 (1971).

3. The analysis of the ^{210}Po data in W. A. Lanford's contributed
    paper for this conference is somewhat similar to that in this
    report.

4. T. T. S. Kuo, Nucl. Phys. A122, 325 (1968).

Work performed under the auspices of the U. S. Atomic Energy
Commission.

TABLE 7.—Two-body particle-particle matrix elements (keV).

		J= 0	1	2	3	4	5	6	7	8
$(1d_{3/2})^2$	$^{34}Cl^{2,b}$	-2870	-2223	160	-2720					
$1d_{3/2}1f_{7/2}$ T = 0	$^{34}Cl^a$			-2980	-1710	-1610	-2060			
$1d_{3/2}1f_{7/2}$ T = 1	$^{34}S^b$			- 390	550	970	- 20			
					± 150	± 150				
$(1f_{7/2})^2$	$^{48}Sc^c$	-2127	-2114	- 814	-1044	61	- 868	276	-2278	
$1f_{7/2}2p_{3/2}$	$^{50}Sc^d$			- 644	- 572	- 140	- 901			
$(1g_{9/2})^2$	$^{90}Nb^e$	-1835	-1316	- 671	- 628	- 40	- 442	159	- 594	225 / -1489 (J=9)
$1g_{9/2}2d_{5/2}$	$^{92}Nb^f$			- 567	- 416	- 223	- 345	-202	- 702	
$1h_{9/2}3p_{3/2}$	$^{208}Bi^g$				- 345	- 126	- 96	-217		
$1h_{9/2}2f_{5/2}$	$^{208}Bi^g$			- 257	- 56	- 52	- 134	- 32	- 283	
$1h_{9/2}2f_{7/2}$ T = 0	$^{208}Bi^h$, $^{210}Po^i$		-1276	- 926	- 509	- 488	- 330	-387	- 250	- 633
$1h_{9/2}2f_{7/2}$ T = 1	"		- 38	90	113	88	116	39	152	25
$1h_{9/2}1i_{13/2}$ T = 0	"			-2913	-1364	- 918	- 439	-614 / -607 (J=9)	- 338 / -147 (J=10)	- 291 / -998 (J=11)

TABLE 7.—Continued.

	J= 0	1	2	3	4	5	6	7	8
$1h_{9/2}1i_{13/2}$ $T=1$ "			- 195	- 12	66	27	126	28	149
							13 (J=9)	187 (J=10)	- 180 (J=11)
$1h_{9/2}2g_{9/2}$ $^{210}Bi^{j}$	- 606	- 653	- 333	- 306	- 152	- 220	- 106	- 220	- 72
									- 375 (J=9)
$(2g_{9/2})^2$ T=1 $^{210}Pb^{k}$	-1231		- 436		- 140		- 38		42

[a] J. R. Erskine, D. J. Crozier, J. P. Schiffer, and W. P. Alford, Phys. Rev. C3, 1976 (1971).

[b] D. J. Crozier (to be published).

[c] H. Ohnuma, J. R. Erskine, J. P. Schiffer, J. A. Nolen, Jr., and N. Williams, Phys. Rev. C1, 496 (1970).

[d] H. Ohnuma, J. R. Erskine, J. A. Nolen, Jr., J. P. Schiffer, and P. G. Roos, Phys. Rev. 177, 1695 (1969).

[e] R. C. Bearse, J. R. Comfort, J. P. Schiffer, M. M. Stautberg, and J. C. Stoltzfus, Phys. Rev. Letters 23, 864 (1969).

[f] J. B. Ball and M. R. Cates, Phys. Letters 25B, 126 (1967) for spins; R. K. Sheline, C. Watson, and E. W. Hamburger, Ibid. 8, 121 (1964) for energies.

[g] J. R. Erskine, Phys. Rev. 135, B110 (1964).

[h] W. P. Alford, et al., Ref. 2.

[i] W. A. Lanford, W. P. Alford, and H. W. Fullbright, Bull. Am. Phys. Soc. 16, 493 (1971).

[j] J. R. Erskine, W. W. Buechner, and H. A. Enge, Phys. Rev. 128, 720 (1962).

[k] J. H. Bjerregaard, O. Hansen, O. Nathan, L. Vistisen, R. Chapman, and S. Hinds, Nucl. Phys. A113, 484 (1968).

DOES AN EFFECTIVE E2 OPERATOR HAVE A TWO-BODY PART?

F. Khanna, M. Harvey, D. W. L. Sprung and A. Jopko

Chalk River Nuclear Laboratories, Atomic Energy of

Canada Ltd. and McMaster University, Hamilton, Ontario

The question of calculating the effective charge for E2 transitions has been quite annoying to the theoretical physicists. In recent years a general theory for the effective transition operators as needed in the nuclear shell model has been formulated (1). The basic idea is to define a model space of dimension d and an operator P $(= \sum_{i \in d} |\phi_i\rangle\langle\phi_i|)$ that projects onto this space. Then we can define an operator Q that projects onto the excluded space $Q = 1-P = 1 - \sum_{i \in d} |\phi_i\rangle\langle\phi_i| = \sum_{i \notin d} |\phi_i\rangle\langle\phi_i|$ where $|\phi_i\rangle$ are the eigenfunctions of the model Hamiltonian H_0. As Bruce Barrett mentioned this morning, an effective Hamiltonian can be defined as

$$H_{eff} = H_0 + P\nu P$$

where

$$\nu = V \Omega \text{ and } \Omega = 1 + \frac{Q}{E-H_0} \nu$$

H_{eff} has the same eigenvalue spectrum as the original total Hamiltonian H. Now the Moller operator Ω can be used to define a general effective transition operator $\tilde{t}$ as

$$\tilde{t} = \Omega^+ t \Omega$$

$$= (1 + \nu \frac{Q}{E-H_0})t(1 + \frac{Q}{E-H_0} \nu).$$

$\tilde{t}$ ought to be used for calculation of transition rates if the wave function is restricted to stay in the model space P. As in the case of an effective interaction, $\tilde{t}$ can be expressed in terms of the two body potential, V, which has got a very strong repulsion at small inter-particle separation. However following the Brueckner method a reaction matrix $G = V + V \frac{Q}{E-H_0} G$ can be defined and then t is expanded as a perturbation theory in G. It may be remarked that for t, a one body operator and V a two body operator, t will in general have a one body (t_1), a two body (t_2), a three body (t_3) and a four body (t_4) component.

Restricting to the case of one particle or one hole in a closed shell nucleus, the one body effective charge has been calculated in three separate approximation schemes: i) lowest order in G, ii) summation in TDA series and iii) summation of RPA series. Using a realistic interaction and single particle energies obtained from experiments, it is found that for ^{17}O and ^{17}F, the effective charge for the extra neutron (e_n) and for the extra proton (e_p i.e. total charge is $1+e_p$) can be explained quantitatively in the RPA approximation scheme. The lowest order and TDA approximation schemes yield too small a value for e_n and e_p. However in the region of ^{40}Ca, the RPA calculations have shown some instabilities which are discussed in the paper by Khanna et al.[1].

In this communication we would like to ask whether the transition rates in A=18 system can be quantitatively estimated by including only the effective charge calculated for A=17 system with small corrections due to Pauli exclusion principle. Benson & Flowers[2] (Nucl. Phys. A126(1969)332) have observed that in calculating the E2 transition rates in A=18 system with Saxon-Woods functions, the transitions between T=1 states require an effective charge of 0.5 - 0.7 while the transition rates between T=0 states require no effective charge. This observation suggests some special property about the core polarization effects for T=1 and T=0 states. To understand the nature of this property we have calculated the two-body part ($\tilde{t}_2$) of the E2 operator to look for features that will lead to a cancellation for T=0 states but will give enhancement for T=1 states. We have used two separate approaches for this study.

a) We have parameterized the coupling operator (v) as

$$v = -\kappa \, Q_2 \cdot Q_2 \, [A^{31}, A^{13}, A^{11}], \quad Q_2 = 4 \, \sqrt{\pi/5} \, r^2 Y^2 \text{ and } A^{2T+1,2S+1}$$

are exchange parameters. Estimating these parameters from the effective charge in ^{17}O, ^{17}F, ^{15}O and ^{15}N, we found that the matrix elements of t_2 were as large as 40% of the matrix elements for the one-body part (t_1) of the effective E2 operator. There are three relevant observations: (i) for T=1 states the contributions from t_1 and $\tilde{t}_2$ are in phase; (ii) for T=0 states $\tilde{t}_1$ and t_2 give contributions that are in-phase or out-of-phase depending on the details of the structure of the states and (iii) for T=1 states $\tilde{t}_2$ gave large enough contributions to bring B(E2: $2^+ \to 0^+$) in ^{18}O and ^{18}Ne into fair agreement with experiment (^{18}O:8.25 (8.8) $e^2 fm^4$, ^{18}Ne:69±16

(65) $e^2 fm^4$, the bracketed numbers are calculated). b) The large magnitude of the two body effect with a simple parameterization of ν suggested that a detailed calculation with realistic inter-actions is needed to establish the presence of t_2 in the E2 transition rates. First we calculate the transition rates for low lying T=1 states in A=18 system using shell model functions corresponding to $(sd)^2$. It is well known that a 4p-2h component has a signifi-cant admixture even in the ground state of ^{18}O or ^{18}Ne. Benson & Flowers have estimated that there is a 17% admixture of 4p-2h component in the ground state (0^+) and the first excited state (2^+). The p-h space chosen by Benson & Flowers is the $1\hbar\omega$ space while for the calculation of t_2 as well as for t_1 we use the $2\hbar\omega$ space. So different pieces of the excluded space Q are being con-sidered in our calculation and in the calculation of Benson & Flowers. There is no overcounting of diagrams in the calculation of the two body part of the effective operator.

We have constructed the reaction matrix for a non-local potential as given by Kahana, Lee & Scott and have used matrix elements obtained directly from phase shifts by the method of Srivastava, Jopko and Sprung.[8] The matrix element of the operator ν are obtained by including i) one bubble, ii) TDA series and iii) RPA series. Then the first eight terms given in equation 3.10 of paper I by Harvey and Khanna[1] are calculated to get the matrix elements of t_2. The three approximation schemes are used so that we can compare the two body matrix elements of t_1 and $\tilde{t}_2$ in similar resummation of diagrams. For the transition between the 4p-2h component of 0^+ and 2^+ states in A=18 system, we use the experi-mental value for B(E2: $2^+ \to 0^+$) in ^{20}Ne of 56 $e^2 fm^4$. The results for B(E2: $2^+ \to 0^+$) in ^{18}O and ^{18}Ne in the RPA approximation scheme are given in the table. The first line gives the transition rate for shell model states $(sd)^2$ with $e_p = 0.6$ and $e_n = 0.5$. The second line gives the effect of including 17% of the deformed state as suggested by the wave functions given by Benson & Flowers. The third line includes the effect of the two body part. Note that with two neutrons in the sd shell there is no contribution of the two body part to the transition rate in ^{18}O. The last line gives the experimental numbers. The results given in brackets are obtained by using $e_p = 0.7$. The experiments at Oxford[4] and at Chalk River[4] suggest that a two body part of t is needed.

Similar procedures can be used to calculate the transition rates in A=42 system. It has been shown by Benson & Flowers that the ground state (0^+) of ^{42}Ca has an admixture of $\sim 21\%$ of 4p-2h com-ponent while the first excited state (2^+) has an admixture of 58% of 4p-2h component. With these wavefunctions and $e_n = 1.0$, Benson & Flowers get for B(E2: $2^+ \to 0^+$) a value of 102 $e^2 fm^4$ while experi-ments give a value of 81 $e^2 fm^4$ (6). Using the same wavefunctions for ^{42}Ti, one can deduce a value for B(E2: $2^+ \to 0^+$) of 110 $e^2 fm^4$. Recently Forster, Ball and Davies (7) have measured this B(E2) to have the magnitude ~ 200 $e^2 fm^4$. As in the case of A=18 system, there is a strong indication that a two body part of the E2 operator is needed to explain this B(E2) in ^{42}Ti. We are presently doing calculations with renormalized reaction matrix elements to compare the contributions of the two body and the one body part of the effective E2 operator in A=42 system.

The following additional observations can be made with regard to the two body part of the E2 operator: i) For transitions between T=1 states, the matrix elements of $\tilde{t}_1$ and $\tilde{t}_2$ are generally in phase and E2 transitions for low lying states are enhanced by 10-15%. However for some matrix elements the two body part is quite large (and even of opposite sign), for example

$$<2s_{1/2} \ 1d_{3/2} \ J=1 || \tilde{t}_2 || 2s_{1/2} \ 1d_{5/2} \ J=3>$$

$$= -1.285 \ <2s_{1/2} \ 1d_{3/2} \ J=1 || \tilde{t}_1 || 2s_{1/2} \ 1d_{5/2} \ J=3>$$

with RPA renormalized G-matrix elements. The net result is that the B(E2) will be reduced by a factor of ~ 12. ii) For transitions between T=0 states, the matrix elements of $\tilde{t}_1$ and $\tilde{t}_2$ are out of phase in general. iii) In the calculation of the effective charge it was observed that in going from one-bubble to TDA to RPA renormalized G-matrix elements, the effective charge increased steadily. But for the matrix elements of $\tilde{t}_2$ no such regularity has been observed. Actually many of the matrix elements of $\tilde{t}_2$ have smaller magnitude (and even opposite signs) for RPA than for one-bubble renormalized reaction matrix elements. This indicates that the off-diagonal matrix elements of ν are substantially different from the bare reaction matrix elements and this will have implications with regard to the exchange character of the effective interaction. iv) the magnitudes of the matrix elements of $\tilde{t}_2$ are quite similar for the two different potentials.

From the study of the effective operators with realistic interactions, it appears that the two body part of the effective E2 operator can be quite large as compared to the one body part of the effective E2 operator. It can perhaps give some dramatic effects like the sharp decrease in the effective charge needed for ^{18}F. This really brings us to the important question as to the behavior of the effective charge as we move away from the closed shell nuclei. In the sd shell an interesting case will be a study of the E2 transitions in ^{20}Ne which we are undertaking presently.

TABLE.--$B(E2: 2^+ \to 0^+)$ in $e^2 fm^4$

Operator	^{18}O	^{18}Ne
one body $(\tilde{t}_1)$ $(e_p=0.6,\ e_n=0.5)$	4.2	43 (48.2)
one body with 17% deformed (4p-2h) state*	8.8	45 (49.3)
one body with 17% deformed state + two body $(\tilde{t}_2)$	8.8	54 (60)
Experiment	8.25±0.5	69±12

*For deformed state, we use $B(E2: 2^+ \to 0^+)$ in ^{20}Ne which has a value
value of $56e^2fm^4$.

REFERENCES

1. B. H. Brandow, Rev. Mod. Phys. 39(1967)771; M. Harvey and
    F. C. Khanna, Nucl. Phys. A152(1970)588; A155(1970)337;
    F. C. Khanna, H. C. Lee and M. Harvey, Nucl. Phys. A164
    (1971)612. (The last reference contains a complete list of
    references to previous work).

2. H. G. Benson and B. H. Flowers, Nucl. Phys. A126(1969)332.

3. S. Kahana, H. C. Lee and C. K. Scott, Phys. Rev. 180(1969)956.

4. B. C. Robertson, R. A. I. Bell, J. L'Ecuyer, R. D. Gill and
    H. J. Rose, Nucl. Phys. A126(1969)431; T. K. Alexander, O.
    Hausser, unpublished.

5. B. H. Flowers and L. D. Skouras, Nucl. Phys. A126(1969)332.

6. R. Hartmann and H. Grawe, Nucl. Phys. A169(1971)209.

7. J. S. Forster, G. Ball and W. Davies, unpublished.

8. M. K. Srivastava, A. M. Jopko and D. W. L. Sprung, Can. J.
    Phys. 47(1969)2359.

COMMENTS

B. R. Barrett: Have you included the vertex renormalization in calculating the effective charge?

F. Khanna: No we have not done it so far. However I would like to consult with you regarding this question later on. I would remark at this point that the vertex renormalization no doubt cancels the collective effect introduced by RPA for the case of an effective interaction and we would expect quite similar effects for the calculation of core polarization effects. However we find on comparing the ratio of the matrix element of t_2 to the two body matrix elements of t_1 that the magnitude of this ratio is similar for the case of one bubble renormalized, TDA or RPA approximations. Though the one body effective charge i.e., the one deduced from t_1 will be smaller after including vertex renormalization, I expect that the relative magnitudes of the two body matrix elements of t_2 and t_1 will remain quite similar to the ones I have used here.

THE COUPLING OF PARTICLE-HOLE STATES TO VIBRATIONS [*]

L. Zamick

Rutgers State University and Brookhaven National

Laboratory

We wish to consider the two-particle-two hole states of a closed shell nucleus such as ^{40}Ca, in a formalism in which the one-particle one hole vibrational states, e.g., 3^-, 5^- enter explicitly. The two particles are in the f-p shell (negative parity) the two holes in the s-d shell (positive parity).

Let $\phi^{I_A T_A}$ be a one particle-one hole vibrational state which can be expanded as

$$\phi^{I_A T_A} = \sum_{p_A h_A} (x_{p_A h_A} [a^{+p_A} a^{h_A}]^{I_A T_A} - y_{p_A h_A} [a^{p_A} a^{+h_A}]^{I_A T_A}) |0>$$

$$H\phi^{I_A T_A} = \omega_{I_A T_A} \phi^{I_A T_A} .$$

Let ψ^{JT} be the (2p-2h) wave function of the system

$$H\psi^{JT} = \omega\psi^{JT}$$

Define the amplitude $A^{JT}((p_L h_L)^{I_L T_L} \phi^{I_A T_A})$ as the overlap

$$= <\psi^{JT} [[p_L h_L]^{I_L T_L} \phi^{I_A T_A}]^{JT} 0> .$$

The amplitudes obey the following equation

$$[\omega - \omega_{I_A T_A} - (\varepsilon_{p_L} - \varepsilon_{h_L})] A^{JT} ((p_L h_L)^{I_L T_L} {}_\phi {}^{I_A T_A})$$

$$= <JT [V, (a^{+p_L h_L}_\alpha {}^{I_L T_L}]_\phi {}^{I_A T_A}>$$

where V is the potential energy. By evaluating the commutator the following equation for the amplitudes can be derived, the solution of which will be obtained by matrix diagonalization, just as in the more conventional shell model calculations:

$$(\omega - \omega_{I_A T_A} - (\varepsilon_{p_L} - \varepsilon_{h_L})) A^{JT} ((p_L h_L)^{I_L T_L} {}_\phi {}^{I_A T_A}) = \sum_{p_R h_R J_R T_R I_B T_B}$$

$$<p_L h_L {}^{I_L T_L} {}^{I_A T_A} | M | p_R h_R {}^{I_R T_R} {}^{I_B T_B}> A^{JT} ((p_R h_R)^{I_R T_R} {}_\phi {}^{I_B T_B}).$$

Here M is a <u>symmetric</u> matrix, which can be written in four parts

$$M = M \text{ (particle-hole)} + M \text{ (particle-vibration)}$$
$$+ M \text{ (hole-vibration)} + M \text{ (backward)}$$

$$<L| M \text{ (p-h)} | R> = < (p_L h_L^{-1})^{I_L T_L} V (p_R h_R^{-1})^{I_L T_L}>$$

$$\delta_{I_L I_R} \delta_{T_L T_R} \delta_{I_A I_B} \delta_{T_A T_B}.$$

Here we have written the particle-hole matrix element, which is equal to

$$- \sum_{I_x T_x} \begin{Bmatrix} p_R & h_L & I_x \\ & & \\ p_L & h_R & I_L \end{Bmatrix} \begin{Bmatrix} \frac{1}{2} & \frac{1}{2} & T_x \\ & & \\ \frac{1}{2} & \frac{1}{2} & T_L \end{Bmatrix} (-1)^{p_L + p_R + h_L + h_R}$$

$$< (h_L p_R)^{I_x T_x} V (h_R p_L)^{I_x T_x}>_A$$

where the subscript A always denotes an <u>unnormalized</u> (this is important because there will be cases where p_L and h_L are the same state) <u>antisymmetrized</u> particle-particle matrix element

$$<L| M \text{ (particle-vibration)} | R>$$

$$= \frac{1}{2} \delta_{h_L h_R} \sum_{st} ((2I_L+1)(2T_L+1)(2I_R+1)(2T_R+1)(2I_B+1)(2I_y+1)(2T_y+1))^{\frac{1}{2}}$$

$$(-1)^{I_A + T_A + J + T + h_L + p_L + 1}$$

$$
\begin{Bmatrix} I_R & I_Y & I_L \\ P_L & h_L & P_R \end{Bmatrix}
\begin{Bmatrix} I_R & I_y & I_L \\ I_A & J & I_B \end{Bmatrix}
\begin{Bmatrix} T_R & T_y & T_L \\ \frac{1}{2} & \frac{1}{2} & \frac{1}{2} \end{Bmatrix}
\begin{Bmatrix} T_R & T_y & T_L \\ T_L & T & T_B \end{Bmatrix}
$$

$$
< (p_L p_R^{-1})^{I_y T_y} y_{V(st^{-1})}^{I_y T_y} y >
$$

$$
x\Delta(s,p) < \phi^{I_B T_B}_{[a^{+s}a t]} {}^{I_y T_y}_{y} {}^{I_A T_A}_{\phi A} {}^{I_B T_B}_{B} >
$$

where $\Delta(s,p) = 1$ if s (and hence t) has the same parity as p_L
(or p_R); $\Delta(sp) = 2$ if s has the opposite parity to p_L.

The last matrix element is easy to evaluate if the coefficients
$x_{ph}{}^{\alpha}$ and $y_{ph}{}^{\alpha}$ are known.

$<L|M \text{ (hole-vibration)}|R>$

$$
= -\frac{1}{2}\delta_{P_L,P_R} \sum_{st} ((2I_L+1)(2T_L+1)(2I_R+1)(2T_R+1)(2I_B+1)(2T_B+1)(2I_y+1)
$$

$$
(2T_y+1))^{\frac{1}{2}} (-1)^{P_L+h_L+I_L+I_R+I_A+J+I_y} (-1)^{1+T_L+T_R+T_A+T+T_Y}
$$

$$
\begin{Bmatrix} I_R & I_y & I_L \\ h_L & P_L & h_R \end{Bmatrix}
\begin{Bmatrix} I_R & I_y & I_L \\ I_A & J & I_B \end{Bmatrix}
\begin{Bmatrix} T_R & T_y & T_L \\ \frac{1}{2} & \frac{1}{2} & \frac{1}{2} \end{Bmatrix}
\begin{Bmatrix} T_R & T_y & T_L \\ T_A & T & T_B \end{Bmatrix}
$$

$$
< (h_R h_L^{-1})^{I_y T_y} y_{V(st^{-1})}^{I_y T_y} y >
$$

$$
\Delta(s,h) < \phi^{I_B T_B}_{[[a^{+s}a t]} {}^{I_y T_y}_{y} {}^{I_A T_A}_{\phi A]} {}^{I_B T_B}_{B} >
$$

M (backward) involves the amplitudes in which p_R is a positive
parity state and h_R is a negative parity state. These will not
be considered here - we restrict p_R to the f-p shell and h_R to the
s-d shell.

The solution to these equations is obtained by matrix diagonalization just as in the more usual shell model calculations.

The representation used for the amplitude forms an overcomplete
non-orthonormal set. Hence some of the eigenvalues of our matrix

diagonalization will be associated with spurious states.   (On the other hand we can avoid to a very large extent the spuriousity associated with centre of mass motion by simply not including the spurious $J = 1^-$ T=0 state $\vec{R}|0>$).

Application: If the Effective Interaction method, in its simplest form, is applied to obtain the lowest 2p-2h energy levels with T=0, 1 and 2, then a large discrepancy results.   One really doesn't know where the T=0 state of this multiplet is because of the presence of 4p-4h states at a very low energy so we shall concentrate upon the T=1 and T=2 states.

In the effective interaction method one assigns the configuration

$$[(f_{7/2}^2)_0^{J_0=0\ T_0=1}(d_{3/2}^{-2})_0^{J_0=0\ T_0=1}]^{OT}$$

and one gets the particle-hole matrix elements from the negative parity spectrum of Ca^{40}.   Assuming that $\varepsilon_{f_{7/2}} - \varepsilon_{d_{3/2}} = 7.2$ MeV these are:

T=0		T=1	
J	V	J	V
2^-	-0.45	2^-	1.3
3^-	-3.5	3^-	0.5
4^-	-1.6	4^-	0.45
5^-	-2.7	5^-	1.35

One takes the particle-particle and hole-hole matrix elements from Ca^{42}, Sc^{42}, A^{38} and K^{38}.

Experimentally, the T = 1 state is at 9.3 MeV in Ca^{40} (the analog of the 1.6 MeV state in ^{40}K) and the T = 2 state is at 12.00 MeV.   But the above technique leads to a much wider spacing between these states; the T = 2 state is well predicted, but the T = 1 is much lower.   In ^{40}K it is predicted to be below the 4^-, T=1 ground state.   This result is clearly seen if one uses the monopole force a+b t_1t_2 for a particle-hole interaction and one fits b to the difference in center of gravity of the T = 1 and T = 0 states.   This leads to b $\approx 2.5 \leftrightarrow 3$ and the T = 2, T = 1 splitting is $\frac{b}{2}[T_1(T_1+1)-T_2(T_2+1)] = 2b \approx 5 \leftrightarrow 6$ MeV.   The experimental separation is only 2.7 MeV.

In order to get results that are better than this simple effective interaction method, we must explicitly display the configuration mixing which causes the collective states to lie so low;

this is done in our formalism.

We employ a surface delta interaction $V= -4\pi A_T \delta(r_1-R)\delta(r_2-R)\delta(\Omega_{12})$. We compare three calculations which are as follows:

1) $A_{T=0}=1.07$ $A_{T=1}=0.49$. These are Glaudemans' parameters. All vibrations are assigned the configuration $f_{7/2}d_{3/2}^{-1}$. With the above force the energies of the T = 0 J = 2^-, 3^-, 4^- and 5^- states are 7.54, 6.87, 7.28 and 5.93 MeV compared with the empirical values 6.75, 3.74, 5.60 and 4.49. For T = 1,the calculated values 7.98, 7.52, 7.50 and 8.50 compare better with the empirical values 8.46, 7.70, 7.66 and 8.55.

2) $A_{T=0}=1.07$ $A_{T=1}=0.49$ as before. The lowest 3^- T = 0 vibration and only this one is made collective in the T.D.A. approximation with the above force. The calculated energy of the 3^- vibration is now 2.99 MeV; its wave function is strongly mixed over many particle-hole states.

3) With the above force it was noted that the particle-particle and hole-hole effective matrix elements were too weak especially in T = 1 states. For example the J = 0 T = 1 $<f_{7/2}^2 V f_{7/2}^2>$ (normalized) matrix element is only -2 MeV whereas the effective value from experiments is 3.2 MeV. It is true that correlations such as particle-particle scattering and core polarization could increase this value but since in this calculation we have not put these correlations in we felt justified in making the T = 1 particle-particle and hole-hole force stronger. Hence we have $A_{T=0}=1.07$ $A_{T=1}=0.89$ for p-p and h-h; but as in cases 1) and 2) for p-h interaction. It should be emphasized that the parameter $\omega_{I_A T_A}$ (see page 2) was always taken from experiment, i.e., lowest negative parity state (except for the 2^- T = 0 which was chosen to be 6.75 MeV because the lower state is a 3p-3h state).

It was further decided that in some of the calculations the 2^-, 4^- and 5^- T = 0 matrix elements would be modified by adding a diagonal shift to the surface delta particle hole matrix elements of -0.79 MeV, -1.67 MeV and -1.44 MeV respectively; this shift is the differ- ence between the surface delta results and experiment. We designate by SHIFT or NO SHIFT whether this was done or not.

Note that there is not a great deal of difference between the SHIFT and NO SHIFT calculations.

As a result of making the 3^- T = 0 collective, the 9.12 MeV (mainly double-octopole T = 0) state goes down from 9.12 MeV to 7.67 MeV. But for T = 1 the 9.71 (mainly double octopole) state goes up in energy to 11.16 MeV so that another state (a mixture of double 4^- and 5^- vibrations) becomes the lowest state.

Energies of 2p-2h J = 0 States in ^{40}Ca.

	I NO COLLECTIVITY NO SHIFT	II 3⁻ COLLECTIVE		III 3⁻ COLLECTIVE SHIFT
		NO SHIFT	SHIFT	
T=0	9.12	7.67	7.67	6.47
	10.82	9.82	8.57	7.52
T=1	9.71	10.40	10.33	8.58
	10.93	11.16	11.14	10.99
T=2	14.11	14.11	14.11	12.06
	16.81	16.81	16.81	16.81

In calculations I and II the T = 2 state is about 2.1 MeV
too high because the particle-particle and hole-hole force was too
weak. In calculation III in which we make this force more attractive it comes down to 12.06 MeV; note that the T = 0 state does
not come down as much as the T = 2 state, because this double
octopole state is less sensitive to the p-p, h-h correlations.

Finally, the results for the T = 2 – T = 1 splitting are
much better than was obtained in the simple effective interaction
approach, although there is still a discrepancy of about 0.7 MeV.
It was necessary to explicitly display the collectivity of the 3⁻
vibration for this to be so. We predict the lowest T = 0 2p-2h
state at 6.47 MeV. We no longer obtain a T(T+1) spectrum as one
would get from the configurational assignment $(f_{7/2}^2)^{J=0} (d_{3/2}^{-2})^{J=0}$.

Work supported, in part, by U.S. Atomic Energy Commission while
a Summer Visitor to Brookhaven National Laboratory, 1971, and in
part by National Science Foundation.

COMMENTS ON THE ISOSPIN DEPENDENCE OF THE TWO-BODY EFFECTIVE
INTERACTION AS DEDUCED FROM NUCLEAR SPECTRA

W. A. Lanford
University of Rochester, Rochester, New York 14627

A multipole expansion of the two-body effective interaction
in the $h_{9/2} \times 2f_{7/2}$ and $h_{9/2} \times i_{13/2}$ shell model configurations has
been made using recent data on the proton-proton and proton-neutron
(hole) multiplets in ^{210}Po and ^{208}Bi. The experimental multipoles
have been compared with the multipoles of charge independent scalar
interactions. It is seen that an interaction of the form $V_{12} =$
$V_0(1+1/10\sigma_1\cdot\sigma_2) \exp[-r/r_0)^2]-V_0^\tau 1/2(1+\tau_1\tau_2)$ gives good agreement
with the data.

EXPERIMENTAL EVIDENCE OF THE GOODNESS OF SENIORITY IN j=9/2 ORBITS

W. A. Lanford
University of Rochester, Rochester, New York 14627

The goodness of the seniority quantum number in the $g_{9/2}$,
$h_{9/2}$, and $2g_{9/2}$ shell model orbits has been investigated by check-
ing the mixing of the $(9/2)^3 J=9/2$ v=1 and the $(9/2)^3$ J=9/2 v=3
states. This mixing is calculated to be less than 1/10 of one
percent. An upper limit of 3% on this mixing in the $g_{9/2}$ orbit
has been set by ^{92}Mo(^{4}He,t)^{93}Tc and ^{92}Mo(^{3}He,d)^{93}Tc reactions. A
residual interaction which exactly preserves seniority is seen to
give two-body matrix elements which have a r.m.s. deviation from
the experimental spectra of less than 16 keV for the present cases.

PROBLEMS OF sd-SHELL CALCULATIONS FOR ^{24}Mg

P. Manakos, H. Feldmaier and T. Wolff
Institut für Theoretische Kernphysik, der Technischen
Hockschule Darmstadt, Germany

Using the supermultiplet and SU(3) classification scheme, a
systematic method for the choice of configurations has been applied
to T=0 positive parity states of ^{24}Mg. It consists in diagonalizing
the Hamiltonian in small vector spaces carrying the set of quantum
numbers [λ], (λμ), L, S, T, J and taking these energies as a
criterion for the states to be truncated. Very recent calculations
including states of symmetry [44] [431] and [422] show that in
addition to central forces, the T=1 S=1 tensor force is needed to
reproduce the experimental spectrum below 10 MeV. Then K=0 and K=2
bands have prominent SU(3) components (8,4) and (9,2) of symmetries
[44] and [431] respectively.

EXCITATION STRUCTURE OF NEGATIVE PARITY STATES IN ^{17}O

A. Müller-Arnke
Institut für Theoretische Kernphysik, der Technischen
Hochschule Darmstadt, Germany

A shell model calculation including all nonspurious $1\hbar\omega$ excitations and certain nonspurious $3\hbar\omega$ excitations, which according to their SU(4)-SU(3) structure are closely related to the first rotational band of ^{16}O, has been performed to decide whether the lowest negative parity ($T=\frac{1}{2}$) states can be of 4p3h structure. From an examination of the dependence of the spectrum on the particle hole energy it is concluded that the lowest negative parity states are of 2p1h type while the second $1/2^-$ and higher $3/2^-$ states presumable contain large 4p3h components.

AN IMPROVED PRESCRIPTION FOR CALCULATING THE NEGATIVE PARITY SPECTRUM OF DOUBLY-CLOSED SHELL NUCLEI

F. Petrovich, R. Schaeffer and R. Trilling
Michigan State University

It is noted that T.D.A. and R.P.A. calculations for the negative parity states in doubly-closed shell nuclei with N=Z generally underestimate the splitting between the T=0 and T=1 members of a given multiplet. It is pointed out that if the monopole components of the particle-hole matrix elements are fixed from the nuclear symmetry potential the results of these calculations can be significantly improved. Results are shown for the levels in ^{40}Ca which belong essentially to the $1f_{7/2}-1d\frac{\bar{3}}{2}^{\frac{1}{2}}$ multiplet.

MONOPOLE POLARIZATION WITH A SURFACE DELTA INTERACTION

L. Zamick
Rutgers University

The monopole vertices $<jV[j(ph)^{J_0=0,T_0}]^j>$, $T_0=0$ and $T_0=1$ are constructed with a surface delta interaction $-4\pi G_T^0 \delta(\Omega_{12})$ using $G_{T=0}=1.07$, $G_{T=1}=0.49$. The results are

$$-(G_0+G_1)\,\frac{3}{2\sqrt{8}}\,\sqrt{2(2h+1)} \text{ for } T_0=0 \text{ and } (3G_0-G_1)\,\frac{1}{2}\sqrt{\frac{3}{8}}\,\sqrt{(2h+1)}$$

for $T_0=1$. We consider 4 problems in which the above enter: Nolen-Schiffer Anomaly, Isotope shift, Monopole exchange between two particles and between a particle and hole. Taking ΔE_{ph} in Ca^{40} = 30 MeV, the

Nolen-Schiffer anomaly (700 keV) is more than explained, but bad things happen to the other things: a large isotope shift is predicted. It is shown independent of the signs or magnitudes of G_0 and G_1 that the T=1 particle-particle states e.g. $(f_{7/2}^2) = 0$, 2, 4, 6 must come down in energy, whereas we know the J=6 is repulsive. For the $f_{7/2} d_{3/2}^{-1}$ multiplets with T=0 and T=1 we find that the T=0 multiplet must be pushed up also, but by a smaller amount. However, it may be that the G_0, T_1 we chose is about a factor two too large; then the bad effects will be reduced by a factor of four.

RENORMALIZED OPERATORS AND COLLECTIVE PARTICLE-HOLE EXCITATIONS IN Ca^{40}

M. Dworzecka and H. McManus
Michigan State University

Collective particle-hole excitations in Ca^{40} are calculated with Sussex matrix elements in Tamm-Dancoff approximation. Calculations are done both with bare and renormalized operators. Renormalization is accomplished by including core polarization effects and ground state correlations by perturbation in both the two-body interaction and transition moments. The results are similar to those obtained in random-phase approximation by using forces which contain core polarization screening.

THE INFLUENCE OF THE TWO-BODY SPIN-ORBIT INTERACTION ON EFFECTIVE CHARGE

G. F. Bertsch
Michigan State University

Valence particles normally carry an effective charge larger than their bare charge for electric transitions, due to the attractive nuclear interaction. However for the highest multipolarity transitions the particles have enough momentum that the two-body spin-orbit force can also play a substantial role. In the case of the E6 $f_{7/2} \rightarrow f_{7/2}$ effective charge, the spin-orbit interaction interferes with the central interaction and yields a correction to the proton effective charge close to zero. This is consistent with a recent calculation of proton scattering in the $g_{9/2}$ shell, which showed that the spin-orbit interaction dominated the central interaction for L=8.[1] Our calculation fails to account for a recent experimental result in the $f_{7/2}$ shell[2], which requires an effective charge of magnitude 0.3-0.4[3].

1. W. G. Love, Physics Letters 35B(1971)371.

2. J. N. Black, W. C. McHarris, and W. H. Kelly, Phys. Rev.
    Letters 26(1971)451.

3. B. H. Wildenthal, private communication.

THE MULTIPLET STRUCTURE OF ^{210}Bi

R. A. Eisenstein, T. R. Canada, C. Ellegaard, and P. D.
Barnes
Carnegie-Mellon University

Detailed studies of the level structure of ^{210}Bi can be expected
to provide information about the (p,n) interaction in the lead
region. Using a 9.0 MeV deuteron beam from the University of
Pittsburgh Tandem Accelerator, we have made a study of the reaction
^{209}Bi(d,p)^{210}Bi. A total of 51 levels below 3.4 MeV excitation
were seen. Probable spins were assigned on the basis of observed
branching ratios and by assuming that all states could be described
by available shell model configurations. The observed energies
and branching ratios are compared to calculations of Kim and
Rasmussen and Kuo and Herling; overall agreement of both quantities
is found with the results of Kuo and Herling.

THE DESCRIPTION OF INELASTIC (p,p') SCATTERING

Richard Schaeffer

C.E.N. Saclay and Michigan State University

INTRODUCTION

The need for additional and more precise information on the excited states of nuclei has been paralleled in recent years by the development of microscopic theories both of the structure of these states and of inelastic scattering reactions leading to them. We shall be mainly interested here in assessing the reliability of information on these states obtained by inelastic proton scattering.

The distorted-wave-Born-approximation (DWBA) is now-a-days commonly used to describe the (p,p') reaction, assuming, basically, that the inelastic cross-sections are much smaller than the elastic cross sections. One then obtains the elastic scattering wave-functions by fitting experimental elastic scattering cross sections and calculates inelastic scattering to first order in the coupling potential:

$$\sigma_{in} \sim |\langle \chi_f \, \psi_e |V| \chi_i \, \psi_g \rangle|^2 \tag{1}$$

Here χ_i and χ_f are optical wave functions describing the elastic scattering of the projectile and ψ_g, ψ_e are the wave functions of the ground and excited states of the target, respectively. Inelastic scattering provides directly a measure of the difference between the ground state and the excited state considered. More precisely, since V is a one-body operator, one gets information on the transition density. The first order assumption (1) needs some experimental confirmation. Nevertheless, by simply examining the second order correction

$$\sum_n \frac{\langle f|V|n \rangle \, \langle n|V|i \rangle}{E-E_n+i\varepsilon} \tag{2}$$

to the scattering amplitude, some "hand waving" arguments can be
given to say why one expects the DWBA approximation to be good. For
f=i, that is for elastic scattering, expression (2) reads

$$\sum_n \frac{|<i|V|n>|^2}{E-E_n+i\epsilon} \tag{3}$$

and one may expect the sum over the intermediate states n to be
important since all terms contribute coherently. For inelastic
scattering, one may use a "random phase" argument for the product

$$<f|V|n> \ <n|V|i>$$

and therefore expect some cancellation when summing over n. This
is of course not true in all cases. For instance if f and i belong
to the same rotational band they are built out of the same intrinsic
state and it is well known that the first order assumption is deficient.
Similarly when the first order term is strongly inhibited by some
selection rule, the second order contributions are expected to show
up. I shall restrict myself to cases where such features do not
occur, for instance vibrational states or almost pure single particle
states.

MODELS FOR NUCLEAR STRUCTURE AND REACTIONS

The assumptions one makes for the usual vibrational model are
rather simple. Defining the effective potential that the projectile
which experiences the inelastic transition sees, as the "form factor"
F(r), one can write:

$$<\chi_f \ \psi_e |V| \chi_i \ \psi_g> = <\chi_f |F| \chi_i>$$

$$F(r) \sim \beta_{pp'} \ \frac{\partial V_{opt}}{\partial R} \ . \tag{4}$$

This form factor shape is derived from the optical potential and the
strength $\beta_{pp'}$ characterizes the excited states and is fixed by
fitting (the) experiment. A nice feature of this model is that it is
not really a first order theory, since $\partial V_{opt}/\partial R$ may contain contributions from higher order terms. However, the model is strictly
valid only for the low lying vibrational states, and it is most
puzzling that it happens to work also for weakly excited states which
are by no means vibrational. There is therefore the suspicion that
this widely used model is successful in predicting good angular
distributions merely because it provides the correct geometrical
features of the (p,p') reaction and not because the underlying
dynamical assumptions (i.e. a vibrational excitation) are correct.

All the dynamics are contained in the fitted parameter $\beta_{pp'}$, and an error (if any) in the assumed transition operator may be hidden if it affects only the strength of the cross-sections. A more annoying feature is that there is no clear connection between the strength $\beta_{pp'}$ and the microscopic structure of the excited states $|\psi_e\rangle$, in contrast with the situation for an electromagnetic transition where the corresponding parameter β_{EM} is nicely related to the difference between ground and excited states:

$$\beta_{EM} \sim \sum_{j'j} \langle\psi_e|a_{j'}^+, a_j|\psi_g\rangle\langle j'|r^J|j\rangle \tag{5}$$

(protons)

The right hand side of equation (5) is the product of the probability amplitude for the transition from shell j to shell j' and of a typical matrix element for a single particle transition, summed over all possible proton excitations. (The wave functions $|\psi_g\rangle$ and $|\psi_e\rangle$ are assumed to be the true wave functions, so we don't need effective charges).

The commonly admitted equivalence $\beta_{pp'} = \beta_{ee'}$ is merely an assumption, especially for nuclei where neutrons and protons play a different role, since (p,p') scattering is sensitive to neutron excitations in nuclei whereas (e,e') is not.

Clearly, if one wants to learn something new about the bound state wave functions from (p,p') scattering, a fully microscopic model is needed. In the DWBA approximation, the scattering amplitude

$$\sim \sum_{j'j} \langle\psi_e|a_{j'}^+, a_j|\psi_g\rangle\langle\chi_f^{(-)}j'|v|\chi_i^{(+)}j\rangle \tag{6}$$

has strong similarities with expression (5), but also two important differences: i) the single particle transition matrix element is now a matrix element of the two body residual interaction, taken between shell model bound states and optical model scattering states and ii) the sum runs now over both proton and neutron excitations.

The largest uncertainty in the microscopic model is due to the lack of knowledge of the interaction v that should be used in (6), particularly its strength, since the choice of the range affects mostly the shape of the angular distribution and is not so crucial. For this reason, the approach taken by Satchler[1] was to determine an effective force strength V_{eff} by fitting experimental (pp') data for some simple transitions. The general conclusions of this review[1] were that the effective strength is much larger than any prediction using realistic forces. More important, V_{eff} is energy dependent, decreasing with increasing bombarding energy. V_{eff} is also state dependent.   Satchler noticed the tendency of V_{eff} to increase when higher angular momenta are transferred.   This makes spectroscopy with the microscopic model quite difficult, since one still does not know accurately enough which strength to use in a given calculation.

EXCHANGE EFFECTS

The sensitivity of the (p,p') scattering amplitude expression (6), to the choice of the two-body force v can be seen in a slightly different way.   For the T=0 vibrational states, where only the spin and isospin independent part V_{00} of v contributes, a Rosenfeld mixture gives a scattering amplitude about 50 times smaller than a Serber mixture, reflecting the corresponding change in V_{00}.   The two forces give however rather similar results in a bound state calculation, since the relevant matrix elements differ on the average only by a factor of 2.   However the bound state matrix elements are antisymmetrized whereas expression (6) is not.   Moreover, in the case of the exchange contribution for a zero range force (such an estimate is good in the limit of two interacting particles with zero relative energy) it is found that this contribution increases the scattering amplitude by a factor of 2 for the Serber force and by more than a factor of 60 for the Rosenfeld mixture.   Exchange effects therefore appear to be very important in (p,p') scattering.

The first calculation including these effects was done by Amos, Madsen and McCarthy.[2]   By a clever choice of the (p,p') transition ($g_{9/2}$ $p_{1/2}$, 4^-) in order to obtain some inhibition of the direct term, they demonstrated that the exchange contribution can be quite large in some cases, in contrast with estimates made earlier[3] which tended to show that exchange effects are generally small.   However, from the arguments given above, one might expect that for a short range force exchange effects are important in every case and increase the cross section by roughly a factor of 4 when a Serber mixture is used, at least as long as the wave length of the incoming particle is larger than the range of the two-body force (this is the case in the energy range where the DWBA is valid).   The aim of the calculation done by Agassi and myself[4] was to check this point.   The code we used was written using the helicity formalism of Raynal.[5] Instead of expression (6), we used:

$$\tilde{\sum_{j'j}} <\psi_e|a_{j'}^+, a_j|\psi_g><\chi_f^{(-)} j'|v \ |\chi_i^{(+)} \ j> - \ |j \ \chi_i^{(+)}> \ . \tag{7}$$

This expression represents only part of the terms that arise in the scattering amplitude when the DWBA approximation is done on fully antisymmetrized wave functions of all particles in the system.   In the limit where the optical and shell model wave functions are calculated with the same potential well, however, these terms vanish. They can therefore be assumed to be small in the real case, except for some pathological situations.

   A good case to check the accuracy of the approximations is the excitation of the lowest 3^-, $T=0$ collective level of ^{40}Ca. The wave function for this state, calculated by Gillet and Sanderson[6] gives good results for electron scattering (i.e. without any effective charge) implying that the proton excitations are well described. Since this state is $T=0$, the neutron excitations can also be assumed to be given correctly. For a good model of (p,p') scattering one should therefore get the experimental magnitude of the cross sections using these wave functions. The best choice for the two-body inter-action is probably a Serber mixture with about 40 MeV strength. Wong and Wong[7] have shown that the G-matrix describing the inter-action of two particles at the nuclear surface is very close to the free nucleon-nucleon interaction at low energies, as described for instance by the Blatt-Jackson force[8] whose parameters are almost identical to the Serber force we use. With this choice of para-meters at 55 MeV, the calculated cross-sections are a factor of 2 too low (Table 1) when exchange is not included. Instead of the factor of 4 expected from the zero-range estimate of exchange, the realistic calculation gives a factor of 2 which is just what is needed in order to reproduce the magnitude of the experimental cross-section. This factor of 2 enhancement is about the same whether one uses a single particle wave function or the strongly configuration mixed one calculated by Gillet and Sanderson[6]. This shows that the exchange term in expression (7) is roughly as coherent as the direct one, as expected from the zero-range estimate. The lowest 3^-, collective state, of ^{208}Pb (See Table 1) as well as the lowest 3^- of ^{16}O and 2^+ of ^{12}C display exactly the same feature: one gets the experimental magnitude of the cross-section without any adjustable parameters.

   Although we had shown that, after including exchange effects, it was possible to obtain a quantitative description of (p,p') scattering without any adjustable parameter, we were not aware of the systematic dependence of exchange effects on the energy of the incoming particle and on the transferred spin. This was shown by Atkinson and Madsen[9] for the $g_{9/2}$-$g_{9/2}$ transition in ^{90}Zr. The calculation for the $f_{7/2}$-$f_{7/2}$ transition in mass 40 presented in Figure 1 is essentially equivalent to the original[9] one. At low energy (20 MeV), the increase of the ratio σ_{D+E}/σ_D of the total cross sections with and without exchange is drastic, and for large angular momentum transfer reaches almost a factor of 20 (compared to 4 for a zero-range force). At higher energies (100 MeV), the maximum increase is a factor of 3. This is a typical finite range effect, and can be easily understood by the rapid decrease of the Fourier transform $v(q)$ of a 1.4 fm range force with increasing q. If k and k' are the relative momenta of the two interacting particles at the nuclear surface before and after the collision, the direct term is proportional to $v(k-k')$, whereas the exchange term varies as $v(k+k')$. For small angular momentum transfer, one has

TABLE 1.--Differential cross-section $d\sigma/d\omega(\theta)$ in units of mb/sr for R.P.A. Taken from Ref. 4.

Force	40Ca E_p=55 MeV (a) E(3⁻)=3.83 MeV				(a) 208Pb E_p=24.55 MeV (a) E(3⁻)=2.6 MeV			
	[D+E]25° (b)	[D]25°	[D+E]85°	[D]85°	[D+E]25°	[D]25°	[D+E]85°	[D]85°
Exp. [5]	13.		0.25		2.8		0.48	0.22
Serber	11.6	6.3	0.24	0.09	1.85	1.0	0.38	0.22
CAL (c)	4.	1.0	0.11	0.015	0.7	0.2	0.14	0.05
Rosenfeld	2.2	0.01	0.1	0.0001				

(a) E_p = incident proton energy, E(3⁻)=excitation energy of 3⁻ state.

(b) [D] denotes the direct term result, [D+E] denotes the direct plus exchange result and the subscripted number corresponds to the scattering angle.

(c) Force used for calculating the corresponding RPA vector in Ref. 6.

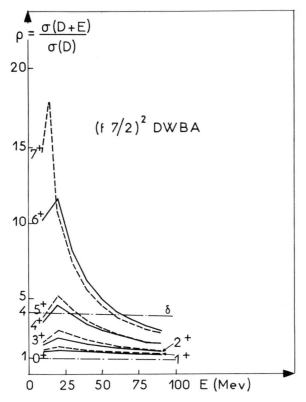

Fig. 1.--Ratio of total (p,p') cross sections for an $f_{7/2}\, f_{7/2}$ transition with various spin transfers for a mass 40 target. Taken from ref. 10.

$|\vec{k}-\vec{k}'| < |k+k'|$ and therefore the direct contribution is larger than the exchange one. However, for large angular momentum transfer, $\vec{k}$ and $\vec{k}'$ have opposite directions, and if their magnitude is not too large (i.e. at low energy), the sum $\vec{k}+\vec{k}'$ cancels somewhat and $v(\vec{k}+\vec{k}')>v(\vec{k}-\vec{k}')$, as is the case at 20 MeV for large angular momentum transfer. Of course, at zero energy, both k and k' are small and one gets back to the zero range estimate, except for Coulomb effects which force the low energy incident proton to remain outside the barrier and which reduce further the exchange contribution.

In summary, one has quite a nice quantitative description of
(p,p') scattering including the energy and angular-momentum-transfer
dependence either using the Blatt-Jackson interaction as I have
described here (see also ref. 12) or the Kallio-Kolltveit force as
described by Petrovich et al.[11] Some detailed results are shown
in Fig. 2-5. It can be seen that exchange increases the cross
section for calculation of the lowest 3^- and 5^- collective T=0 states
of ^{40}Ca by a factor which ranges from 2 for the 3^- at 55 MeV to 7
for the 5^- at 20 MeV. After exchange has been included, in all
four cases, the theory predicts exactly the magnitude of the
experimental cross-sections. This means that the exchange contribution brings just the increase needed in order to remove the
normalization discrepancy observed in the calculation with the
direct term only. The same situation is true for the 2^+ collective
state of ^{12}C (Fig. 6). The Kallio-Kolltveit and Blatt-Jackson
forces give very similar results. This is not astonishing since
according to Wong and Wong[7] they both represent the same G matrix
at the nuclear surface. The asymmetry of the outgoing particle
for the 3^- state in ^{40}Ca at 20 MeV is shown in Fig. 7. As was the
case for the shape of the angular distribution (Figs. 4,5), the
fits for the asymmetry are only qualitative (Fig. 7).

More recently Love et al.[13] have used an interaction derived
from the Hamada-Johnston potential. They introduced a separation
distance for the central part and suppressed the odd-state interaction. After correction for the effects of the second order tensor
contribution, they obtained a Serber type force which is consistently
weaker (by about 20%) than the Blatt-Jackson or Kallio-Kolltveit
interactions. This procedure[13] is very crude for obtaining the
G matrix and may not be accurate to within 20%. However, an
estimate[14] of the strength of a two-body interaction which fits
the Sussex matrix elements also gives a potential slightly weaker
than those used in the earlier[4,9,11,12] calculations.

THE APPROXIMATE TREATMENT OF EXCHANGE

I want also to say a few words on the approximate treatment
of the exchange term in the DWBA matrix element. This approximation
was originally introduced by Petrovich et al.[11] and used in their
calculation[11] with the Kallio-Kolltveit interaction. The antisymmetrized matrix element in expression (7) can be written as a
direct one with an effective, non-local force:

$$<\chi_f^{(-)} j' | v | \chi_i^{(+)} j> - | j \chi_i^{(+)}> \ = \ <\chi_f^{(-)} j' | v_{eff} | \chi_i^{(+)} j> \qquad (8)$$

where $v_{eff} = v(1 - P_\sigma P_\tau P_x)$,

and the operators P_σ, P_τ and P_x are the permutation operators

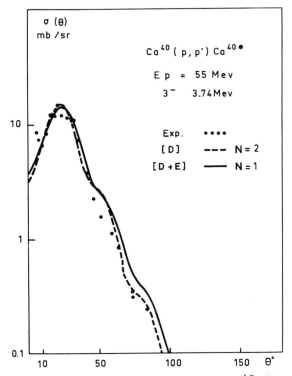

Fig. 2.--Excitation of the lowest 3^- state of ^{40}Ca by 55 MeV protons.
The theoretical curves were obtained by a microscopic DWBA
calculation with the nucleon-nucleon Blatt-Jackson potential.
The cross sections are calculated including exchange contributions (solid line) and without (dashed line). The excited
state is desribed by RPA[8] microscopic wave functions. In
order to compare the angular distributions, the theoretical
cross sections are multiplied by the indicated (N) normalization factors. Taken from ref. 12.

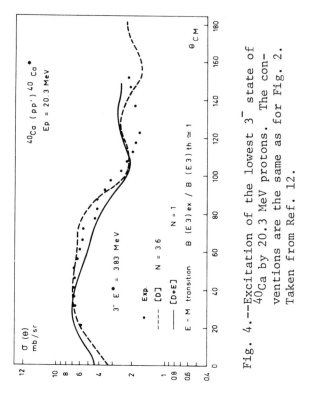

Fig. 4.—Excitation of the lowest 3⁻ state of
40Ca by 20.3 MeV protons. The conventions are the same as for Fig. 2.
Taken from Ref. 12.

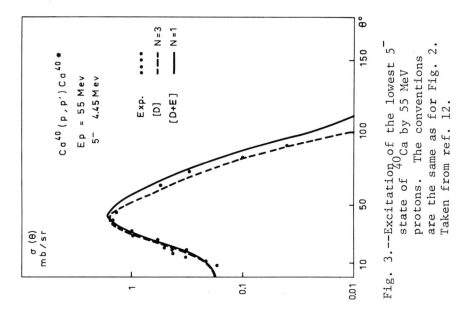

Fig. 3.—Excitation of the lowest 5⁻
state of 40Ca by 55 MeV
protons. The conventions
are the same as for Fig. 2.
Taken from ref. 12.

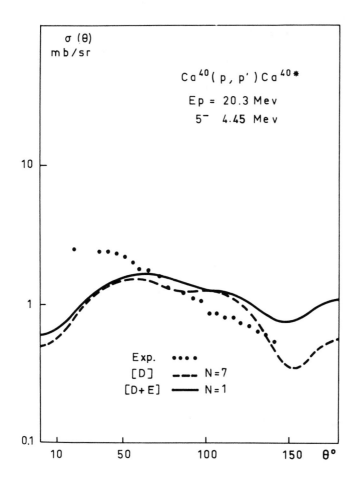

Fig. 5.--Excitation of the lowest 5⁻ state of ⁴⁰Ca by 20.3 MeV
    protons. The conventions are the same as for Fig. 2.
    Taken from ref. 12.

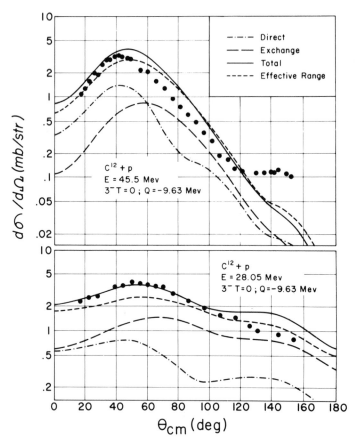

Fig. 6.--Differential cross sections obtained with the K-K and
effective range interaction for the L=3 transition in
the ^{12}C(p,p') reaction at 28.05 and 45.5 MeV. A
decomposition of the cross sections into direct and
exchange components is shown for the K-K interaction.
Taken from ref. 11.

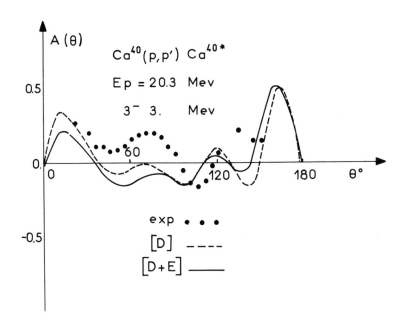

Fig. 7.--Asymmetry relative to the excitation of the lowest 3⁻
state of ⁴⁰Ca by 20 MeV protons. The theoretical curves
were obtained by a microscopic DWBA calculation with the
nucleon-nucleon Blatt-Jackson potential. The cross sections
are calculated including exchange contributions (solid
line) and without (dashed line). The excited state is
described by RPA[8] microscopic wave functions. Taken
from Ref. 10.

respectively of the spin, isospin and space coordinate. The P_σ and P_τ operators just change the force mixture. For P_x, Petrovich et al.[11] derived a local approximation, following a method due to Perey and Saxon[15]:

$$v(r) \ P_x \sim v(K) \ \delta(r) \tag{9}$$

where K is a constant, suitably chosen momentum.

The main assumption is that k+k' is about constant, or more precisely that one can choose K such that the quantity (K- k+k') is small. We denote by λ the range of the non-locality of $v(r)p_x$. This range is roughly the range of the force v(r) when the wave functions of the interacting particles overlap, but it may be much larger at low energy for incoming protons which are obliged by the Coulomb barrier to stay far from the target nucleus. Consistent with the hypothesis of bound particles having low momenta at the nuclear surface, and assuming that the optical distortions don't change the absolute value of the momentum too much, a good choice for K is the momentum of the incoming proton. It can be noted that, in this approximation, the exchange term is constant as the transferred angular momentum L increases, since one has neglected the variations of k+k' around K. Using $|k-k'| = q \sim (\frac{L}{R})$ (q is the momentum transfer and R the radius of the nucleus), the strong increase of the ratio

$$\sigma_E/\sigma_D \ \frac{v^2(K)}{v^2(L/R)}$$

is obtained in this model by the rapid decrease of the direct cross-section for increasing L transfer. A comparison between exact and approximate calculations for exchange effects (Fig. 8) shows that the approximation is rather good, except at low energies when the interacting particles stay far from each other. In the expression for σ_E/σ_D, it can be seen that in the local approximation this ratio is maximum at low energies whereas in the exact calculation σ_E/σ_D decreases, the exchange operator becoming more non-local.

Treating the exchange effects approximately has however some advantages because of its extreme simplicity. It is possible to use more complicated nucleon-nucleon forces which don't have a Yukawa or a Gaussian dependence, as was done by Petrovich et al.[11] (The approximation can however not be extended simply to non-central forces). Another application was[16] to estimate exchange effects in the inelastic scattering of composite projectiles (h, t or α particles).

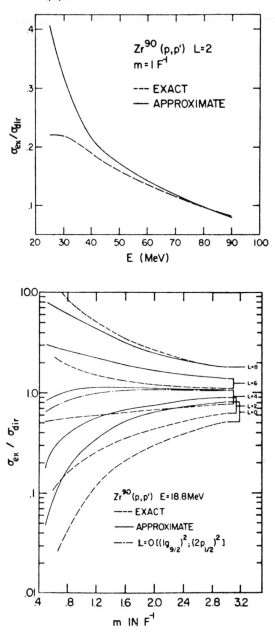

Fig. 8.--Comparison of approximate and exact results, for the
variation with energy and interaction range, of the
ratio of the exchange to the direct total cross sections
for several multipoles in the ^{90}Zr(p,p′) reaction.

COMPARISON OF PROTON AND NEUTRON EXCITATIONS IN NUCLEI

The main interest in having a quantitative theory of (p,p')
scattering is that one can obtain specific information on the
excited state wave functions. A quantity widely used to test the
microscopic wave functions is the electromagnetic transition rate
to the ground state. However this latter quantity provides infor-
mation only on the amplitudes $X_{j'j}^{(p)} = <\psi_e | a_{j'}^+, a_j | \psi_g>$ for proton
excitations, as can be seen in expression (5). Inelastic proton
scattering provides additional information on the similar amplitudes
$X_{j'j}^{(n)}$ for neutron excitation.

The procedure chosen [17,18] is to consider some model for the
bound state wave function and to renormalize the amplitudes $X_{model}^{(p)}$
by a factor λ_p in order to fit the experimental electromagnetic
transition rate. Using this same factor λ_p in expression (7), it
is then possible to fit a similar renormalization factor λ_n for
$X_{model}^{(n)}$ in order to reproduce the experimental (p,p') cross section
strength. Since the (p,p') reaction mechanism is described correctly,
any discrepancy in the predicted strength of the (p,p') cross section
can only be due to the model used for the bound states. There is
however still an approximation in our procedure, which is to assume
that the model for the bound states gives correctly the shape of the
transition density, but not its strength. In any case, the present
description of the (p,p') reaction is still too crude to give much
information on the shape of the transition density, but does give a
reasonably accurate measure of its strength.

Since the quantity

$$\sum_{prot} X_{j'j}^{(p)} <j'|r^J|j>$$

is widely used for testing microscopic wave functions, we have
extracted from experiment the same quantity for neutrons and consider

$$\rho_{mod}(n/p) = \frac{\sum_{neut} X_{mod}^{(n)} <j'|r^J|j>}{\sum_{prot} X_{mod}^{(p)} <j'|r^J|j>}$$

and also

$$\rho_{extr}(n/p) = \frac{\sum_{neut} X_{extr}^{(n)} <j'|r^J|j>}{\sum_{prot} X_{extr}^p <j'|r^J|j>} = \frac{\lambda_n}{\lambda_p} \rho_{mod}(n/p).$$

Most interesting are of course the nuclei with a different
number of protons and neutrons. For simplicity, we have considered

nuclei where either the proton or neutron shell is closed, the other being open. The wave functions of Gillet, Giraud and Rho[19] were available in this case. We have considered the lowest 2^+, collective state of the Ni isotopes with a closed proton shell and the ^{88}Sr, ^{90}Zr isotones with a closed neutron shell.

For a typical shell-model wave function, where the active particles are supposed to be those in the valence orbitals of the open shell, ρ_{mod} (n/p) is 0 or ∞. Gillet, Giraud and Rho[19] have included the core excitations of both kinds of particles. Typical values of ρ_{mod} (n/p) being then 1/3 or 3, the open shell particles contributing about three times more to the transition than the closed shell particles. When ρ_{mod}(n/p) is corrected in order to obtain agreement with experiment, the new ratio ρ_{extr}(n/p) is about 1 in both cases, as can be seen in Table 2. This shows that proton and neutron behave much more symmetrically than predicted by the current nuclear models.

The findings of Bernstein[20], who compares (α,α') scattering to the electromagnetic transition rates are very similar to ours, i.e. he also finds very small isovector transition rates for the low lying, collective states.

CONCLUSION

In conclusion, exchange effects have to be included for (p,p') scattering. They explain both the angular momentum transfer and the energy dependence of the observed cross sections, whereas a DWBA calculation with the direct contribution, fails to explain either. More important, if exchange effects are included, and realistic forces are used, the DWBA appears as a quantitative, parameter free, theory of (p,p') scattering, which can be used in order to extract new spectroscopic information on the excited states of a nucleus.

The trend towards more realistic two-body interactions should also be pursued. This is however a difficult problem, since the Moshinsky transformation cannot be used for scattering states which are described by optical wave functions. Moreover, there is still a basic inconsistency which hampers any realistic calculation, namely the necessity of using fitted optical parameters. One needs a good microscopic calculation of elastic scattering with a correct treatment of the absorption, before one can do a fully realistic calculation.

Nevertheless, some indications can be obtained from a phenomenological theory for example, concerning the importance of the two-body tensor and spin-orbit force. Most interesting in this study will be the unnatural parity states of doubly closed shell nuclei.

TABLE 2.--Factors λ_p and λ_n which display the extent of the agreement of respectively the proton and neutron part of the microscopic wave functions.

	$^{90}Zr3^-$ 20.3	$^{90}Zr3^-$ 12.7	$^{60}Ni3^-$ 40	$^{60}Ni3^-$ 13	$^{90}Zr2^+$ 20.3	$^{90}Zr2^+$ 12.7	$^{60}Ni2^+$ 40	$^{60}Ni2^+$ 13	$^{62}Ni2^+$ 11
λ_p	1.4	1.4	1.9	1.9	2.0	2.0	3.0	3.0	2.6
λ_n	1.2	1.3	0.9	1.5	5.5	5.0	0.9-1.3	1.4	1.1
$\rho_{mod}(n/p)$	0.8	0.8	1.5	1.5	0.36	0.36	2.6	2.6	3
$\rho_{extr}(n/p)$	0.7	0.75	0.8	1.2	1	0.9	0.8-1.1	1.2	1.3

For accurate wave functions, these factors would be unity. The relative ratio ρ of neutron and proton excitation is given (i) as predicted by the microscopic wave functions (ρ_{th}) which are made of two quasiparticle excitations of the outer shells and 1p-1h excitations of the core, and (ii) as extracted from the ee' and pp' experiments (ρ_{extr}). Since the calculated angular distributions do not fit exactly the experimental ones. there is some uncertainty for λ_n, the largest being $\pm 20\%$ in the case of the 2^+ state of ^{60}Ni as indicated. Taken from Ref. 14.

Configuration mixing for these states is much smaller than for the collective levels, and the non-central components of the two-body force are therefore not averaged out when summing over all possible excitations. These terms are also expected to show up in the calculated asymmetries or polarizations, and may explain the remaining discrepancy seen in Fig. 7.

REFERENCES

1. R. Satchler, Nucl. Phys. A95(1967)1.

2. K. A. Amos, V. A. Madsen and I. E. McCarthy, Nucl. Phys. A94 (1967)103.

3. See discussion in ref. 2.

4. D. Agassi and R. Schaeffer, Phys. Letters, 26B(1968)703.

5. J. Raynal, Nucl. Phys. A97(1967)572.

6. V. Gillet and A. Sanderson, Nucl. Phys. A91(1967)292.

7. C. W. Wong and C. Y. Wong, Nucl. Phys. A91(1967)433.

8. J. M. Blatt and J. D. Jackson, Phys. Rev. 76(1949)18.

9. J. Atkinson and V. A. Madsen, Phys. Rev. Letters, 21(1968)295.

10. R. Schaeffer, Thesis 1969, Rapport CEA R-4000.

11. F. Petrovich, H. McManus, V. A. Madsen and J. Atkinson, Phys. Rev. Letters 22(1969)895.

12. R. Schaeffer, Nucl. Phys. A132(1969)186.

13. W. G. Love, L. W. Owen, R. M. Drisko, G. R. Satchler, R. Stafford, R. J. Philpott and W. T. Pinkston, Phys. Letters 29B(1969)478.

14. F. Petrovich, private communication.

15. F. Perey and D. Saxon, Phys. Letters, 10(1964)107.

16. R. Schaeffer, Nucl. Phys. A158(1970)321.

17. J. Picard, O. Beer, A. El Behay, P. Lopato, Y. Terrien, G. Vallois and R. Schaeffer, Nucl. Phys. A128(1969)481.

18. R. Schaeffer, Nucl. Phys. A135(1969)231.

19. V. Gillet, B. Giraud and M. Rho, private communication.

20. A. Bernstein, Phys. Letters, $\underline{29B}$(1969)332; A. Bernstein, Advances in Nucl. Phys. Vol. 3, Plenum Press, New York.

CORE POLARIZATION EFFECTS IN (p,p') REACTIONS*

H. McManus

Michigan State University, East Lansing, Michigan

Let me remark first of all that, regardless of the status of its theoretical description at any moment, core polarization is a well established experimental phenomenon, or rather is well established in the interpretation of experimental phenomena in terms of a microscopic model, i.e. the shell model, in an extended version or otherwise. A typical example is the interpretation of electromagnetic transitions, in terms of effective charge. The effective charge is the bare charge of a nucleon, +1 for a proton, 0 for a neutron, plus the polarization charge. Typically for an E2 transition, as the shell model basis becomes larger, and the calculation increases in complexity, the polarization charge needed to fit experiment decreases but usually more and more slowly, until when the limits of present computing are reached, it still remains at ~0.5e. Figure 1 shows such an example taken from calculations in the s-d shell.[1]

Here the polarization charge Δe required to match experimentally observed E2 transitions in ^{33}S is shown as a function of the average number of basis states, N, in the wave functions corresponding to three models III, II and I in which the shell model basis is successively enlarged (A). Figures B and C show the same effect in a different representation. Curve 1 plots the polarization charge Δe versus neglect of components in the wave function with amplitudes less than x and curve 3 plots the average number of configurations corresponding to this neglect. It is seen that with only a few basis states, with a corresponding neglect of components with amplitudes less than 0.1, the polarization charge is ~1-1.2. Increasing the number of basis states to ~100 brings down the polarization charge required to 0.4, with the corresponding inclusion of a large number of amplitudes of the order ~1%.

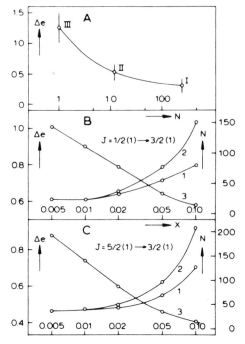

Figure 1.--Plots of quadrupole polarization charge required to fit
experiment versus complexity of shell model states used in calculation.[1]

        All the basis states considered were in the s-d shell. It is
clear that within this basis, one is running out of steam, and that
to get rid of the residual polarization charge ~0.4 one has to go to
states outside this shell.

        This simple example illustrates the essential feature of core
polarization. Core polarization is the effect of the excluded part
of the vector space. One tries to describe a system with many degrees
of freedom by a system with a much smaller number of degrees of freedom, say a few valence nucleons. The excluded degrees of freedom can
be taken into account via core polarization, more precisely by defining effective operators.

        Of course there are situations where for particular purposes the
extra degrees of freedom are adequately represented in some approximate calculational scheme, such as pairing theory, or the R.P.A. in
the case of the excitation of closed shell nuclei. Another case, in
which the term core-polarization can be taken literally, is where
weak coupling of valence nucleons to the actual collective excitations
of the core is of predominant importance. In general however it
remains true that one has to deal with effective operators which take

into account the degrees of freedom not explicitly described.

Thus when we deal with a shell model description, we are dealing with quasi-particles, whose interactions, as compared with particles, are modified. In the nuclear case, we are in any case dealing with quasi-particles to begin with. First the meson degrees of freedom have been eliminated to give a nucleon-nucleon interaction, in fact usually obtained empirically. We might have to correct operators for neglect of these degrees of freedom, i.e. the effect of exchange currents. Development of soft pion theorems lead us to believe that these effects are usually small though not always negligible. Then as the nucleon-nucleon interaction is strong and singular, to get to a shell model basis, as explained in Bethe's talk, we use the Brueckner Hartree-Fock theory to get rid of the corresponding short range two nucleon correlations, and replace the singular interaction by a smooth well behaved one, the bare G-matrix. This gives for instance the binding energy of nuclear matter and finite nuclei. Again operators should be corrected for the truncation of the vector space, in this case those parts giving short range correlations.

What we are concerned with, in core polarization, are not these corrections to transition operators, which are assumed, not always correctly, to be negligible in the long wave length limit, but with effects near the fermi surface i.e. long-range correlations between nucleons which are important in the low-excitation spectra and in the structure of transition operators for relatively low lying excitations and in the case of inelastic scattering for momentum transfers up to momentum transfer $q \sim k_F \sim 1.5^{-1}f$. Here starting from the Brueckner-Hartree-Fock theory we still cannot solve anything by direct diagonalization--the vector space would be too large. So we have to resort to effective interactions, different from the bare G-matrix of nuclear matter or that used for the calculation of the binding energy and single particle states of finite nuclei, and effective operators.

This defines an effective force $\tilde{V}$ in the truncated shell model space, written schematically

$$\tilde{V} = G + G \ Q/e \ \tilde{V} \tag{1}$$

where G is the bare G-matrix of nuclear matter or finite nuclei Q is a projection operator projecting out of the truncated space actually used into a wider, but still limited, shell model space. This limitation is to avoid intrusion into the space already used in creating G, the bare G-matrix. Equation 1 is of course similar in form to the Bethe-Goldstone (B-G) equation. The energy denominator e, however is quite different. The limitation of the wider shell model space to be used in 1, and the comparative independence of the

two truncations, the first putting the effects of short-range
correlations in the effective operators, and the second, core
polarization, putting in the effect of long range correlations, have
been discussed at this conference by G. E. Brown in his remarks after
Barrett's talk. As with the B-G equation we also have

$$G\psi = V\phi ,\tag{2}$$

where ϕ is our truncated space and ψ the extended shell model
space. This then defines effective transition operators, $\tilde{M}$, such
that

$$<\psi_B|M|\psi_A> = <\phi_B|\tilde{M}|\phi_A>$$

$$\tilde{M} = Z_B^{-\frac{1}{2}}(M+\tilde{V}\frac{\phi}{e}M + M\frac{\phi}{e}\tilde{V} + \tilde{V}\frac{Q}{e}M\frac{Q}{e}\tilde{V})Z_A^{-\frac{1}{2}}\tag{3}$$

where M is the "bare" operator, and the other terms, including the
normalization factors Z_B, Z_A which come from the fact that we are
only using the projection of the "true vectors" onto the truncated
space ϕ, contain the core polarization effects, like polarization
charge.

The theory of these effective operators is gone into in some
detail by M. Harvey, M. H. Macfarlane and others.[2] The theory will
not be discussed here, but the difficulties of actually interpreting
and solving such equations have been discussed at length by Barrett
at this conference.

I would however again like to reiterate what I said at the
beginning, that the existance of "effective" rather than "bare"
operators, is amply documented by analysis of experiments. I want
to emphasize the following points.

1. If a shell model basis is adopted and the matrix elements
of the effective interaction found by fitting to experimental spectra,
these matrix elements differ very considerably from those of the
bare G-matrix--a point sufficiently emphasized by Schiffer at this
conference.

2. In interpreting electromagnetic transitions it is necessary
to introduce quantities like polarization charges.

3. Simple lowest order perturbation treatment of equations
1 and 3 usually leads to effective interactions and operators which
are much closer to experiment than the bare operators. The most
striking example of this is the effective interaction between like
particles. As first noted by Bertsch and studied in detail by Kuo

and Brown for nuclei in the vicinity of ^{16}O, ^{40}Ca, ^{48}Ca, ^{50}Ni, ^{88}Sr and ^{208}Pb, core polarization gives rise to a strong pairing effect which is the major feature of the observed spectra.[3]

Also effective electromagnetic operators calculated in this simple way go very much in the direction of experiment, i.e. electric transitions tend to be enhanced, magnetic transitions hindered, as if the nuclear medium were diamagnetic. This was noted long ago by Arima[4] in the 50's. But for some reason this type of calculation was not revived until a decade later. Of course the successes of this simple treatment are limited--the monopole component of the effective force is very poorly given, and attempts to justify the procedure by looking at all orders have not been successful as demonstrated by Barrett in his talk. However with G. E. Brown's remarks after that talk in mind, one may proceed for the moment with an empirical, rather than a theoretically justified, approach.

As then the effects of core polarization are easily observed in the spectra and electromagnetic transitions of nuclei, they should also be evident in inelastic scattering, and in fact they are provided we can make some assumptions about inelastic scattering. I will confine my remarks to the inelastic scattering of nucleons, in fact protons, though much of the discussion could be extended to other projectiles. It is convenient to make the following 3 assumptions:

1. It is assumed that the reaction mechanism is adequately given by the distorted-wave-Born-approximation DWBA. This makes the scattering amplitude linear in the interaction between projectile and target nucleons and the target single nucleon transition density. Obviously it is easier to deal with a linear theory. If one had to deal with anything more complicated like coupled channels, it would be much harder to extract information.

2. It is assumed that the "bare" force between the incident nucleon and the target nucleons is known. This is necessary to set a scale factor for inelastic scattering, i.e. to define the equivalent of a Weisskopf unit to measure enhancement factors.

3. It is assumed that the projectile nucleon acts like a one-body operator as far as the target is concerned.

The first two assumptions have been discussed in detail by Schaeffer in the preceeding talk. As far as the second is concerned, the effective bare interaction is expected to vary only slowly as a nucleon is raised above the fermi surface, so that for relatively low energy projectiles ≤ 40 MeV it should be very much the same as the effective interaction between nucleons below the fermi surface,

i.e. the bare G-matrix used in Hartree-Fock calculations. At any
rate, such interactions give a good account of the real part of
the optical potential,[5] and are successful as detailed by Schaeffer
in predicting the cross sections for excitation of nuclei in those
cases, such as the excitation of collective normal parity T=0
excitations in T=0 nuclei where proton and neutron transition
densities are the same, and the proton transition densities can be
obtained from electron scattering experiments. The main point is
that all plausible methods of calculating or rather estimating
the interaction tend to give very much the same results as far as
inelastic scattering is concerned.

This, however, only applies to the central parts of the interaction. The non-central parts are still uncertain as we will
mention later. As pointed out by Schaeffer in his talk the main
features of the central part of the force are: 1) that it is
approximately a Serber mixture; 2) that the effects of antisymmetrization of the projectile nucleon with the nucleons in the target in
the DWBA are very important. In the interest of brevity, the part
of the DWBA matrix element that comes from antisymmetrization is
referred to as "exchange", which is a source of confusion to people
who deal with Hartree-Fock problems. If we say for the excitation
of some given state that "exchange" is unimportant we do not mean
that the amplitude is insensitive to the exchange character of the
forces. It simply means that we have separated out the DWBA matrix
elements in a particular way which is in fact dictated by angular
momentum coupling and that in this case the particular matrix element
coming from antisymmetrization happens to be numerically small.
The relative importance of this "exchange" component of the amplitude,
due to antisymmetrization compared to the "direct" term (and remember
that "direct" here does include exchange components of the force)
changes rapidly with angular momentum transfer and bombarding energy.
Its importance falls off with energy but it dominates high angular
momentum transfer.

On the third point the assumption that the projectile nucleon
acts like a one-body operator as far as the target is concerned
is not necessary but is certainly convenient and goes together with
the first assumption, the use of DWBA. It enables us to compare
inelastic scattering with other one-body operators, i.e. electromagnetic transitions and electron scattering, β decay, μ-capture,
etc. Because of the effects of antisymmetrization which introduces
components corresponding to two-body operators, it is only an
approximation but seems to be a good one. The DWBA amplitude
(Schaeffer's equation 6) can be written with an interaction v
between projectile and target nucleon.

$$T_{DWBA} \sim \sum_{j'j} <\psi_e | [a_{j'}^+, a_j]^J | \psi_g> <\chi_f^- j' | v | \chi_i^{(+)} j> \tag{4}$$

i.e. a product of the target transition density expressed as spectroscopic factors for single particle transitions $j \to j'$ with angular momentum transfer J, and the weighting of these components by the interaction v. If ψ_e, ψ_g are shell model wave functions in the truncated space, then we have to take core polarization into account and renormalize the amplitude. Clearly it does not matter whether we interpret this as a renormalization of the transition density or the projectile-target interaction v. What is actually done in most calculations, however, is to introduce explicit variables, collective or single particle, corresponding to the core polarization, with amplitudes determined either theoretically or empirically. This is looking at changes in the transition density and leaving the interaction v alone to act as a scale factor. With the one-body assumption, the DWBA amplitude can be factorized again

$$T_{DWBA} \sim \sum_{j'j} <\psi_e | [a_{j'}^+, a_j]^J | \psi_g> <j' | t_J | j> <\chi_f^- t_J | \chi_i^+>$$

where t_J are the multipole components of the interaction, v, which as discussed before are supposed to be known (roughly the multipole components of the bare G-matrix). Then the first two factors are the matrix elements of a single particle operator t_J, and the effect of core polarization can be taken into account by computing the diagrammatic expansion of (3) for the effective single particle operator $\tilde{t}_J$. Within this approximation this is not an infinite but a limited set of diagrams. The first is the effect of the "bare" operator, or the effect of the valence nucleons in the truncated space; the second diagram is one of the diagrams giving core polarization effects. Note that in the excitation of the core

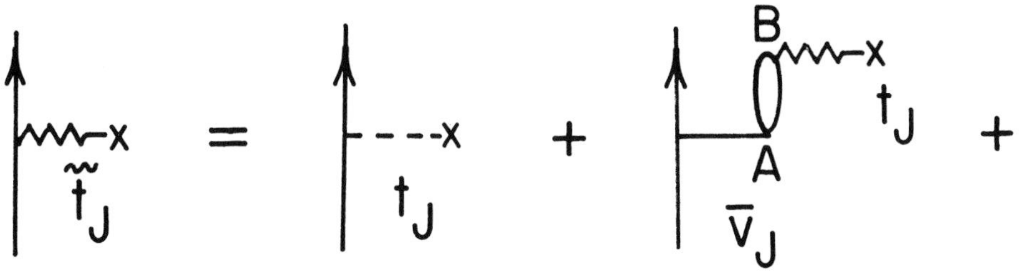

Figure 2

polarization at vertex A, the effective interaction $\bar{V}$ between valence and core comes in, not the bare G-matrix. Clearly there is plenty of room for phenomenology at vertices A and B, with for instance a macroscopic model used for the excitation and de-excitation of the core.

Within this simple framework we can estimate, following an argument due to Madsen,[6] whether in cases where there is known to be polarization charge, core polarization effects in p-p' scattering are going to be large or not. The components of this argument are the following:

1. The interaction v is a Serber mixture.

2. Whatever the core polarization is, it is an isoscalar excitation.

3. Only normal parity excitations are considered, in which the spin flip components of v are unimportant.

4. Transition densities $\rho_{tr}^{(r)} = <\phi_e | [a_j^+, a_j]^J | \phi_g > [\phi_j, (\vec{r}) \phi_j (\vec{r})]^J$

have a characteristic shape peaked near the nuclear surface. It is assumed that this shape is independent of the orbitals involved. For the DWBA we need components of the interaction folded into the transition densities, i.e. the quantities

$$\int v^{(\alpha)} (r-r') \rho_{tr} (\vec{r}') dr'$$

It is assumed that as the interaction v is short-ranged, then this does not appreciably alter the shape of the transition density, i.e.

$$\int v^\alpha (\underline{r}-\underline{r}') \rho_{tr} (\vec{r}') d\vec{r}' \sim v^\alpha \rho_{tr} (\vec{r}')$$

where v^α is a constant characteristic of the particular component of the force. With assumption (3) that the spin dependent parts of the interaction are unimportant, we have only two components coming in

$$V = v_{00}(r) + T_1 T_2 v_{01}(r) \quad \text{or} \quad v_{pp} = v_{00} + v_{01}$$

$$v_{pn} = v_{00} - v_{01}$$

and hence two constants V_{00}, and V_{01}. Or alternatively $V_{pp} = V_{00} + V_{01}$ $V_{pn} = V_{00} - V_{01}$. With assumption (1), that the force is a Serber mixture

$$V_{01} = -\frac{1}{3} V_{00}$$

Now consider the case of valence protons outside a closed shell with protons incident. The transition amplitude due to these valence protons alone is $\rho_v(r)$. The electromagnetic transition amplitude, is then $\propto \rho_v(r)$ and the p-p' scattering amplitude $\propto V_{pp}\rho_v(r)$.

However the transition is characterized by a polarization charge δe. Hence the proton transition density with core polarization is $\rho_v + \delta e\rho_v$. If we make assumption (3), isoscalar excitation of the core, then the neutron core-polarization transition density is just $\delta e\rho_v(r)$, and the p-p' scattering amplitude including core polarization is then

$$(V_{pp}(1+\delta e) + \delta e\ V_{pn})\ \rho_v(r).$$

Thus the ratio of the p-p' scattering amplitude with core polarization to the amplitude with valence protons only is

$$(V_{pp}(1+\delta e) + \delta e\ V_{pn})/V_{pp} = (V_{00}(1+2\delta e) + V_{01})/(V_{00} + V_{01})$$

and with the Serber mixture $V_{01} = -\frac{1}{3}V_{00}$, this becomes $1+3\delta e$.

Thus the effect of core polarization is amplified in this case. A polarization charge of 1 which would amplify an electromagnetic transition by a factor of 4, would amplify the p-p' cross section by a factor of 16. So core polarization effects should be clearly seen in the scattering of protons from valence protons.

The reason for the amplification is simple. The proton neutron force is twice as strong, for a Serber mixture, as the proton-proton force and the scattering therefore is much more sensitive to neutron polarization in the core then to valence protons.

For the case of protons incident on valence neutrons, the situation is not quite so favorable. For polarization charge δe, which of course gives all the electromagnetic transition amplitude of the neutron, the ratios for p-p' amplitudes with and without core polarization are $(V_{pn}(1+\delta e) + \delta e\ V_{pp})/V_{pn} = 1 + 3/2\ \delta e$. Thus for a polarization charge $\delta e \sim 1$, appropriate to say neutrons in the Pb region, the cross section amplification for p-p' would be ~ 6.

In fact the first clear evidence of core polarization in inelastic scattering was discovered by Love and Satchler[7] several years ago in analyzing the scattering of protons from ^{90}Zr. Here on the simple shell model we have 2 valence protons outside closed neutron and proton shells. The ground state which is 0^+ is a mixture of $p_{1/2}^2$ and $g_{9/2}^2$ orbitals: there is a band of states 2^+, 4^+, 6^+

and 8^+ which are taken to be $g_{9/2}^2$. It is known that core polarization is important for these states because:

1. The $2^+ \rightarrow$ ground electromagnetic transition requires a large
polarization charge, $\Delta e \sim 1.5$;

2. The relative positions of this band of states are not given
by the bare G-matrix which predicts a very compressed spectrum.
However Kuo[3] has shown that the calculation of core polarization
(perturbatively) brings the spectrum into agreement with experiment.

With a polarization charge of 1.5, Madsen's simple estimate would
give an amplification of the pp' cross section to the 2^+ state due
to core polarization of ~30. Satchler[7] found the contribution of the
valence protons was, in fact, very small. Using the measured
polarization charge to determine the parameters of a collective
model of core polarization, i.e. for the bubble diagram of Fig. 2,
he found cross-sections in agreement with experiment. A more recent
experiment and analysis, both done by the Oak Ridge Group[8] on the
inelastic scattering of 62 MeV protons from this nucleus is shown
in Figure 3. Here for the valence interaction essentially the
Kuo-Brown bare G-matrix interaction was used. Curves marked V are
the contribution of the valence nucleons, C the contribution of
core polarization and V + C, the total cross section.

It is seen that for the excitation of the 2+ state, L=2, the
contribution of the valence nucleons is quite negligible. Most
of the cross-section comes from core polarization, with an amplification reasonably close to the estimate above, even though in this
case, the core polarization transition density, given by the
collective model, was rather different from the valence transition
density. For the excitation of states of higher spin, there is no
measurement of polarization charge, so the contribution of core
polarization is a free parameter, which can be used by fitting pp' to
estimate a polarization charge for the transitions. It is seen that
core polarization is necessary for all states. However this conclusion is weakened if we consider non-central forces as we shall
see later. A very thorough study of the reliability of our present
knowledge of the central parts of the effective interaction as a
scaling factor in such analyses is given by Love and Satchler[9] and
a detailed and very general study of the use of collective coordinates for core polarization in inelastic scattering has been
made by the same authors.[10]

Figure 4a shows some calculations done by F. Petrovich[11]
analyzing the excitation of the same states at a different bombarding energy, using a completely microscopic description of core
polarization, i.e. the effect of the bubble diagram was calculated

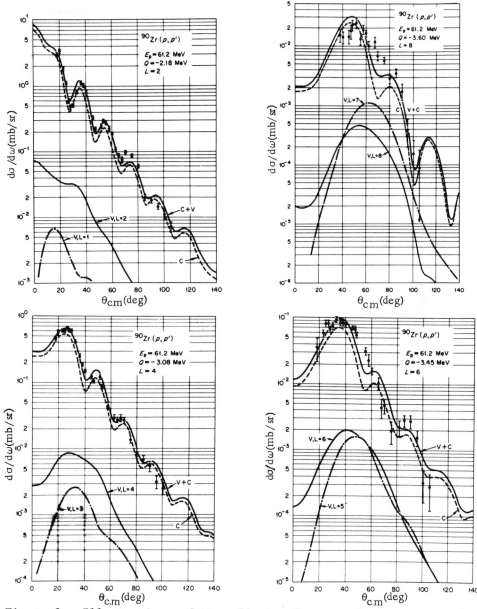

Figure 3.--Illustrations of the effects of core polarization on the inelastic scattering of 61.2 MeV protons for ^{90}Zr exciting the 2^+(L=2), 4^+(L=4), 6^+(L=6) and 8^+(L=8) states. The curves marked V,L give the contribution of the valence protons to the cross section for various multipoles using central forces only for the interaction of the incident proton with the target. The dotted curve is the contribution due to core polarization alone and the solid curves marked C+V the total cross section, valence plus core. For the 2^+ state, the core polarization contribution is estimated from the measured polarization charge. For the other states it is fitted to experiment.

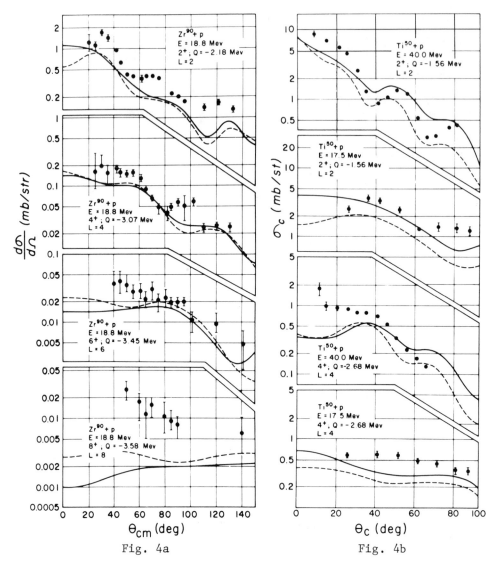

Fig. 4a

Fig. 4b

Figure 4a.--Comparison of theory (solid curves) with experiment for the excitation of the 2^+, 4^+, 6^+ and 8^+ states of ^{90}Zr by 18.8 MeV protons. Here core polarization was calculated on a microscopic model and central forces only were used for the interactions between incident proton and target nucleons.

Figure 4b.--The same comparison of theory with experiment as in 4a for the excitation of 2^+ and 4^+ states of ^{50}Ti by 17.5 MeV and 40 MeV protons.

The dotted curves in 4a and 4b are a collective model calculation with theoretically calculated parameters.

in the same way as Kuo and Brown. There were no free parameters in the calculation. Again core polarization gives a much larger contribution to the cross section than the valence protons. The calculation predicts that core polarization remains important as the spin of the excited state increases. Note that despite the large contribution from antisymmetrization for such a high spin state, the theoretical core polarization is totally insufficient to account for the excitation of the 8^+ state.

Also for the excitation of the 2^+, the microscopic calculation does not do as well as the Oak Ridge calculation[8] which uses information on core polarization from the measured effective charge. I emphasize this because I want to make the point that the experimental data on level structure, effective charge, and p-p' are quite consistent with each other, at least in cases where non-central forces are relatively unimportant. All that has to be assumed is that the mass polarization is strongly correlated with the charge polarization, which is obvious, and that the mass polarization for these normal parity transitions is mostly isoscalar, which is what, as Schaeffer notes, is empirically observed in most cases. In fact, with central forces, p-p' is rather insensitive to isovector admixtures. It should also be pointed out that, as mentioned in connection with equation (4), the microscopic and macroscopic calculations are very similar to each other. In both cases explicit variables are introduced for the core polarization, in the one case collective variables of the core, in the other, for this particular case, 3 particle, 1-hole components of the wave-function. The calculations are in fact independent of any assumption about whether the interaction between projectile and target acts like a one-body operator, though this assumption is convenient in interpreting the results.

What is particularly interesting about this type of analysis is the possibility of getting information about the mass and hence indirectly the charge polarization for high multipoles which are inaccessible by other means. Thus the quadrupole polarization charge δe_2 is well measured electromagnetically and a few isolated values of δe_4 are becoming available via ee' but nothing is known about δe_6 and δe_8. The theoretical calculation used in Fig. 4a predicts, (via mass polarization), $\delta e_2 \sim 1.5$, $\delta e_4 \sim .62$, $\delta e_6 \sim .3$, $\delta e_8 \sim .16$, i.e. a steady fall off with increasing multipole but with non-negligible contributions for the higher multipoles. As Fig. 4a shows these are quite insufficient to account for the excitation of the higher multipoles using central forces only for the interaction of projectile and target.

Figure 4b shows the results of calculations similar to those illustrated in Fig. 4a for the excitation of states in ^{50}Ti, treated as two $f_{7/2}$ protons outside a closed ^{48}Ca core, and therefore entirely

analogous to ^{90}Zr. As expected core polarization gives the bulk
of the cross-section. Again the microscopic calculation somewhat
underestimates its effect. Recently electron scattering measurements exciting the 2^+ and 4^+ states in this nucleus have become
available.[12] They have been analyzed by G. Hammerstein[13] and give
polarization charges (for transitions to the ground state) of 0.71
and 0.43 respectively. The first is in agreement with other measurements, the other has not been measured before but is in rough
agreement with the theoretical prediction. Both values are quite
consistant, on a collective model of core polarization, with the
p-p' results. A measurement[14] of the polarization charge for the
$6^+ \rightarrow 4^+$ level transition gives a quadrupole polarization charge of
0.62 in fair agreement with the $2^+ \rightarrow$ ground transition, again in
accord with the expectations of a simple core polarization picture.
The same calculation has been successfully applied by Preedom et al.[15]
to the analysis of inelastic scattering from ^{51}V and ^{52}Cr, with
three and four valence protons outside a ^{48}Ca core. The spectra of
these nuclei are also reproduced by an $(f^{7/2})^n$ configuration with
effective forces as calculated by Kuo and Brown, plus a little
configuration admixing to give the seniority mixing observed in
^{52}Cr (in fact observed in p-p'[15]).

Figure 5a shows the identical calculation for the $p_{1/2} \rightarrow g_{9/2}$
single particle transition in ^{89}Y. This can go by an M4 or an E5
transition. Core polarization increases the strength of the E5 by
a factor of about 10 and decreases the M4 by a factor of 40%.
Momentum transfer considerations make the electromagnetic decay of
the $g_{9/2}$ state entirely M4. Both contribute to inelastic scattering, the E5 dominating at large momentum transfer.

Figure 5b shows the corresponding calculation of single neutron
hole transitions in ^{207}Pb. Here the effect of core polarization is
much less as expected from Madsen's simple argument. This theoretical
calculation probably underestimates the amount of core polarization.
It corresponds to polarization charges of ~0.5 whereas a phenomenological analysis[16] gives polarization charges of ~1.

Figure 6 shows the analysis of inelastic scattering of 40 MeV
protons from ^{209}Bi [17] for the single particle proton transition
$1h_{9/2} \rightarrow 1i_{13/2}$. Here practically the entire cross-section is due to
the coupling of the protons to the 3^- collective state of ^{208}Pb.
The parameters used were predicted by Mottelson.[18] M2 and E3
transitions are both allowed. In contrast to ^{89}Y, both γ-decay
and inelastic scattering go by E3.

This is typical of a case in which the concept of polarization
charge is inappropriate until the relevant degrees of freedom of the
system are included i.e. the valence proton and the physical collective
3^- state of the ^{208}Pb core. The effect of other degrees of freedom
contributing to the mass and charge polarization outside the space are

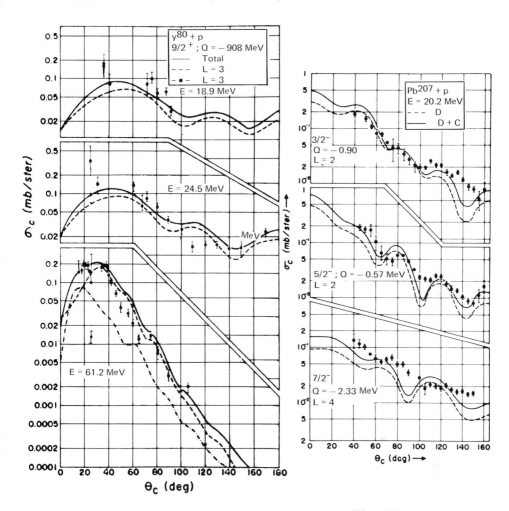

Fig. 5a   Fig. 5b

Figure 5a.--Comparison of theory with experiment for the excitation
of the $9/2^+$ state in ^{89}Y by 18.9, 24.5, and 61.2 MeV protons. The
dashed curve is the contribution of the E5 multipole. The dashed
dot the contribution of the M4 and the solid curve the total
theoretical cross section.

Figure 5b.--Comparison of theory with experiment for the excitation
of the $3/2^-$, $5/2^-$ and $7/2^-$ states in ^{207}Pb by 20.2 MeV protons.
The dashed curve is the contribution of the valence neutron hole.
The solid curve the total theoretical cross section including core
polarization.

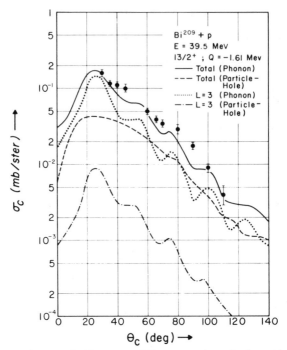

Figure 6.--Comparison of theory with experiment for the excitation of the 13/2[+] state of [209]Bi by 39.5 MeV protons. The dashed curve shows the contribution of all core multipoles coming from high-lying excitations of the core. The dotted line shows the contribution of the collective 3[-] state of the [208]Pb core which should be contrasted with the contribution of high-lying 3[-] excitations. (Dash-dot curve) The solid line is the total theoretical cross section.

rather small as shown by the difference between the dotted (labelled L=3 (Phonon curve) which is the effect of the relevant degrees of freedom, and the solid curve, which includes also all the small core-polarization effects.

So things look relatively straight forward. The picture that emerges is that core polarization effects as measured by inelastic scattering decrease with multipole but remain appreciable even for high multipoles. Referring back to Fig. 4a calculations with theoretical core polarization but central forces only failed completely to describe the excitation of the 8[+] state. However, non-central forces have been left out, in particular the two-body spin orbit. Love[19] has done a recent calculation of the effects of the spin orbit force on the excitation of the various states in [90]Zr. Its effect increases rapidly with multipole. The results for the 6[+] and 8[+] states are shown in Figure 7a. Here only valence nucleons are considered but non-central forces are included in the

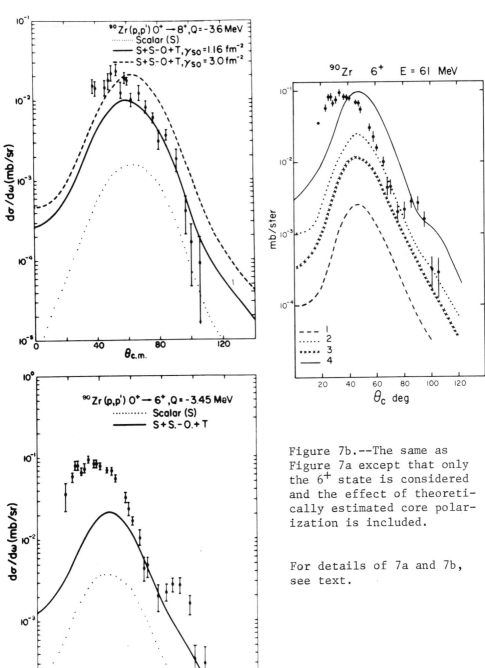

Figure 7b.--The same as Figure 7a except that only the 6⁺ state is considered and the effect of theoretically estimated core polarization is included.

For details of 7a and 7b, see text.

Figure 7a.--Comparison of theory without core polarization, but with non-central force, with experiment for the excitation of 6⁺ and 8⁺ states of ^{90}Zr by 60.2 MeV protons.

interaction with the incident protons. The dotted line is the contribution for central forces only, the solid line includes the effect
of the non-central forces (spin orbit and tensor). The dashed line
for the 8^+ is the same using different spin orbit parameters. As
this calculation shows the two-body spin orbit goes a long way to
remove the discrepancy in the excitation of the 8^+ state of ^{90}Zr,
in fact no core polarization seems necessary in this case. Figure
7b shows the results of a similar calculation for the 6^+ excitation
using a microscopic calculation of core polarization and a two-body
spin orbit force. The core polarization corresponded to $\delta e_6 \sim 0.3$,
and the parameters used for the spin-orbit force were slightly
different from those of Love. The bottom (dashed) curve (1) includes
valence nucleons only and central forces. The next curve up (3)
(crosses) has valence contribution only but central and spin orbit
forces. The dotted curve (2) includes core polarization but with
central forces only, and the solid curve (4) includes core polarization and non-central forces.

Including all these effects, core polarization and the spin-
orbit interaction between incident proton and target nucleons,
gives a peak cross-section in rough agreement with experiment, but
hardly what would be considered a satisfactory fit as far as differential cross section is concerned. The microscopic calculations
suggesting $\delta e_6 \sim 0.3$ were made with only central forces between valence
nucleons and core. George Bertsch has estimated that if one includes
the two-body spin orbit in this interaction then core polarization
is reduced. For the transition connecting the state of highest
spin possible to the ground state the spin-orbit gives a contribution of opposite sign to the central part of the force, resulting
in almost complete cancelation of the contribution of core polarization so that $\delta e_8 \sim 0$. But what the spin-orbit takes away with one
hand, it gives back with the other, and referring back to fig. 7a,
the cross-section is more than restored by the spin-orbit interaction
between the projectile and the valence nucleons. Unfortunately the
spin-orbit interaction is not known well enough to test this point
as will be discussed in detail by Sam Austin in the next talk. It
is usually chosen to give a reasonable prediction of the optical
model spin orbit, but this only gives information about its isospin
average. However, Love informs me that his latest analysis, relying
on shape as well as peak cross-section, does require appreciable core
polarization for excitation of the 8^+. Clearly a more precise scale
for the two-body spin orbit is required to answer this interesting
question.

More work needs to be and is being done on the non-central
parts of the interaction, both theoretically and by phenomenological
analysis. Unnatural parity excitations and charge exchange reactions
are sensitive to these components of the interaction, but of course
core polarization comes in too, sometimes very strongly,[20] and
Madsen will have something to say about this.

So perhaps this is the point to stop and listen to what Sam Austin has to tell us about empirical values of the effective interaction.

In conclusion I would like to thank Fred Petrovich and Gary Love for supplying me with information for the talk--do not blame them for the result!--and the typist, Mrs. Julie Perkins, and principal organizers Sam Austin and Gary Crawley, for their patience.

REFERENCES

1. P. R. de Kock and P. W. M. Glaudemans, Phys. Lett. $\underline{34B}$(1971) 280.

2. M. Harvey, AECL-3366, June, 1969; M. Harvey and F. C. Khanna, Nucl. Phys. $\underline{A152}$(1970)588; $\underline{A155}$(1970)337; M. H. Macfarlane, "Enrico Fermi" Course 40, Varenna, Italy, p. 457 (Academic Press, 1969); M. Gmitro, A. Rimini, Theory of Nuclear Structure, Trieste Lectures 1969, 697, 713 Vienna, 1970.

3. G. F. Bertsch, Nucl. Phys. $\underline{74}$(1965)234; T. T. S. Kuo and G. E. Brown, Nucl. Phys. $\underline{85}$(1966)40; $\underline{A92}$(1967)181; $\underline{A114}$(1968) 241; T. T. S. Kuo, Nucl. Phys. $\underline{A90}$(1967)199; $\underline{A103}$(1967)71.

4. H. Horie and A. Arima, Phys. Rev. $\underline{99}$(1955)778.

5. D. Slanina and H. McManus, Nucl. Phys. $\underline{A116}$(1968)271.

6. J. Atkinson and V. A. Madsen, UCRL-71809, Phys. Rev. $\underline{C1}$(1970) 1377.

7. W. G. Love and G. R. Satchler, Nucl. Phys. $\underline{A101}$(1967)424.

8. A. Scott, M. L. Whitten, G. R. Satchler, Bull. Am. Phys. Soc. $\underline{15}$(1970)498, Nucl. Phys. $\underline{A}$ to be published.

9. W. G. Love and G. R. Satchler, Nucl. Phys. $\underline{A159}$(1970)1.

10. G. R. Satchler and W. G. Love, Nucl. Phys. $\underline{A172}$(1971)449.

11. F. Petrovich, Ph.D. Thesis, Michigan State University, 1970.

12. J. Heisenberg, J. S. McCarthy and I. Sick, Nucl. Phys. $\underline{A164}$ (1971)353.

13. G. R. Hammerstein, private communication.

14. T. Nomura, C. Gil, H. Saito, and T. Yamazaki, Phys. Rev. Lett. $\underline{25}$(1970)342.

15. B. M. Preedom, C. R. Gruhn, T. Kuo and C. Maggiore, Phys. Rev. C2(1970)166.

16. C. Glasshausser, B. G. Harvey, D. L. Hendrie, J. Mahoney, E. A. McClatchie and J. Saudinos, Phys. Rev. Lett. 21(1968) 918.

17. W. Benenson, S. M. Austin, P. J. Locard, F. Petrovich, J. R. Borysowicz and H. McManus, Phys. Rev. Lett. 24(1970)

18. B. R. Mottelson, J. Phys. Soc. Japan, Suppl. 24(1968)96.

19. W. G. Love, Phys. Lett. 35B(1971)371.

20. V. A. Madsen, V. R. Brown, F. Becchetti, and G. W. Greenlees, Phys. Rev. Lett. 26(1971)454.

*Supported in part by the U. S. Atomic Energy Commission.

THE EFFECTIVE TWO-NUCLEON INTERACTION FROM INELASTIC PROTON SCATTERING*

Sam M. Austin

Michigan State University

Since inelastic scattering cross-sections depend on a nuclear matrix element containing both the wave functions of the nuclear states involved and the effective interaction V_{eff} between the projectile and the target nucleons,[1] it is necessary to have an a priori value for V_{eff} before one can use inelastic scattering as a tool for nuclear spectroscopy. An approach to determining V_{eff} which I will explore in detail in this paper, is entirely empirical in nature. One examines transitions where the wave functions are well known and adjusts the strength of the two-body interaction to fit the observed cross-sections. If the results for a representative sample of cases are consistent one has obtained a workable interaction for spectroscopic studies.

This approach was tried by Satchler[2] in his early studies of inelastic scattering, but was unsuccessful for reasons which we now understand fairly well. The major difficulty was that the states involved were collective in nature so that the cross sections were greatly enhanced. On the other hand, only simple wave functions were used in the theoretical calculations, so single-particle-size cross sections were predicted, and when the two-body potential was adjusted to fit the measured cross sections, the resulting potential was unphysically large. Since the collective enhancement depends on the nucleus, the state involved, and the angular-momentum transfer (L) of the transition, the effective interaction obtained was also nucleus, state and L dependent. In addition the calculations neglected exchange contributions which depend on L and the bombarding energy,[1] and again the empirical V_{eff} mirrored this dependence. For these reasons it was not possible to find a consistent effective-interaction for the strong, spin independent part of the force. Studies[2-9] of L=0 transitions involving spin-flip or isospin-flip were more successful since both collective enhancements and exchange effects are relatively small in these cases.

Faced by this difficulty several investigators[10-12] introduced "realistic" effective interactions derived in a more-or-less plausible fashion from the free two-body force. As Schaeffer and McManus have shown at this symposium, this approach has been relatively successful and has led to an understanding of the importance of collective enhancement and of the contribution of exchange processes. However, the derivation of these forces is not free of uncertainties. For example, in approaches[10,12] using the Moszkowski-Scott[13] separation method it is necessary to neglect the odd-state forces since the method cannot be applied to repulsive forces. The effect of this omission is not clear and is particularly serious for the parts of the force responsible for spin-flip and isospin-flip transitions since they are given as the small differences of large numbers. It is worth noting that the validity of the realistic forces has been checked in detail only for the spin-isospin-independent part of the force. It is also difficult to evaluate the accuracy of approximations made in obtaining the interaction and elsewhere in the theory. In the empirical approach, on the other hand, the consistency of the results obtained provides an immediate measure of the accuracy of the calcula-tion, with the additional advantage that inaccuracies which affect all transitions in the same fashion are automatically corrected through a renormalization of the effective force.

For several reasons the time seems ripe for a return to the empirical procedure. First, we now understand the importance of the exchange process and can do calculations including exchange; secondly, we have a good understanding of when wave functions are adequate and in many cases electron scattering data is available as a check; and thirdly, enough data are available so we have some chance of isolating individual parts of the two-body interaction. The next section of this paper contains a brief outline of the theoretical formalism. This is followed by a review of experimental information on the central (or scalar) part of V_{eff}. Finally, in the last part of the paper, recent work on the tensor and spin-orbit parts of the force is discussed and available information is summarized.

THEORETICAL CONSIDERATIONS

In the distorted wave approximation (DWA) neglecting exchange, the cross-section for a reaction $A(a,b)B$ is proportional to the square of a transition amplitude

$$T_{ba} = \int \chi_b^{(-)} \langle \psi_f | V_{eff} | \psi_i \rangle \chi_a^{(+)} d\vec{r}, \tag{1}$$

where the χ's are distorted waves generated in an appropriate optical potential. The form factor $<\psi_f|V_{eff}|\psi_i>$ depends on both the wave functions ψ of the states involved and the effective interaction.[14] It is assumed that V_{eff} can be written as the sum of the two-body interactions between the projectile "p" and the target nucleons "i",

$$V_{eff} = \Sigma V_{ip} \tag{2}$$

where the sum is over the valence nucleons.

For the central part of the force we have

$$V_{ip}(r)=V_{00}(r)+V_{10}(r)\vec{\sigma}_i\cdot\vec{\sigma}_p+V_{01}(r)\vec{\tau}_i\cdot\vec{\tau}_p+V_{11}(r)(\vec{\sigma}_i\cdot\vec{\sigma}_p)(\vec{\tau}_i\cdot\vec{\tau}_p). \tag{3}$$

The subscripts on the V_{ST} are the spin and isospin transferred in the reaction. For mneumonic purposes we will often write $V_o=V_{00}, V_\sigma=V_{10}$, $V_\tau=V_{01}$, $V_{\sigma\tau}=V_{11}$, since for example, V_{11} is responsible for a reaction involving both spin and isospin flip.

The selection rules describing the scattering then determine which of the V_{ST} are important in any particular reaction. For the direct (non-exchange) process these are[14]

$$\vec{J}=\vec{J}_f-\vec{J}_i \qquad\qquad \vec{T}=\vec{T}_f-\vec{T}_i$$

$$\vec{S}=\vec{s}_i-\vec{s}_f \qquad\qquad \pi_i\pi_f=(-1)^L. \tag{4}$$

$$\vec{L}=\vec{J}-\vec{S}$$

Here $\vec{J}$, $\vec{S}$ and $\vec{L}$ are the total, spin and orbital angular momenta transferred in the reaction and T is the transferred isospin. The subscripted quantities refer to initial(i) and final(f) states, π denoting the parity of these states. For the case of spin-one-half projectiles we have

$$S=0,1$$
$$T=0,1. \tag{5}$$

As an example we apply these rules to a transition from a $J_i=1^+$, $T_i=0$ state to a $J_f=0^+$, $T_f=1$ state (denoted as $(1^+,0)\rightarrow(0^+,1)$). We find $J=1$; $L=0,2$; $S=1$; $T=1$, which means that only $V_{ST}=V_{11}$ can contribute to the cross-section. Table I shows transitions which isolate other parts of the force. The important case of the excitation of a collective natural parity state, $(0^+,0)\rightarrow(2^+,0)$ for example, does not precisely isolate V_o since both $S=0$ and 1 are allowed. However, the $S=1$ amplitude is not enhanced by collective effects and is usually negligible so only V_o is important.

TABLE I.--Transitions Isolating V_o, V_σ, V_τ, $V_{\sigma\tau}$.

V_{ST}	Transition[a]	Reaction
V_o	$(0^+,0) \rightarrow (0^+,0)$	(p,p')
V_σ	$(0^+,0) \rightarrow (2^-,0)$	(p,p')
V_τ	$(0^+,1) \rightarrow (0^+,1)$	(p,n) to analog state
$V_{\sigma\tau}$	$(1^+,0) \rightarrow (0^+,1)$	(p,p')

[a]The notation is $(J_i^\pi, T_i) \rightarrow (J_f^\pi, T_f)$.

Of course the selection rules apply strictly only to the direct part of the scattering amplitude, and in the general case all V_{ST} can contribute to the exchange amplitude. However, if the effective interaction acts only in even states (or only in odd states) the force contributing to the exchange amplitude is the same (within a sign) as that in the direct amplitude.[12,15,16] We will find that the force obtained from our analysis is quite similar to a Serber force (V_o: V_σ: V_τ: $V_{\sigma\tau}$=-3:1:1:1) which is an even state force, so we are still able to isolate single V_{ST}. Angular momentum transfers not satisfying $\pi_i\pi_f=(-1)^L$ are also allowed in exchange, but in most cases studied to date these amplitudes are small[15] and we shall not consider them here.

Although the discussion has been carried out for inelastic scattering reactions, we have used an isospin formalism, so the calculations for (p,n) reactions are formally the same, only isospin quantum numbers being changed.

EXPERIMENTAL INFORMATION ON THE V_{ST}

A large number of analyses of inelastic scattering and charge exchange data are available in the literature. However, the great bulk of these were done with inadequate wave functions and neglecting exchange, or have other defects which prevent the inclusion of their results in this review.

The assumptions and criteria used to select data were:

1) The interaction has a Yukawan radial dependence with a range μ of 1.0F, or else the reaction has L=0 so we can obtain an estimate of the strength of an equivalent 1.0F Yukawa by matching the volume integrals of the forces. Calculations in a few special cases have shown that this procedure is not accurate for L>0.[16,17] The choice of 1.0F as the standard range was made primarily because it is the

most commonly used range and shapes of angular distributions are not strongly affected by changes of μ, at least for μ between 0.7 and 1.4F.[17] However, Table II shows that this value of the range also yields a mean-square radius for the potential which is roughly equal to that for the long-range part of the Hamada–Johnston potential.[18]

2) The transition isolates a single one of the V_{ST}. This effectively restricts us to self-conjugate nuclei (whose ground states have T=0) since otherwise both V_{S0} and V_{S1} can contribute to the cross section.

TABLE II.--Yukawas Matched to the Hamada–Johnston Potential[a].

ST	00	01	10	11
V_{ST} (MeV)[b]	24.8	12.8	4.6	8.3
μ_{ST} (F)[b]	1.06	0.98	1.18	1.06

[a] These Yukawas $\dfrac{e^{-r/\mu}}{r/\mu}$ have the same volume integral and value of of $<r^2>$ as the long range part of the Hamada–Johnston potential.

[b] Adapted from Table I of Ref. 12.

3) In the case of V_0, where collective effects enhance the relevant cross sections, we require that the wave functions properly describe the electromagnetic transitions between the states involved. In some cases electron scattering data were directly used in the analysis[19] and in some other cases effective charge techniques[15] were used to renormalize the inelastic scattering amplitudes.

4) Exchange effects were small (L=0), or the calculation included exchange (L≠0). In the L=0 case, values of V_{ST} obtained were decreased by 20% (15% above 30 MeV), corresponding to the average contribution[12,15] of exchange to such cross sections. This requirement unfortunately eliminated the large number of analyses using macroscopic core-polarization techniques. In a few cases data were reanalyzed to include exchange effects using the code DWBA 70.[20]

5) A Serber exchange mixture was assumed in the analysis. This choice is roughly consistent with what one expects from the Hamada–Johnston force (see Table II) and with the results of the present review.

Empirical optical model fits to elastic scattering data were also used to provide estimates of V_0 and V_τ. The real part U of the optical model potential for a self-conjugate nucleus is given in first order by folding the nuclear matter density $\rho(r')$ with the two-body interaction between the projectile and the target nucleons[21]

$$U(r) = \int \rho(r') V_o(|\vec{r}-\vec{r}'|) d\vec{r}'. \tag{6}$$

One can then show that

$$\int V_o(r) d\vec{r} = \frac{1}{A} \int U(r) d\vec{r} \tag{7}$$

where A is the atomic number. Thus from the volume integral of U one can obtain the volume integral of V_o and hence V_o itself. A similar relationship connects the optical model symmetry potential and V_τ. An energy dependent correction of between -11 and -26%[21,22] was applied to these values of V_o and V_τ to account for the contribution of exchange processes.

The results are shown in Figs. 1-4, where the laboratory bombarding energy is plotted along the abscissa and the strength of the 1.0F Yukawa potential along the ordinate. The relative signs of the V_{ST} have not all been fixed by experiment, and the examples studied in this review shed no light on them, since only a single V_{ST} is important in each case. The signs chosen are those universally predicted by the realistic models, namely, V_o is attractive while V_σ, V_τ, and $V_{\sigma\tau}$ are repulsive. Also shown on each figure as lines labelled HJ, KK or KKD are comparisons with the so-called realistic forces. These lines denote the Yukawa with the same volume integral as the long range parts of the Hamada-Johnston force (HJ),[18] of the Kallio-Kolltveit force (KK),[23] and of the density-dependent[24] Kallio-Kolltveit force averaged over the lead nucleus (KKD). The numbers near the points are the mass numbers of the targets. Some comments on individual V_{ST} follow.

V_o: The spin-isospin independent part of the effective interaction appears to be independent of bombarding energy and of L, and is in essential agreement with the theoretical estimates. It is particularly reassuring that the results obtained from inelastic scattering and from the optical model are in such close agreement. The two values for the ^{90}Zr point[15] correspond to different experimental values of the B(E2) used to fix the effective charge.

V_σ: There is rather little solid information available here. The two lowest energy points from the ^{16}O(p,p')^{16}O(2$^-$) reaction[17] are upper limits because of compound nucleus contributions. It now seems likely, as we shall discuss below, that the spin-orbit force contributes to this cross section, so the other points may also be upper limits. Both V_σ and $V_{\sigma\tau}$ contribute to the spin-flip part of the cross section for the ^{89}Y point; it was assumed in the analysis[25] that $V_\sigma = V_{\sigma\tau}$.

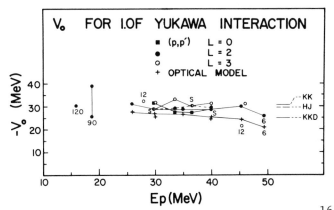

Figure 1.--Values of V_0. Points not numbered are from ^{16}O. S means two points are superimposed.

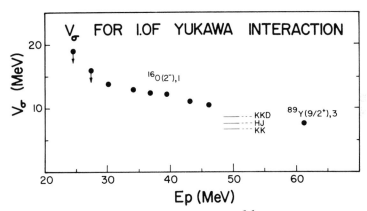

Figure 2.--Values of V_σ. The annotation $^{16}O(2^-),1$ means the data is from $^{16}O(p,p')^{16}O(2^-)$ and that L=1.

V_τ: Only a few analyses are available above 20 MeV and the results of these are not particularly consistent. The points labelled <lp> are the average lp-shell results of Clough et al.,[8,9] and have been used to argue that V_τ is strongly energy dependent between 30 and 50 MeV. Recent experiments, analyzed using a macroscopic-model form factor, indicate that any energy dependence is much less dramatic than indicated by the <lp> points. One can see, by comparing the lines labelled KK and KKD, that V_τ is particularly sensitive to any density dependence in the two-body force. It is not clear how important such effects should be in charge exchange reactions, but the effect would tend to damp contributions from the high density interior region and lead to a more strongly surface-peaked form factor.

$V_{\sigma\tau}$: There is a substantial amount of data available from both (p,p') and (p,n) reactions. $V_{\sigma\tau}$ seems fairly well determined and independent of energy, but all these transitions are dominated by L=0, so we have no information on a possible L-dependence.

To assess the accuracy to be expected in analyses such as this, one must have some idea of the sensitivity of the calculations to the input assumptions, in addition to those already considered. In the light nuclei considered here, the optical model parameters are suspect. This problem has not been thoroughly studied, but the shapes of the angular distributions do not seem to be sensitive to reasonable changes in the optical model.[17] However, the magnitude of a predicted cross section has been observed to change by as much as 30% when different sets of optical model parameters were used,[17] thus introducing a 15% change in V_{ST}. No damping (due to non-local potential effects, for example) was applied to the distorted waves in the quoted calculations.

Although harmonic oscillator (HO) wave functions were used in most of the calculations quoted here, Woods-Saxon (WS) wave-functions were used occasionally. The shapes and magnitudes of the cross sections are not usually very sensitive to either the range of the HO potential or to the choice of a HO or WS radial dependence.

The sources of information for the figures are tabulated in Table III.

One can summarize these results by stating that V_0 and $V_{\sigma\tau}$ are fairly well determined, are nearly independent of energy, and have the values V_0=27±5 MeV and $V_{\sigma\tau}$=12±2.5 Mev, while V_σ and V_τ are poorly determined. Very few transitions isolate V_σ and other spin-dependent forces may contribute to these, so it will be difficult to get precise numbers for this small part of the force. On the other hand new (p,n) data is becoming available in the 20-50 MeV range and it seems likely that reliable estimates of V_τ will be available soon.

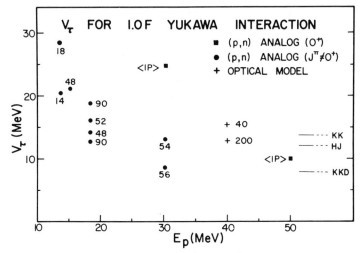

Figure 3.--Values of V_τ.

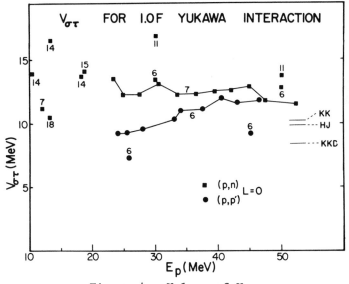

Figure 4.--Values of $V_{\sigma\tau}$.

TABLE III.--Sources of information on V_{ST}.

Inter-action	Nucleus(L)[b]	Ep(MeV)	Reference	Modifications[a]
V_o	^{6}Li(OM)	25.9-49.5	22	E
	^{6}Li(2)	25.9-49.5	22	E
	^{12}C(3)	28.05,45.5	10	
	^{16}O(0,2,3)	29.8-40.1	19	E
	^{16}O(3)	30.1,46.1	17	E
	^{58}Ni to ^{208}Pb(OM)	30,40	21,26,27	E
	^{90}Zr(2)	18.8	15	
	^{120}Sn(2)	16	15	
V_σ	^{16}O(1)	24.5-46.1	17	E
	^{89}Y(3)	61	25	
V_τ	<1p>	30,50	8,9	R,E
	^{14}C(0)	13.7	7	R,E
	^{18}O(0)	11.9,13.5	7	R,E
	^{48}Ti(0)	15.25,18.5	7,2	R,E
	^{52}Cr(0)	18.5	2	E
	^{54}Fe(0)	30.2	28	R,E
	^{56}Fe(0)	30.2	28	R,E
	^{90}Zr(0)	18.5	2,7	R,E
	A=40,200(OM)		21,29	E
$V_{\sigma\tau}$	^{6}Li(0)	30,50	8,9	R,E
	^{6}Li(0)	25.9,45.1	22	E
	^{6}Li(0)	24.3-46.5	6	E
	^{7}Li(0)	12.0-52.3	4,6	E
	^{11}B(0)	30,50	8,9	R,E
	^{14}C(0)	10.4,13.3,18.3	7,30	R,E
	^{15}N(0)	18.8	7	R,E
	^{18}O(0)	13.2	7	R,E

a) E or R means the results of the reference were modified to account for the effects of exchange or a different range, respectively.

b) In the case of (p,p') or (p,n) reactions, L is the momentum transfer. For optical model studies L=OM.

So far we have neglected the contribution of the tensor and spin-orbit parts of the effective interaction. In most cases this is a reasonable assumption. However, we shall find that in certain circumstances either the long-range tensor force or the short-range spin-orbit force can dominate an inelastic scattering reaction. We examine first the evidence for the tensor force.

THE TENSOR FORCE

In the simplest case, the one-pion-exchange-potential (OPEP), the tensor force has the form

$$\text{OPEP:} \quad V_T(r) = V_{T\tau} \vec{\tau}_1 \cdot \vec{\tau}_2 \, S_{12} f(\alpha) \tag{8}$$

$$f(\alpha) = (1 + \frac{3}{\alpha r} + \frac{3}{(\alpha r)^2}) \frac{e^{-\alpha r}}{\alpha r},$$

where S_{12} is the usual tensor operator and $\alpha^{-1} = (\frac{h}{m_\pi c}) = 1.415F$ is the Compton wavelength of the pion. Because of the strongly divergent form of the OPEP as $r \to 0$, alternative shapes have usually been chosen. The most common of these are the regularized OPEP or ROPEP form [30,31]

$$\text{ROPEP:} \quad V_T(r) = (V_T + V_{T\tau} \vec{\tau}_1 \cdot \vec{\tau}_2) S_{12} [f(\alpha) - \frac{\beta^3}{\alpha^3} f(\beta)], \tag{9}$$

and the r^2-Yukawa form [20]

$$r^2\text{-Y:} \quad V_T(r) = (V_T + V_{T\tau} \vec{\tau}_1 \cdot \vec{\tau}_2) S_{12} r^2 \frac{e^{-\alpha r}}{\alpha r}. \tag{10}$$

In Eq. 9, β is taken to be substantially larger than α so the potential resembles the OPEP at large r, while in both these forms a term V_T which can act in no-isospin-transfer cases is allowed.

The first two terms in the expansion of the Fourier transform of the tensor and spin-orbit forces are proportional to the integrals[32]

$$J_4 = \int r^4 V(r) dr$$

$$J_6 = \int r^6 V(r) dr. \tag{11}$$

Schaeffer and I[38] estimated the strength and range for the r^2-Y form by matching J_4 and J_6 for the r^2-Y form of Eq. 10 to those for the OPEP or the Hamada-Johnston potential. In the latter case we find $V_T \sim 0.05 \, V_{T\tau}$, so one would expect the tensor force to contribute strongly only to transitions in which the isospin changes.

Some special selection rules apply to the tensor force, namely[30,33,34]

$$\begin{aligned} S &= 1 \\ L &= \lambda, \lambda \pm 2. \end{aligned} \tag{12}$$

The first equation states that only spin-flip transitions are possible. The second reflects the fact that the nature of S_{12} allows the two angular momentum transfers involved in the problem, that to the valence nucleon (λ) and that to the projectile (L), to differ by ± 2. For the central-force case $\lambda = L$.

Most of the available information on $V_{T\tau}$ comes from analyses of the $^{14}C(p,n)^{14}N(g.s.)$ reaction, which is the analog of the strongly inhibited β^- decay of ^{14}C. Since this is a $(0^+,1) \to (1^+,0)$ transition we have S=T=1; L=0,2 and for a central force one would expect the L=0 amplitude to dominate the reaction. However, it has been shown by Madsen[3] that the L=0 amplitude is approximately proportional to the β-decay matrix element which is very small in this case. One then expects the central-force contribution to the cross-section to be small and to have an L=2 shape. Adding a tensor force relaxes these selection rules. The amplitude with angular momentum $\lambda=0$ transferred to the nucleus is still small but the amplitude with ($\lambda=2$, L=0), to which only the tensor force can contribute, is now expected to be important.

Measurements and calculations of cross sections and polarizations for $^{14}C(p,n)$ have been performed by Wong, et al.[30] at a variety of energies below 20 MeV. Inclusion of a tensor force of the ROPEP form greatly improved the fits to cross sections, but perhaps because of compound nucleus effects or poorly known optical model parameters, quantitative fits were not generally obtained. Fits to the polarization data were very poor either with or without a tensor force. Exchange processes were not included in this analysis, but they should not be large for this L=0 transition.

The analogous transition in inelastic proton scattering leaves ^{14}N in its first excited state (J=0$^+$,T=1) at 2.311 MeV. Crawley, et al.[31] measured the cross section for this transition at Ep=24.8 MeV and analyzed it in terms of a force of ROPEP shape, again without including exchange processes. A good fit to the data was obtained for $\theta < 80°$ by including a tensor force of roughly OPEP magnitude, but not with a central force alone. The data at back angles were not fitted, even qualitatively, but the cross sections were so small (~30 µb/sr) that it was not possible to exclude compound nucleus processes in such a light nucleus.

Because this reaction promises perhaps the cleanest measure of $V_{T\tau}$, Fox, et al.[35] at Michigan State have measured cross sections at 29.8, 36.6, and 40.0 MeV. The angular distribution at 29.8 MeV and a preliminary analysis are shown in Fig. 5. Theoretical cross sections including exchange processes were calculated using the code DWBA 70.[20] The central force ($V_{\sigma\tau}$) had a strength consistent with the results of Fig. 4; it is clear from the curve labeled "central (exchange)" that the central force alone cannot describe the data. The tensor force had an r^2-Y form. The range of $V_{T\tau}$ was obtained by matching the integrals of Eq. 11 to the OPEP potential as described

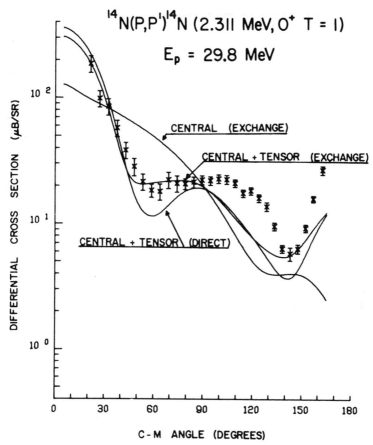

Figure 5.--Cross sections for the ^{14}N(p,p')^{14}N(2.311 MeV, 0^+,T=1) reaction at 29.8 MeV.

earlier and its strength was adjusted to fit the data. The calculations including central and tensor forces are labeled "Central + OPEP". The curve labeled "direct" contains only the direct amplitude, while that labeled "exchange" contains both direct and exchange amplitudes. The fit is good out to $\theta=90^\circ$ and even the back angle dip is qualitatively predicted. An additional calculation including a spin-orbit force of reasonable size produced an overall 10% decrease in the cross section, but did not change the shape. The curves shown were calculated for $V_T=0$, but the results are not sensitive to this assumption. It should be noted that the asymmetries predicted by the present calculation are not in good agreement with the measurements of Escudie, et al.[36] at 24 MeV.

Love and Parish[34] found that the $^{12}C(p,p')^{12}C(15.1$ MeV, 1^+, T=1) reaction at 45.5 MeV was sensitive to tensor force contributions, and that a good fit to the angular distributions was obtained with a tensor force of reasonable magnitude. This calculation included the exchange amplitude. The first calculations to do so were performed by Love et al.[33] on the $^{14}N(p,n)^{14}O(g.s.)$ and $^{13}C(p,n)$ $^{13}N(g.s.)$ reactions at 12.2 MeV, but for reasonable values of the force, the predicted cross sections were much too small.

The results discussed above are summarized in Table IV. The parameters are defined in Eqs. 8-11, and the data are compared using the J_4 integral of Eq. 11. The results all lie fairly close to the value J_4=318 MeV-F^5 for the OPEP, and to the estimates[38] obtained from the Hamada-Johnston potential. We may conclude that although the tensor force is not yet well determined from the data, its value is consistent with one's qualitative expectations.

THE SPIN-ORBIT FORCE

We next examine the spin-orbit part (V_{LS}) of the effective interaction. One might qualitatively expect that this part of the force would be important in the study of spin-dependent phenomena such as asymmetries, in transitions to high-spin states, and at high energies. We shall examine these cases in turn and find that there is as yet no firm empirical information on the LS interaction, although there are promising possibilities for further investigation.

Raynal[39] has made a fairly extensive study of the effect of a microscopic LS force on the nature of the predicted asymmetries in inelastic proton scattering. He finds that inclusion of V_{LS} improves the fits, but that one does not come close to a quantitative fit. Consequently it is difficult to obtain a measure of the strength of the force from asymmetry measurements.

The situation is more encouraging in the case of high spin states. Love[40] has found that an LS force is important in the excitation of 6^+ and 8^+ states in $^{90}Zr(p,p')$ at 61.2 MeV. In the case of the 8^+ state a form of V_{LS} obtained by Gogny, et al.[41] in a fit to the free-two-body system, is sufficient by itself, to reproduce the experimental cross section. The difficulty here is not a lack of sensitivity to V_{LS} as it apparently dominates the transition. Rather one needs a better understanding of the importance of collective enhancements in this high multipolarity transition[42] so that the magnitude of the cross sections can be used to empirically determine the spin-orbit force.

We now examine the energy dependence of cross sections dominated by various parts of V_{eff}. Using the code DWBA 70,[20] I have calculated total cross-sections σ_T, for the $^{16}O(p,p')^{16}O(8.87$ MeV, 2^-, T=0)

TABLE IV.--Values of the tensor force.[a]

Determination	$V_{T\tau}$ (MeV)	$\alpha(F^{-1})$	$\beta(F^{-1})$	J_4 (MeV-F^5)
$^{14}C(p,n)^{14}N(g.s.)$ [b]				
Ep=10.4 MeV	5.4	0.707	4.0	444
12.7	5.1	0.707	4.0	420
13.3	5.1	0.707	4.0	420
18.3	3.9	0.707	4.0	321
$^{12}C(p,p')^{12}C(15.1\text{MeV},1^+,1)$ [c]				
Ep=45.5 MeV	2.35	0.707	-	200*
$^{14}N(p,p')^{14}N(2.311\text{MeV},0^+,1)$ [d]				
Ep=24.8 MeV	3.9	0.707	2.0	290
29.8	14.6	1.23	-	421*
1p-shell, two body [e] matrix elements	5.1	0.707	4.0	420
OPEP				318
HJ [f] (r_c=1.0F)				288
(r_c=0.6F)				294

a) Only numbers marked (*) include exchange.
b) Ref. 30. ROPEP form.
c) Ref. 34. OPEP form.
d) 24.8 MeV: ref. 31, ROPEP form. 29.8 MeV: ref. 35, r^2-Y form.
e) Determined by Schmittroth, ref. 37, from Cohen-Kurath 1p-shell
   two-body matrix-elements involving the 1^+, T=0 and 0^+, T=1 states
   only.
f) Ref. 38. For the part of the Hamada-Johnston potential with
   $r \geq r_c$.

reaction when mediated by Yukawan central forces (V_σ) with ranges
of 1.0F and 1.4F, by a r^2-Y tensor force (V_T), and by a Yukawan
spin-orbit force. Figure 6 shows the ratio of the cross section at
Ep to that at 20 MeV for the direct amplitude only and for the
direct-plus-exchange amplitudes. The contribution of V_{LS} increases
with energy while that due to other components decreases, the relative contribution of V_{LS} compared to V_σ(1.0F) changing by a factor
of five between 20 and 60 MeV. This result allows one to hope that
one can disentangle the effects of V_{LS} by studying inelastic scattering as a function of energy.

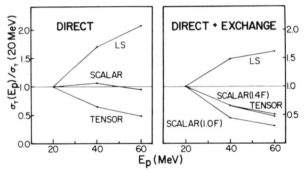

Figure 6.--Energy dependence of cross sections mediated by central, tensor and spin-orbit forces.

The excitation of an unnatural parity state, such as the 2^- state at 8.87 MeV in ^{16}O, would seem a likely place to look for spin-orbit effects, since the competing central interaction, V_σ, is relatively small. Data on this reaction has recently become available[17] at energies between 23.4 MeV and 46.1 MeV, and the analysis of this data in terms of central forces yielded a puzzling anomaly. At low energies relatively good fits to the data were obtained as is shown in Figure 7. As the energy increased, however, the fits became worse, the DWA calculations continuing to predict an L=1 shape while the experimental angular distribution began to resemble that for an L=3 transfer. The character of the discrepancy is shown in Figure 8, where one should compare the data points and the curve labeled "Central". We have used the code DWBA 70[20] to perform calculations with exchange, and an effective interaction including V_σ, a tensor force obtained[38] from the Hamada-Johnston potential[18] as discussed earlier, and a spin-orbit force somewhat larger than that used by Love[40] in his studies of high spin states. The spin-orbit force was expressed as the sum of two Yukawas with ranges of 0.392F and 0.595F, and its strength was adjusted to roughly fit the data. The curve labeled "Central + tensor + LS" shows the results of this calculation displaced upward 20% for display purposes. It is clear that a good fit is obtained for $\theta < 90°$. The asymmetry data of Benenson, et al.[43] at Ep=40.5 MeV are also fairly well described at forward angles by the same forces. It should be pointed out that one can obtain a cross section peak at roughly the correct angle without an LS force if one includes a tensor force with $|V_T| \simeq |V_{T\tau}|$. There seems little justification for such an approach, since the Hamada-Johnston potential gives $V_T \simeq 0$. The tensor force used in the present calculation had very small V_T, and hence contributes only to

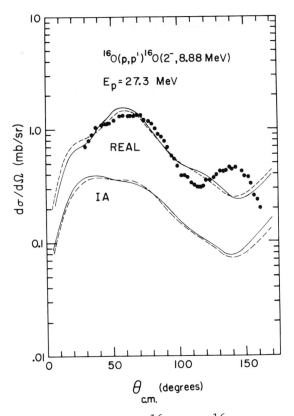

Figure 7.--Cross-sections for the $^{16}O(p,p')^{16}O(8.88$ MeV, 2^-,T=0)
reaction at Ep=27.3 MeV. The upper two curves are for a Yukawa form
of V_σ with a range of 1.0F and two different optical model potentials.
The two lower curves should be ignored.

the exchange amplitude, giving the forward peaked cross-section
shown in Figure 8.

A summary of the forces which have been used in various calculations is shown in Table V and compared with the results implied
by the empirical optical model potential.

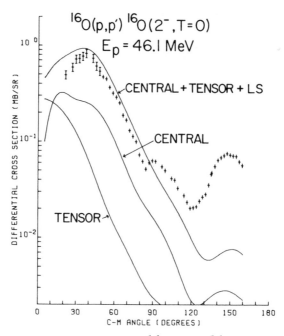

Figure 8.--Cross sections for the $^{16}O(p,p')^{16}O$(8.88 MeV, 2^-, T=0) reaction at 46.1 MeV.

THE IMAGINARY PART OF THE EFFECTIVE INTERACTION

Finally, I will discuss briefly a recent and interesting development. Satchler[44] has suggested that the microscopic effective interaction may contain an important imaginary component and has given a recipe for determining it from the imaginary part of the optical model potential. In the specific case of a transition to the 3^- state at 3.73 MeV in ^{40}Ca, inclusion of this imaginary effective interaction substantially improved the fit to the cross-section. Recently Howell and Hammerstein[45] have examined the effect on cross sections, asymmetries and spin-flip fractions of adding an imaginary form factor to the usual real microscopic form factor. They conclude that an imaginary component is definitely required, but that a simple collective model is more reliable than the Satchler[44] prescription.

TABLE V.--Values of the spin-orbit force.[a]

Determination	J_4(T=0) (MeV-F^5)	J_4(T=1) (MeV-F^5)
Optical Model V_{so}=6.7MeV[b])	−80	−
Raynal-Asymmetries[c])	−61.5	−35.8
Love-^{90}Zr(p,p')[d])	−37.6	−15.2
^{16}O(p,p')^{16}O(8.87,2$^-$,0)[e])	−50.8	−32.2
HJ[f](r$_c$=1.0F)	− 7.3	− 6.5
(r$_c$=0.6F)	−27.7	−13.7

a) The spin-orbit force has the form $[V_{LS}(T=0)+V_{LS}(T=1)\vec{\tau}_1\cdot\vec{\tau}_2]\vec{\ell}\cdot\vec{s}$
b) Ref. 26.
c) These are twice the numbers quoted in ref. 39, but correspond to values actually used (ref. 32).
d) Ref. 40.
e) See the text of this paper.
f) Ref. 38. From the part of the Hamada-Johnston potential (Ref. 18) with r>r$_c$.

SUMMARY

The present situation is summarized in Table VI. It seems likely it will be difficult to determine V_σ accurately, the present results being in the nature of upper limits. High quality (p,n) data are now available from Colorado[46] at 23 MeV and from MSU[47] at 22, 30, and 40 MeV. Careful analysis of these data should lead to reliable values of V_τ. Though the tensor force appears to be important in only a few cases, its effects are distinctive and more detailed fits to existing data should pin down this part of the force reasonably well. Almost nothing is presently known about V_{LS}, but circumstances have been found in which this part of the force is important and the situation looks hopeful.

Since ^{40}Ca is the heaviest stable N=Z nucleus and its wave functions are well known, the recent high resolution ^{40}Ca(p,p') experiments of Gruhn, et al.[48] at Ep=25, 30, 35, and 40 MeV may prove particularly useful in future studies of V_{eff}.

TABLE VI.--Our present knowledge of V_{ST}, V_T, V_{LS}.

Force	Value (for 1.0F range Yukawa)
V_O	27±5 MeV
V_σ	poorly determined
V_τ	poorly determined, hopeful
$V_{\sigma\tau}$	12±2.5 MeV
$V_{T\tau}$	consistent with OPEP
V_{LS}	poorly determined, hopeful

REFERENCES

1. R. Schaeffer, Gull Lake Symposium, September, 1971.
2. G. R. Satchler, Nucl. Phys. A95,1(1967).
3. C. Wong, J. D. Anderson, J. McClure, B. Pohl, V. A. Madsen and F. Schmittroth, Phys. Rev. 160,769(1967).
4. P. J. Locard, S. M. Austin and W. Benenson, Phys. Rev. Lett. 19,1141(1967).
5. S. M. Austin and G. M. Crawley, Phys. Lett. 27B,570(1968).
6. S. M. Austin, P. J. Locard, W. Benenson and G. M. Crawley, Phys. Rev. 176,1227(1968).
7. J. D. Anderson, S. D. Bloom, C. Wong, W. F. Hornyak and V. A. Madsen, Phys. Rev. 177,1416(1969).
8. A. S. Clough, C. J. Batty, B. E. Bonner, C. Tschalär, L. E. Williams and E. Friedman, Nucl. Phys. A137,222(1969).
9. A. S. Clough, C. J. Batty, B. E. Bonner and L. E. Williams, Nucl. Phys. A143,385(1970).
10. F. Petrovich, H. McManus, V. A. Madsen and J. Atkinson, Phys. Rev. Lett. 22,895(1969).
11. R. Schaeffer, Nucl. Phys. A135,231(1969).
12. W. G. Love and G. R. Satchler, Nucl. Phys. A159,1(1970).
13. S. A. Moszkowski and B. L. Scott, Ann. Phys. 11,65(1960).
14. See for example, G. R. Satchler, Nucl. Phys. 77,481(1966).
15. Jay Atkinson and V. A. Madsen, Phys. Rev. C1,1377(1970).
16. F. Petrovich, Ph.D. Thesis, Michigan State University, 1970 (unpublished).
17. S. M. Austin, P. J. Locard, S. N. Bunker, J. M. Cameron, J. R. Richardson, J. W. Verba and W. T. H. van Oers, Phys. Rev. C3,1514(1971).
18. T. Hamada and I. D. Johnston, Nucl. Phys. 34,382(1962).
19. D. Bayer, Ph.D. Thesis, Michigan State University, 1970 (unpublished).
20. R. Schaeffer and J. Raynal, unpublished.
21. D. Slanina and H. McManus, Nucl. Phys. A116,271(1968).
22. K. H. Bray, M. Jain, K. S. Jayaraman, G. Lobianco, G. A. Moss, W. T. H. van Oers, D. O. Wells and F. Petrovich, preprint.

23. A. Kallio and K. Kolltveit, Nucl. Phys. 53,87(1964).
24. A. M. Green, quoted in A. Lande and J. P. Svenne, Phys. Lett. 25B,91(1967).
25. M. L. Whitten, A. Scott and G. R. Satchler, to be published, Nucl. Phys. A.
26. G. W. Greenlees, G. J. Pyle and Y. C. Tang, Phys. Rev. 171, 1115(1968).
27. G. W. Greenlees, W. Makofske and G. J. Pyle, Phys. Rev. C1, 1145(1970).
28. C. J. Batty, B. E. Bonner, E. Friedman, C. Tschalär, L. E. Williams, A. S. Clough and J. B. Hunt, Nucl. Phys. A116, 643(1968).
29. F. D. Becchetti, Jr., and G. W. Greenlees, Phys. Rev. 182, 1190(1969).
30. C. Wong, J. D. Anderson, V. A. Madsen, F. A. Schmittroth and M. J. Stomp, Phys. Rev. C3,1904(1971).
31. G. M. Crawley, S. M. Austin, W. Benenson, V. A. Madsen, F. A. Schmittroth and M. J. Stomp, Phys. Lett. 32B,92(1970).
32. R. Schaeffer, private communication.
33. W. G. Love, L. J. Parish and A. Richter, Phys. Lett. 31B,167 (1970).
34. W. G. Love and L. J. Parish, Nucl. Phys. A157,625(1970).
35. S. H. Fox, S. M. Austin and D. Larson, B.A.P.S. 16,1163(1971).
36. J.-L. Escudie, A. Tarrats and J. Raynal, Polarization Phenomena in Nuclear Reactions (University of Wisconsin Press, Madison, 1970), p. 705.
37. F. A. Schmittroth, Ph.D. Thesis, Oregon State University (1968) unpublished.
38. R. Schaeffer and S. M. Austin, unpublished.
39. J. Raynal, Trieste Lectures, 1971.
40. W. G. Love, Phys. Lett. 35B,371(1971).
41. D. Gogny, P. Pires and R. deTourreil, Phys. Lett. 32B,591(1970).
42. H. McManus, Gull Lake Symposium, September 1971.
43. W. Benenson, P. J. Locard, J.-L. Escudie and J. M. Moss, B.A.P.S. 16,509(1971).
44. G. R. Satchler, Phys. Lett. 35B,279(1971).
45. R. H. Howell and G. R. Hammerstein, preprint.
46. R. F. Bentley, J. D. Carlson, D. A. Lind, R. B. Perkins and C. D. Zafiratos, Phys. Rev. Lett. 27,1081(1971).
47. R. K. Jolly, T. M. Amos, A. Galonsky and R. St. Onge, B.A.P.S. 16,1164(1971).
48. C. R. Gruhn, T. Y. T. Kuo, C. J. Maggiore, H. McManus, F. Petrovich and B. M. Preedom, Preprint (MSUCL-44).

*Research supported in part by the National Science Foundation.

THE ENERGY DEPENDENCE OF THE ISOSPIN PART OF THE OPTICAL POTENTIAL

T. J. Woods

University of Minnesota

The possible energy dependence of the isospin part of the optical potential was suggested by Thurlow[1] in her analysis of the (p,n) angular distributions obtained at a proton bombarding energy of 94 MeV by Langsford, et al.[2] She observed that the depth of the isospin potential at 94 MeV was significantly shallower than the depth obtained from analysis of (p,n) data at lower energies. She concluded that this discrepancy could be accounted for by an energy dependence of the isospin potential. However, it was pointed out later that the shallow isospin potential at 94 MeV could be due to the poor knowledge of the optical parameters at this energy.[3]

Recent analysis of (^{3}He,t) reactions populating isobaric analog states (IAS) indicates that the isospin potential required to fit the data may be energy dependent.[4,5] The purpose of this paper is to show that the isospin part of the optical potential definitely possesses an energy dependence.

At UCLA we recently began a program to measure the (p,n)-IAS angular distributions utilizing the proton (p) decay of the IAS to obtain neutron time-of-flight spectra. With this p timing technique we have measured angular distributions for the ^{208}Pb(p,n) ^{208}Bi(IAS) reaction at proton bombarding energies of 24.8, 30.5, and 38.6 MeV. Fits to these angular distributions as well as the one taken at 50 MeV at the Rutherford PLA[6] were obtained using the DWBA-program DWUCK.[7]

The proton and neutron optical parameters were taken from the analytic expressions derived by Becchetti and Greenlees.[8] The analytic expressions used were the ones labelled "best fit". The isospin form factor was complex, having a real volume and an

307

imaginary surface component, and the isospin strengths was taken
to be four times the coefficients of the (N-Z)/A terms in the
"best fit" expressions.

The fits from the DWUCK calculations were all obtained by
normalizing the predicted cross sections to the data. The normalized
predictions and the experimental data for the four bombarding energies
are shown in fig. 1. The agreement between the normalized predictions and the data is quite good, particularly for the 24.8 and
30.5 MeV data.

The unnormalized predictions were smaller than the experimental data for all four energies. The interesting fact is that
the ratio of the data to the unnormalized predictions tended to
become smaller as the proton bombarding energy increased; but,
due to the large uncertainties for the ratios, particularly for
the 38.6 and 50 MeV data, only qualitative conclusions can be
drawn as regards any possible decrease of the isospin potential
depth with increasing proton bombarding energy. The point to be
made here is that the "best fit" parameters adequately describe the
shapes of the angular distributions for the (p,n)-IAS data, particularly the 24.8 and 30.5 MeV data. If any energy dependence is
introduced into the isospin potential, it must be able to reproduce
the angular distributions as well as the energy independent isospin
potential did in fig. 1.

The observed differential cross section for the production of
$\tilde{p}$ events, $\frac{d\sigma}{d\Omega}(\tilde{p})$, is proportional to the total (p,n) cross section
(σ_τ). The calculation of σ_τ from such a measurement uses the
relation

$$\sigma_\tau = 4\pi \frac{d\sigma}{d\Omega}(\tilde{p}) \frac{\Gamma}{\Gamma_{\tilde{p}}}$$

where the factor 4π is based on the assumption that the $\tilde{p}$ yield
is isotropic and $\Gamma/\Gamma_{\tilde{p}}$ is a correction for the fact that the IAS
might decay through channels other than the observed $\tilde{p}$ channels.
Since this ratio is energy independent, the knowledge of its exact
value is not necessary to obtain the energy dependence of σ_τ. Such
measurements of σ_τ excitation functions using the $\tilde{p}$ decay of the
IAS have been made for various targets and energy ranges.

It should be pointed out that mere possession of the σ_τ
excitation function without any angular distributions allows only
the most qualitative of conclusions as to the energy dependence of
the isospin potential. Without knowing whether the form factor
would also reproduce the (p,n)-IAS reaction angular distribution to
real conclusions can be drawn.

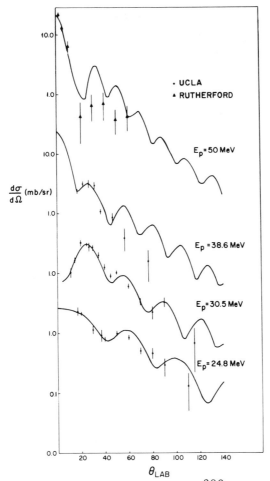

FIGURE 1.--Normalized DWBA fits to the ^{208}Pb$(p,n)^{208}$BI(IAS) data.

The excitation function for the ^{208}Pb$(p,n)^{208}$Bi(IAS) reaction was measured over an energy range from 25.2 to 47.3 MeV.[9] The errors on the data were less than 10%. The same DWBA parameters which were used to obtain the unnormalized fits of fig. 1 were used to calculate the excitation function from threshold to 50 MeV. The data and the calculation (labelled A) are shown in fig. 2. The calculated curve A is normalized to the experimental data at 38.6 MeV.

It is readily apparent that the calculated excitation function A exhibits distinctly non-direct reaction characteristics, in contrast to the data. The data shows a definite decrease in the total (p,n)-IAS cross section over the observed energy range;

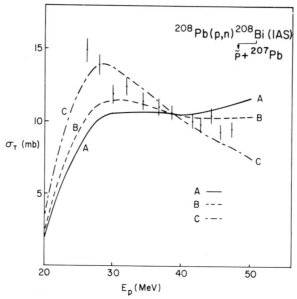

FIGURE 2.--Total cross section excitation function for the ^{208}Pb
(p,n)^{208}Bi(IAS) reaction. The DWBA predictions shown
correspond to form factors which are energy independent
(A) and energy dependent (B and C).

whereas the curve A, using the energy independent isospin form
factor, indicates that σ_τ should still be rising at 50 MeV. It
should also be pointed out that rather good fits to the (p,n)-IAS
angular distributions were obtained using a purely real form factor,
and the calculated excitation function for this form factor showed
the same general features as did calculation A of fig. 2. Furthermore, energy-independent complex isospin form factor predictions
for the (p,n)-IAS reactions on ^{91}Zr, ^{181}Ta, and ^{197}Au, whose
excitation functions have been measured at this laboratory, show
similar disagreement with the data.

To investigate possible limits on the energy dependence of the
isospin potential, two other DWBA excitation functions were calculated in which the isospin potential was assumed to have a linear
energy dependence. These calculations are labelled B and C in
fig. 2. The potentials themselves are listed below.

TABLE 1. $--V_{E \cdot T} = V_R f(X_R) + i4a_I V_I \dfrac{d}{dX_I} f(X_I)$.

	A	B	C
V_R	96	96-0.32E	120-E
V_I	48	48-0.25E	60-0.75E

The resulting excitation functions for these two energy-dependent isospin form factors were also normalized to the measured σ_T at 38.6 MeV (fig. 2). The form factor B, while slowing the increase of the calculated excitation function, still results in an increasing excitation function, albeit slower than A, at 50 MeV. The failure of B to reproduce the data indicates that a 0.3% rate of decrease per MeV of bombarding energy for the isospin potential is not fast enough.

The form factor C exhibits better agreement with the data than does B. For proton bombarding energies less than 28 MeV the calculated excitation function is still increasing, while the data is decreasing from 25 MeV on, thus indicating that a 0.8% decrease per MeV of bombarding energy is too slow. At bombarding energies greater than 43 MeV the calculations decrease faster than the data, indicating that a 0.8% gradient is too large.

One can simply conclude that the energy dependence of the isospin potential is non-linear. There is nothing sacred about linearity, and optical potentials with quadratic energy dependence have been used before. A second conclusion, and at this stage of the game, a more reasonable one, is that the amount of data below 30 MeV is not sufficient to allow very definite conclusions as to whether the energy dependence of C is too large or too small. Lastly, one could conclude that the energy dependence in the region just above threshold could be quite different due to threshold effects, and so would have to be treated separately.

We submit that the energy dependence of the isospin potential is definitely indicated. Good quantitative results for the actual energy dependence as well as the E=0 depth will require measurements of (p,n) angular distributions over a range of energies and targets as well as the excitation functions. At present there is a paucity of (p,n)-IAS data as regards to the number of targets and energy ranges investigated, particularly for A>100 nuclei.

BIBLIOGRAPHY

1. Nola Thurlow, Nucl. Phys. A109, 471 (1968).

2. A. Langsford, P. H. Bowen, G. C. Cox, and M. J. M. Saltmarsh, Nucl. Phys. A113, 433 (1968).

3. G. R. Satchler, Isospin Dependence of Optical Parameters. In: D. H. Wilkinson, ed. Isospin in Nuclear Physics (North Holland, Amsterdam) (1969).

4. W. L. Fadner, R. E. L. Green, S. I. Hayakawa, J. J. Kraushaar, and R. R. Johnson, Nucl. Phys. A163, 203 (1971).

5. F. D. Becchetti, Jr., W. F. Makofske, and G. W. Greenlees, to be published.

6. C. J. Batty, B. E. Bonner, E. Friedman, C. Tschalar, L. E. Williams, A. S. Clough and J. B. Hunt, Nucl. Phys. A116, 643 (1968).

7. P. D. Kunz, University of Colorado, Boulder, private communication.

8. F. D. Becchetti, Jr. and G. W. Greenlees, Phys. Rev. 182, 1190 (1969).

9. G. J. Igo, C. A. Whitten, Jr., Jean-Luc Perrenoud, J. W. Verba, T. J. Woods, J. C. Young, and L. Welch, Phys. Rev. Letters 22, 724 (1969).

STRENGTH OF EFFECTIVE TWO-BODY INTERACTION OBTAINED FROM A STUDY
OF INELASTIC SCATTERING OF 50 MeV PROTONS BY ^{6,7}Li, ^{9}Be and ^{42}Ca

G. S. Mani

Schuster Laboratory, Manchester University

Inelastic scattering to the first two excited states in ^{6,7}Li and to the 2.5 and 6.4 MeV levels in ^{9}Be were measured at 50 MeV. The angular distributions were analyzed using both collective and microscopic form factors. The strength of the effective interaction was obtained with the nuclear wave functions being treated in the LS coupling scheme. In the case of ^{42}Ca the inelastic scattering data to low lying levels were fitted using a microscopic DWBA model. The nuclear wavefunctions were obtained from the calculations of Flowers and Skonvas with a 6p-2h deformed admixture. The strength of the effective interaction obtained agrees well with those obtained from a realistic force if the effective charge correction is made.

THE MICROSCOPIC DESCRIPTION OF INELASTIC SCATTERING

AND CHARGE EXCHANGE WITH COMPLEX PROJECTILES

V. A. Madsen

Department of Physics, Oregon State University

I. MICROSCOPIC FORMALISM

The simplest microscopic picture of inelastic scattering is one in which each nucleon in the projectile interacts with each one in the target in first order with no exchange of particles between them. This picture was formalized several years ago[1,2,3] and has been the theoretical framework within which many analyses of inelastic scattering have been carried out. The nuclear wave functions for the target are obtained from some sort of structure-model calculation, such as a diagonalization in a truncated space, resulting in spherical shell-model description. The distorted-wave approximation gives for the transition amplitude

$$T_{fi} = \left\langle X_f^{(-)}(\vec{r}_0) \; \Phi_{f'}(1'2'...A')\Phi_f(12...A) \middle| V \right.$$

$$\left. \middle| \Phi_{i'}(1'2'...A')\Phi_i(12...A)X_i^{(+)}(\vec{r}_0) \right\rangle \qquad (I-1)$$

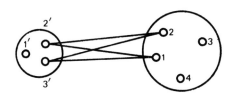

Fig. I-1

where $\vec{r}_0 = \dfrac{\sum \vec{r}_{i'}}{A'} - \dfrac{\sum \vec{r}_i}{A}$ is the coordinate between center of mass of target and projectile, and Φ is an unperturbed internal wave function for target or projectile. For the interaction we assume a sum of local, finite-range, two-body potentials. In the first treatments only a central force was used, but it has been shown[4,5,6,7] that to fit some data on the $(p,n)(p,p')$ and (He^3,t) reactions a tensor term is also required,

$$V = \sum_{ij'} V_{ij'} = \sum_{ij'} [V_{T0} + V_{T1} \tau_i \cdot \tau_{j'}] S_{12}(i,j')$$

$$+ [V_0 + V_\tau \tau_i \cdot \tau_{j'} + \sigma_i \cdot \sigma_{j'}(V_\sigma + V_{\sigma\tau} \tau_i \cdot \tau_{j'})]. \qquad (I-2)$$

As a first step in the evaluation of the amplitude Eq. $(I-1)$ we assume that the projectile internal wave function can be factored[3] into a space-coordinate function times a spin-isospin function,

$$\Phi_{i'}(1'2'\ldots A') = f_{i'}(\xi')\mathcal{S}_{S'T'}, \qquad (I-3)$$

where ξ' represents internal spatial coordinates. We may now carry out the integration of the two-body force over this space coordinate,

$$\overline{V}(r_0, r_1) \equiv \int d\xi' f_{f'}(\xi') f_{i'}(\xi') V_{11'}(\vec{r}_1 - \vec{r}_1'). \qquad (I-4)$$

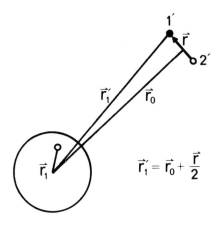

Fig. I-2

Generally speaking, we can expect to be able to write $V_{11'}$ as a sum over spin tensors,

$$V_{11'} = \sum_N S_N \cdot T_N = \sum_N S_N \cdot \frac{1}{L L} [T_L(R_0) T_L(\xi)]_N, \qquad (I-5)$$

where in this form we have also expanded the space coordinate in terms of the center of mass $\vec{r}_0$ and internal coordinates ξ. In the simple case of a deuteron the relationship between $\vec{r}_1{}'$ and $\vec{r}_0$ (Fig. 2) is just $\vec{r}_1{}' = \vec{r}_0 + \vec{r}/2$, and ξ represents the three coordinates of the relative vector $\vec{r}$. If space functions f are pure s-states then the effective interaction $\bar{V}$ will be, term by term, of the same tensor rank as V. If they are different or if there are several terms of the type Eq. (I-3), then the integration can change the tensor rank of the various terms of V. For simplicity we somewhat optimistically restrict ourselves to s-state projectiles in both initial and final state. Then the integration in Eq. (I-4) selects $\bar{L} = 0$, $L = N$, so we have a force of the same type with a different radial shape.

Having folded in the spatial wave functions we may now expand the projectile spin-isospin wave function and the target internal functions in states of a single particle to obtain a result in terms of two-body matrix elements. The resulting cross section for a central-plus-tensor force is[7]

$$\frac{d\sigma}{d\Omega} = \left(\frac{2m}{4\pi\hbar^2}\right)^2 \frac{k_f}{k_i} \frac{1}{(2J_i + 1)(2J' + 1)}$$

$$\times \sum_{II'LM} \frac{|\langle X_f^{(-)} | G_{II'L}(r_0) Y_L^M(\hat{r}_0) | X_i^{(+)} \rangle|^2}{2L + 1} \qquad (I-6)$$

where

$$G_{II'L}(r_0) = \sum_{\tau\rho} C(T'T'\tau; P_i{}' - P_f{}'\rho)(-1)^{T'-P_f{}'} S(J'I'; T'\tau)$$

$$\times \sum_{j_1 j_2 \lambda} (2^{3/2}) \langle j_2 || T_{I(\lambda I')} || j_1 \rangle \, C(T_i T_f \tau; P_i - P_f - \rho) (-1)^{T_i - P_i}$$

$$\times S(J_i J_f I; T_i T_f \tau; j_1 j_2) G_{II'L\lambda\tau}^{j_1 j_2}(r_0), \qquad (I-7)$$

$$G_{II'L\lambda r}^{j_1 j_2} = \int \mathcal{R}_{j_2 l_2}(r_1) F_{II'L\lambda\tau}(r_0, r_1) \mathcal{R}_{j_1 l_1}(r_1) r_1{}^2 dr_1 . \qquad (I-8)$$

$F_{II'L\lambda\tau}$ is the radial multipole of the central-plus tensor force,

$$F_{II'L\lambda\tau} = \delta_{L\lambda} V_{I'\tau} v_\lambda(r_0, r_1)$$

$$+ 5(-1)^I V_{T\tau} \delta_{I'1} W(L\lambda 11; 2I) v_{L\lambda}(r_0, r_1), \qquad (I-9)$$

with v_λ and $v_{L\lambda}$, the radial multipole functions for expansion of the central and tensor forces. The numbers $S(J_i J_f I; T_i T_f \tau; j_1 j_2)$ and $S(J'I'; T'\tau')$ are spectroscopic amplitudes[3] for target and projectile, respectively, for example

$$S(J_i J_f I; T_i T_f \tau; j_1 j_2) = \frac{\left\langle \phi_f || [a_{j_2}^\dagger a_{j_1}]_{I\tau} || \phi_i \right\rangle}{(2I + 1)^{1/2} (2\tau + 1)^{1/2}} \cdot \quad (I-10)$$

These amplitudes contain all the detailed information about the nuclear wave functions of the initial and final states. They are the only quantities which change from transition to transition or from nucleus to nucleus except for the dependence of the radial integrals on nuclear size.

We have heard from Dr. McManus on core-polarization effects in proton-proton scattering. The best known effect is that of enhancement of collective transitions or single-particle transitions through a cooperative phenomenon involving the core. There are many others; a very well-known one is retardation of allowed or unique forbidden β-decay transitions in medium and heavy nuclei. Charge-exchange transitions connect the same states as beta decay. Should transitions retarded in β-decay also be retarded in charge exchange? To answer this question we compare the expression for the nth-order unique forbidden beta decay rate,

$$R = \frac{1}{2J_i + 1} C(T_i T_f 1; P_i, -P_f, P_i -P_f)^2 \left| \sum_{j_1, j_2} S(J_i J_f I; T_i T_f 1; j_1 j_2) \right.$$

$$\times \left\langle j_2 || T_{I(nI')} || j_1 \right\rangle \int_0^\infty R_{j_2}(r) R_{j_1}(r) r^{n+2} dr \Big|^2, \qquad (I-11)$$

with the $\lambda = n$ term of the expression for the charge exchange form factor, Eq. (I-7) with $\tau = 1$. We see that it is identical to the beta decay amplitude except for the radial integral. If it happens that these are roughly proportional to the beta decay radial integrals from one single-particle transition to another, then the $\lambda = n$ term of Eq. (I-7) is expected to be inhibited if the corresponding beta decay is inhibited. Exactly the same considerations apply to quadrupole enhancement or to any other collective effect. The effect of the collective phenomenon on scattering is complicated

by the presence of $\lambda \neq n$ terms in the sum, which are not affected in
any systematic way. In many cases, however, either there is only
the $\lambda=n$ term present or else it dominates a transition. Except in
the long-range-force limit, it is difficult to estimate the effect
of the lack of proportionality of the scattering and beta-decay
radial integrals. In actual calculations the phase is essentially
always right, but there are deviations of the order of 40% in
magnitude. This is not great enough to ruin the correspondence.
In case of an enhancement, where we are dealing with a number of
constructive terms, the proportionality of complete amplitudes is
often very close. For retardations, where different single-
particle terms interfere destructively, the correspondence is only
qualitative. Nevertheless it is always close enough that the use
of wave functions for charge exchange not capable of giving collective retardation in β-decay makes no sense at all.

A recent calculation[8] of $Ni^{58}(He^3,t)$ was made on the basis
of a simple model) in which the two valence nucleons induce core-
polarization components consisting of $(f5/2)(f7/2)^{-1}$ particle-hole
pairs. Using a simple spatially uniform charge-exchange, spin-
exchange force for the valence-core interaction, one can show that
in first order perturbation theory the core-polarization components
always retard the beta decay. Choosing the parameter
$V_{\sigma\tau}/(\varepsilon_{5/2} - \varepsilon_{7/2})$ to fit the beta retardation factor of 20 resulted
in a retardation of about a factor of 5 in the (He^3,t).

It is important to realize that this correspondence we have
shown applies only to ordinary direct scattering, which we do not
necessarily expect to give a very accurate description of reality.
One of the things I wish to do later is to examine the question of
whether selection rules or expected retardations calculated on the
basis of direct reaction theory have any greater generality.

Another related point is that the parity rule, which states
that $(-1)^{\lambda}$ is equal to the parity of the transition[9], applies
only to local interactions. Exchange scattering, for example,
which is a special example of scattering with a non-local operator,
does not satisfy the parity rule.

How well does the theory of direct scattering of composite
particles work? We'll hear more details on analysis of experimental data later this evening from Dr. Roos, but let us compare
forces obtained empirically in nucleon scattering and compute particle scattering.

Table I-1 shows empirical forces obtained from analysis of
some strong transitions produced by complex projectiles. Part A
shows (p,n) forces deduced from (He^3,t) reactions. Generally they
are a little stronger than the corresponding free nucleon

Table I-1. Empirical Interaction Strengths

A. Charge Exchange

Nuclear Trans.	$V_\tau(p,n) = V_{01}$	
$Ti^{48}(He^3,t)0^+$ An.	32	Wesolowski et al[a]
$0^{18}(He^3,t)0^+$ An.	32	Hansen et al[b]
$Ti^{48}(He^3,t)0^+$ An.	$33.8(0^+)$	Saclay[c]
	$59.5(4^+)$	Saclay[c]
$Ti^{48}(p,n)0^+$ An.	20 MeV[a]	
Ave(p,n) Lt. nuclei	18 ± 4	Livermore[d]
KK Effective force	18 MeV	Petrovich and McManus[e]

B. Satchler: Ca^{40}, Pb^{208} Inelastic

Ca	$V_0 \approx 1.1 \times \overline{V}$	(α,α')[f]
Pb	$V_0 \approx 1.2 \times$ "	(α,α')[f]
Ca	$V \approx .7 \times$ "	(h,h')[f]
Pb	$V \approx 1.1 \times$ "	(h,h')[f]

C. Tensor

$$V_T \approx 1.3 \overline{V}$$

Rost-Kunz[g], Schaeffer[h]

a. J. J. Wesolowski, Ervin Schwarcz, R. G. Roos, and C. A.
    Ludemann, Phys. Rev. 169, 878 (1968).
b. Luisa F. Hansen, Marion L. Stelts, Jose G. Vidal, J. J.
    Wesolowski and V. A. Madsen, Phys. Rev. 174, 1155 (1968).
c. P. Kossanyi-Demay, P. Roussel, H. Faraggi, and R. Schaeffer,
    Nucl. Phys. A 148, 181 (1970).
d. J. D. Anderson, S. D. Bloom, Calvin Wong, W. F. Hornyak, and
    V. A. Madsen, Phys. Rev. 177, 1416 (1969).
e. F. Petrovich, H. McManus, V. A. Madsen, and J. Atkinson, Phys.
    Rev. Lett. 22, 895 (1969).
f. J. Y. Park and G. R. Satchler, to be published, G. R. Satchler,
    invited paper presented at the Growth Point Meeting on Nuclear
    Interactions with complex projectiles, University of North
    Carolina, Chapel Hill, April, 1971.
g. E. Rost and P. D. Kunz, Phys. Lett. 30 B, 231 (1969).
h. R. Schaeffer, UCRL 19927 and to be published in Nuclear Physics.

interactions. The strength required to fit a 4^+ transition is
much stronger than the others. This point will be discussed later.
Part B shows effective complex-projectile, target-nucleon interactions compared to that deduced from effective nucleon-nucleon
interactions using the folding procedure, Eq. (I-4). The cross

sections have been roughly corrected to account for experimental B(EL). The agreement is quite good. Finally, part C shows the empirical average tensor strength compared to that deduced from nucleon-nucleon effective tensor forces, the latter being very close to the one-pion force. Again the agreement is fairly good.

What are these empirical forces we are obtaining from DWBA analysis of direct reactions? Although we know that nuclear forces are too strong, we derive reaction formulas as though we were dealing with weak forces, which we may treat in perturbation theory. We have seen from Table I-1 and from Dr. Austin's talk that there is a large amount of consistency among interactions obtained from reactions, and these in turn seem to be consistent with effective interactions used in nuclear structure. I will next discuss this point in a formal way.

II. FORMALISM FOR EFFECTIVE INTERACTION

In this section we use the method of Feshbach[10] and the concepts of effective nuclear interaction as expressed by Macfarlane[11] and by McVoy and Romo[12] to develop the idea of nuclear effective interaction in reaction theory. The aim is to obtain coupled-channel equations which can serve as a starting point for making nuclear reaction calculations.

We begin by defining projection operators P' and Q', such that $P' + Q' = 1$, and P' includes a model space P, which we will deal with explicitly. The space of P' and P are subspaces of complete Hilbert space of the nuclear system in question (See Fig. II-1). The operator Q' will be defined more explicitly later.

Following Feshbach[10] we apply projection operators P' and Q' as defined in Fig. II-1 to the Schroedinger equation

$$0 = P'(E - H)(P' + Q')\psi$$

$$= (E - P'HP')(P'\psi) - (P'HQ')(Q'\psi) \qquad (II-1a)$$

$$0 = Q'(E - H)(P' + Q')\psi$$

$$= (E - Q'HQ')(Q'\psi) - (Q'HP')(P'\psi). \qquad (II-1b)$$

Eqs. (II-1) are coupled-channel equations between the model space P' and the excluded space Q'. We can formally eliminate $Q'\psi$ by solving in terms of $(P'\psi)$ from Eq. (II) to give

$$[E - P'HP' -P'HQ'(E - Q'HQ')^{-1} Q'HP']$$

$$\equiv (E - H')P'\psi = 0 . \qquad (II-2)$$

$$P' = P + Q''$$
$$Q = Q' + Q''$$
$$P + Q = P' + Q' = 1$$

Fig. II-1

H' is an effective Hamiltonian taking into account the effect of the excluded space Q' on the included model space P'. Following Macfarlane[11] and McVoy and Romo[12], we choose Q' to consist of states of very highly excited pairs of nucleons. These intermediate states are important in describing the collisions of two particles involving repulsive cores. If we choose our basis states for the projection operators as eigenstates of H_0 we may write

$$H' = P'\{H_0 + [V + VQ'(E - Q'HQ')^{-1} Q'V]\} P' \qquad (II-3)$$

The operator in the brackets is the effective interaction operator[11,12]. In principle it is a many-body operator because of the multiple-scattering nature of the second and higher order terms in V. However, the main effect of the exclusion of the Q' space is to give us a well-behaved, effective two-body interaction G'. It is calculated by solving the Bethe-Goldstone equation with Q' playing the role of the Pauli operator. Because the Q' states are so high in energy, the G' matrix is expected to be fairly independent of energy and of the details of the nuclear system under consideration. G' is a suitable lowest approximation to the effective interactions used in shell model or reaction calculations. Furthermore, in the Kuo-Brown[13] approximation it is a local interaction.

We now further restrict our model space to P, which in the shell model would include configurations actually to be diagonalized, and, which in nuclear reaction theory, would include channels to be explicitly coupled. Applying P and Q" to Eq. (II-2) and eliminating Q" P'ψ from the resulting coupled equations we have analogously to Eq. (II-2)

$$(E - H'')P\psi = 0. \qquad (II-4)$$

Our new effective Hamiltonian for the model space P,

$$H'' = P[H' + H'Q''(E^+ - Q''H'Q'')^{-1} Q''H'] P \qquad (II-5)$$

is given in terms of the relatively weak, well-behaved reaction matrix G'. Perturbation expansions carried to second order in G' have been used[13] with considerable success in shell-model calculations to include core-excitation effects on the model interaction.

In Eqs. (II-4) and (II-5) we have included a plus superscript on E corresponding to the fact that in open Q" channels the scattered waves in a reaction problem are out-going. The evaluation of the integral implied by the operator $(E^+ - H')^{-1}$ around the pole gives rise to a complex effective Hamiltonian H". As pointed out by Feshbach[10], this is the origin of the imaginary absorptive potential in the optical model, which is obtained formally from Eq. (II-5) by choosing P to include only the ground state of the colliding nucleus and projectile. For complex projectiles we have internal degrees of freedom which can contribute to the sum over intermediate states used for evaluating the effective interaction in Eq. (II-5). This feature, not shared by nucleon projectiles, may be very important in determining the characteristics of the imaginary part of the effective interaction.

For many years, shell-model calculations were done with a simple, local, empirical interaction in H" with considerable success. This is also what we have been doing recently in inelastic-scattering and charge-exchange analysis. We have seen that for simple transitions there is some consistency in the effective interaction required for a range of states, nuclei, and energies. Also, as shown in Table I, there is consistency between forces used in nucleon and complex-projectile reactions. We expect this consistency when it is a fair approximation to use G' only to first order in H", that is when $H" \approx H = H_0 + G'$. Because of the high energy of the important states in Q' it is reasonable to expect that not only is the two-body G' nearly independent of energy and nucleus but also roughly independent of whether one of the nucleons is itself a low-energy projectile or a nucleon in a complex projectile.

III.   COUPLED-CHANNEL EQUATIONS

Starting with Eq. (II-4) we wish to obtain a set of coupled-channel equations in the model space P. We therefore expand in sets of inelastic states of all bound arrangements we regard as important,

$$P\psi = \sum_{n,i}^{N_i} \Phi_n(\xi_i)\, u_n(r_i), \qquad (III-1)$$

where i labels a particular arrangement and n labels the inelastic states within the arrangement. For example r_1 could be center of mass coordinate for a deuteron projectile with respect to the

center of mass of the target and r_2, the coordinate of the proton
with respect to all the other nucleons. Substitution of Eq. (III-1)
into the model-space Schroedinger Eq. (II-4) multiplication by
$\Phi_n(\xi_i)$ and integration over ξ_i yields the coupled-channel
equations[14]

$$[E_n - h''(i) - V_{nn}''] \, u_n(r_i) = \sum_{n'} V_{nn'}'' \, (r_i) \, u_{n'}(r_i)$$

$$+ \sum_{n'i'} \int \Phi_n(\xi_i) \, V_{i'}'' \Phi_{n'}(\xi_{i'}) \, u_{n'}(r_{i'}) \, d\xi_i$$

$$- \sum_{n'i'} \int \Phi_n(\xi_i) \, (E - H_{i'}''(\xi_{i'})) \, \Phi_{n'}(\xi_{i'}) \, u_{n'}(r_{i'}) d\xi_i \qquad (\text{III-2})$$

where E_n is the energy in state n and $h''(i)$ is the unperturbed
center-of-mass Hamiltonian. The last term would vanish if the expansion wave functions were orthogonal; it is a correction term
for the lack of orthogonality among the rearrangement channels.
It may be that in many cases this term is small. In stripping
from a weakly bound deuteron the continuum wave function for the
neutron would be expected to be approximately orthogonal to the
bound-neutron function, so the term would be small. This is a
very rough argument and needs to be examined carefully. Calculations neglecting this term have been made for deuteron elastic
scattering[15] and for the (t,p) reaction[16] using zero-range
forces, which yield a set of coupled equations essentially the
same as for inelastic scattering. The calculations, although ad-
mittedly crude, indicate considerable effect from the coupling of
the rearrangement channel.

   Eqs. (III-2) are a set of integrodifferential equations. If
the effective forces are weak enough, successive approximation can
be used to solve them. An interesting feature of this procedure
is that if distorted waves $u_{n'} = u_0^{(0)} \, \delta_{n'0}$, $u_n = 0$ are used as the
zero-order approximation, the non-orthogonality term vanishes,
because $\Phi_{n'}(\xi') \, u_{n'}^{(0)} \, (r_{i'})$ is in zero-order an eigenstate of H_i''
with energy E, leaving only the interaction term. Solution in
first order using an optical Green's function then yields the
DWBA result for $u_n(r_i)$. Neglect of the non-orthogonality term and
the use of fixed, inelastic, coupled-channel wave functions for
$u_{n'}(r_{i'})$, followed by solution of the resulting inhomogeneous,
inelastic, coupled-channel equations is the source-term method of
Glendenning and collaborators[17]. This method is inconsistent
by order but can be used when the rearrangement coupling strength
is much weaker than the inelastic coupling. However, for example,
it would not give correct results for including exchange effects
in excitation of collective states.

Antisymmetry can be included in Eqs. (III-2) by including ex-change rearrangements and then using an antisymmetric boundary condition; i.e., antisymmetrized plane waves giving the incoming part of the wave function. I will use a simpler alternative pro-cedure which is convenient for calculating anti-symmetrized ampli-tudes to second order.

IV.   INCLUSION OF ANTISYMMETRY

Suppose that we have a reaction initiated by a complex nuclear projectile striking a nucleus, and suppose that we are able to calculate wave functions ψ $(1'2'3'...A'; 1\ 2\ 3...A)$ for processes in which no particles, one particle, two particles, etc., are ex-changed between them, using a theory which treats all the nucleons in each nucleus as identical but distinguishable from those in the other. A correctly antisymmetrized wave function is then obtained by applying an antisymmetrization operator

$$\mathcal{a}\psi(1'2'...A';12...A) = [1 - \sum_{ii'} P_{ii'}$$

$$+ \sum_{\substack{i<j \\ i'<j'}} P_{ii'}\ P_{jj'} - \sum_{\substack{i<j<k \\ i'<j'<k'}} P_{ii'}\ P_{jj'}\ P_{kk'} + ...]$$

$$\times \psi(1'2'...A';12...A) \qquad (IV-1)$$

In words, it consists of the original function minus all one-particle exchanges, plus all two-particle exchanges, etc. The function ψ has the asymptotic forms

$$\psi \xrightarrow[r_0 \to \infty]{} e^{ik \cdot r_0}\ \Phi_i(1'2'...A';12...A)$$

$$+ \sum_n \frac{e^{ik_n r_0}}{r_0}\ \Phi_n(1'2'...A',1...A)F^{(0)}$$

$$\xrightarrow[r_1 \to \infty]{} \sum_n \frac{e^{ik_n r_1}}{r_1}\ \Phi_n(12'...A;1'2...A)F^{(1)}$$

$$\xrightarrow[r_2 \to \infty]{} \sum_n \frac{e^{ik_n r_2}}{r_2}\ \Phi_n(123'...A';1'2'3...A)F^{(2)} \qquad (IV-2)$$

where Φ is a separately antisymmetrized product of target and projectile wave functions, and $F^{(n)}$ is the amplitude for exchange of n particles back and forth, for example,

$$F^{(1)} \propto \Big\langle X_f^{(-)}(\vec{r}_1) \Phi_n(12'...A';1'2...A) \Big| V_f - \hat{U}_f$$

$$\Big| \psi_i^{(+)}(1'2'...A';12...A) \Big\rangle \qquad (\text{IV-3})$$

The contributions to the asymptotic current from the different asymptotic forms Eq. (IV-2) are incoherent for bound Φ. The coherent exchange-scattering effect comes from $\mathcal{A}\psi$; we get contributions to the asymptotic r_0 from rearrangement elements in the exchanged terms of (IV-1). We need consider only one asymptotic coordinate, since they all give the same result. (The fact that there are many different coordinates contributing asymptotically is exactly compensated by the normalization of the incident flux density of $\mathcal{A}\psi$. The complete amplitude is obtained by adding the contributions of each kind together

$$T_{fi} = F^{(0)} - AA'F^{(1)} + \frac{A(A-1)A'(A'-1)}{2!^2} F^{(2)} - \ldots \quad (\text{IV-4})$$

A calculation of the exchange effects based on the first two terms of Eq. (IV-4) has been made by Schaeffer[18] using an expansion about zero range to treat the exchange integrals.

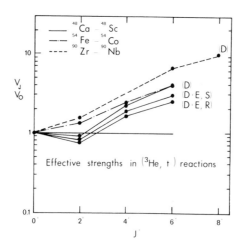

Fig. IV-1. Ratio of the effective strengths V_J and V_0 needed in order to fit the magnitude of the experimental cross-sections. [D] is the calculation neglecting exchange effects and [D+E] the one including them. S stands for a Serber and R for a Rosenfeld mixture for the nucleon-nucleon interaction.

Because the ordinary direct amplitude for orbital-angular-momentum transfer L selects only the L multipole of the two-nucleon potential, and, because these multipoles fall off with increasing L, the calculations using only the direct (first) term Eq. (IV-4) tend to be small for transitions to high J states. This result is in disagreement with data as we have already seen in table (I-1), where a higher interaction strength was required to fit the $0^+ \to 4^+$ transition than the $0^+ \to 0^+$ transition. The exchange amplitude does not share this restriction to the L-multipole with the direct one[19], so it becomes increasingly important for transitions to high J states. Fig. (IV-1) shows the effect of including exchange[18] on the empirical interaction strength determined from fitting (He[3],t) reactions. The inclusion of one-particle exchange improves the consistency slightly, but does not solve the problem.

V. OTHER MECHANISMS

Earlier, in section I, we saw that direct reaction theory with simple local forces has two restrictions on it: it gives the parity rule and it and the various nuclear matrix elements of multipoles of the two-body force are closely tied to collective enhancements or retardations in the corresponding beta or gamma transitions. To what extent are these features retained if we go beyond ordinary direct scattering?

First let me mention antisymmetry effects, which we have just discussed. It was pointed out earlier that the exchange amplitude, being a matrix element of a non-local operator, does not satisfy the parity rule. However, although the proportionality is not so close as for direct amplitudes, one-particle-exchange amplitudes are related in strength to the corresponding beta and gamma transitions. The reason for this is primarily that one is dealing with the same nuclear spectroscopy, namely one-particle-parentage of initial and final states, as in the direct amplitudes.

So far we have been talking primarily about one-step mechanisms. Higher order processes are often important in inelastic scattering and charge exchange. As a first example, consider the excited analogue problem for even nuclear targets.

It was first pointed out by Blair[20] that in the collective model[21] the two two-step processes shown in Fig. V-1 are mathematically of the same order, βV_1 as the one-step process. They must therefore be included for consistency. Early calculations[22] made using the Blair-Austern model for (He[3],t) showed enhancements of factors of about four from inclusion of the two-step processes. Recent coupled-channel calculations[23] show that this result was

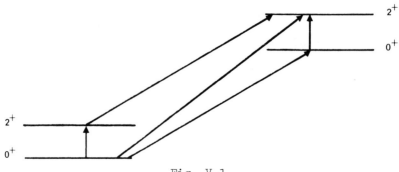

Fig. V-1

wrong and that there is not necessarily any increase. Nevertheless, the two-step processes are not negligible and tend to change
the angular distribution to better agreement with experiment.

Second-order processes involving rearrangement have been considered for (d,d), (He[3],p) and (p,p') reactions[15,16,24]. Since
one-particle transfers in direct reactions are much stronger than
two-step reactions, we should consider intermediate one-particle
stripping or pickup. As an example we consider the (He[3],t) reaction in which both can participate. Charge exchange can take
place by stripping of a proton followed by neutron pickup or the
reverse, Fig. V-2, and the effect is very much like ordinary charge
exchange of a neutron to a proton. In order to calculate the
appropriate amplitude we start with Eq. (IV-4) and recognize that

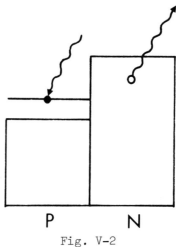

P   N

Fig. V-2

in (He^3,t) the stripping-pickup mechanism is a one-particle exchange amplitude; Eq. (IV-4) gives us the appropriate sign and
normalization. Since we need a second-order amplitude, we drop
the non-orthogonality correction term and solve Eq. (III-2) to
first order in the interaction using the result as an approximation
to ψ_i^+ in Eq. (IV-3). In order to put the interaction in terms of
the stripped state, the intermediate one in this case, another
term like the non-orthogonality correction must be dropped. The
past-prior equivalence does not hold for "in the meantime" due to
lack of energy conservation in the intermediate state. Expanding
the various wave functions in single-particle functions for the
stripped particle, and approximating the interaction by a zero-
range force between stripped particle and intermediate projectile,
we have the amplitude

$$-3A \sum_{n\alpha_1\alpha_2\beta_1\beta_2} \left\langle \Phi_f(1'2\ 3\ldots A)|a_{\alpha_2}|\Phi_n(1'1\ 2\ldots N)\right\rangle$$

$$\times \left\langle \Phi_n(1'1\ 2\ldots N)|a_{\alpha_1}^\dagger|\Phi_i(1\ 2\ldots A)\right\rangle$$

$$\times \left\langle \Phi_t(1\ 2'3')|C_\beta^\dagger|\Phi_d(1'2')\right\rangle \left\langle \Phi_d(1'2')|C_\beta|\Phi_t(1'2'3')\right\rangle$$

$$\times \int X_f^{(-)}(\vec{r})\phi_{\alpha_2}(\vec{r})\mathcal{g}_n^\dagger(\vec{r},\vec{r}')\ \phi_{\alpha_1}(\vec{r}')\ X_i^{(+)}(\vec{r}')\ d^3r'\ dr^3 \qquad \text{(V-3)}$$

where $\mathcal{g}_n^\dagger(r_1 r')$ is the optical Green's function for the center-
of-mass motion of the intermediate deuteron and $\phi_{\alpha_2}(\vec{r})$, $\phi_{\alpha_1}(\vec{r})$
are single-particle bound-state wave functions for the stripped
particle. If we reverse the order of ϕ_{α_2} and ϕ_{α_1} in the integral
we recognize Eq. (V-3) as an ordinary exchange integral for one-
particle scattering with the Green's function playing the role of
a complex two-body finite-range interaction. The process does not
satisfy the parity rule. This fact may be very important for the
(He^3,t) reaction to antianalogue states. The dependence of Eq.
(V-3) on angular momentum transfer is expected to be similar to
one-particle exchange, namely to fall off as a function of increasing L-transfer much less rapidly than the direct. This, of course,
is the sort of thing we need to explain cross sections for (He^3,t)
in high L-transfers. Preliminary calculations along this line
have been made by Bertsch and Schaeffer, and some results will be
reported later this evening by Dr. Bertsch.

The point I wish to make has to do with the nuclear spectroscopy of the pickup-stripping mechanism. Suppose that all of the

intermediate nuclear states for each single-particle state into
which the particle is stripped have about the same energy. We may
then drop the n subscript on the optical Green's function and then
apply closure to the intermediate nuclear states Φ_n. We have

$$\sum_n \left\langle \Phi_f | a_{\alpha_2} | \Phi_n \right\rangle \left\langle \Phi_n | a_{\alpha_1} | \Phi_i \right\rangle = - \left\langle \Phi_f | a_{\alpha_1}^+ a_{\alpha_2} | \Phi_1 \right\rangle$$

which is precisely the nuclear matrix element that enters the
spectroscopic amplitude for direct scattering. Thus one expects
under these conditions the stripping-pickup mechanism to be related in its transition rates to direct charge exchange. If a
similar sum rule were applied for second-order intermediate inelastic processes it would give amplitudes of two creation and two
destruction operators. Of all possible second-order mechanisms
only pickup-stripping may be related to gamma and beta transitions.

VI. SUMMARY

Direct-reaction theory applied to complex projectiles seems
to give empirical forces in rough agreement with nucleon scattering. These forces appear to be given roughly by the bare G matrix
from nuclear structure theory, which is a well behaved, primarily
local interaction. Future calculations will probably have to include second-order transitions proceeding through rearrangement
channels as well as inelastic channels. Of all intermediate
arrangements possible in two-step processes, the one-particle
stripping may be related to nuclear beta and gamma ray transitions
between the same pairs of states.

REFERENCES

1. N. K. Glendenning, Phys. Rev. 144, 829 (1966).

2. G. R. Satchler, Nucl. Phys. 77, 481 (1966).

3. V. A. Madsen, Nucl. Phys. 80, 177 (1966).

4. E. Rost and P. D. Kunz, Phys. Letters 30B, 231 (1969)

5. G. M. Crawley, S. M. Austin, W. Benenson, V. A. Madsen,
    F. A. Schmittroth and M. J. Stomp, Phys. Letters 32B, 92
    (1970).

6. W. G. Love, L. J. Parish and A. Richter, Phys. Letters 31B,
    167 (1970).

7. C. Wong, J. D. Anderson, V. A. Madsen, F. A. Schmittroth
    and M. J. Stomp, Phys. Rev. C 3, 1904 (1971).

8. V. A. Madsen, V. R. Brown, F. Becchetti and G. W. Greenlees,
    Phys. Rev. Lett. 26, 454 (1971).

9. See Eq. (I-7); the reduced matrix element $\left\langle j_2 \left| \left| T_{I(\lambda I')} \right| \right| j_1 \right\rangle$
    vanishes unless the parity rule is satisfied.

10. H. Feshbach, Ann. of Phys. 5, 357 (1958).

11. M. Macfarlane, Proc. of the Enrico Fermi Summer School of
     Physics, Course XL, 1967 (Academic Press, Inc., New York,
     1968).

12. K. W. McVoy and W. J. Romo, Nuclear Physics A126, 161 (1969).

13. T. T. S. Kuo and G. E. Brown, Nucl. Phys. 85, 40 (1966).

14. Norman Austern, Direct Nuclear Reaction Theories, P 253
     (Wiley-Interscience, New York, 1970).

15. G. W. Rawitscher, Phys. Rev. 163, 1223 (1967).

16. A. P. Stamp, Nucl. Phys. 83, 232 (1966).

17. R. J. Ascuitto and N. K. Glendenning, Phys. Rev. 181, 1396
     (1969).

18. R. Schaeffer, Nucl. Phys. A158, 321 (1970).

19. Jay Atkinson and V. A. Madsen, Phys. Rev. C, 1, 1377 (1970).

20. J. S. Blair, Direct Reactions and Nuclear Reaction Mechanisms,
     E. Clementel and C. Villi, Eds. (Gordon and Breach Science
     Publishers, 1962).

21. G. R. Satchler, R. M. Drisko and R. H. Bassel, Phys. Rev.
     136B, 637 (1964).

22. W. E. Frahn, Nucl. Phys. A107, 129 (1968).

23. P. D. Kunz, E. Rost, R. R. Johnson and S. I. Hayakawa, Phys.
     Rev. 185, 1528 (1969).

24. R. Ascuitto, N. Glendenning and B. Sørensen, to be published.

EXPERIMENTAL REVIEW OF THE (^{3}He,^{3}He'), (^{3}He,t) AND (α,α') REACTIONS

P. G. Roos

University of Maryland
Department of Physics and Astronomy

Madsen[1] has discussed the formalism for describing reactions involving complex projectiles, such as ^{3}He and α. I would
like to now discuss primarily the difficulties which have arisen
out of applications of the microscopic theory to experimental
data. Most of the analyses have used the very simple model discussed in the first part of Madsen's talk. Let me first briefly
review this microscopic model.

One assumes that the projectile-target nucleon effective
interaction is made up from a sum of nucleon-nucleon interactions,
where the sum is over the nucleons in the projectile. The nucleon-
nucleon interaction has generally been assumed to be real and central of the type given in equation 1.

$$v_{nn}(\vec{r}_i-\vec{r}_j)=g(|\vec{r}_i-\vec{r}_j|)[V_{00}+V_{10}\vec{\sigma}_i\cdot\vec{\sigma}_j+V_{01}\vec{t}_i\cdot\vec{t}_j+V_{11}\vec{\sigma}_i\cdot\vec{\sigma}_j\vec{t}_i\cdot\vec{t}_j] \qquad (1)$$

By summing and folding this nucleon-nucleon interaction into the
projectile internal wave function (usually assumed to be a 0S
oscillator wave function), one obtains an effective projectile-
target nucleon interaction. This procedure is indicated in eq.(2),

$$v_{0i}(\vec{R}-\vec{r}_i)= \sum_j <\phi_f|v_{ij}|\phi_i> \qquad (2)$$

where ϕ_i and ϕ_f are the initial and final internal wave functions
of the projectile, $\vec{R}$ is the center of mass coordinate of the projectile, and $\vec{r}_i$ is the coordinate of the target nucleon. This
interaction is then placed in the standard DWBA transition matrix
element describing the reaction. An example of such a procedure (2)
for generating a projectile nucleon interaction using a Gaussian
nucleon-nucleon interaction, 0S oscillator projectile wave functions,

and a Serber exchange mixture is given in eq.(3).

$$v_{He^3-i} \simeq [22.5-2.5(\vec{\sigma}_0 \cdot \vec{\sigma}_i + \vec{t}_0 \cdot \vec{t}_i + \vec{\sigma}_0 \cdot \vec{\sigma}_i \vec{t}_0 \cdot \vec{t}_i)]e^{-.2|\vec{R}-\vec{r}_i|^2}$$

$$v_{\alpha-i} \simeq 36e^{-.25|\vec{R}-\vec{r}_i|^2} \tag{3}$$

More often analyses have been performed using a Yukawa shaped
nucleon-nucleon interaction with a range of approximately 1 fm.
It is then a reasonably good approximation to assume the complex
projectile-target nucleon interaction is a Yukawa interaction with
the same range but a modified strength[3]. In addition to the
interaction I have mentioned, one has found a need for including
a tensor force in the effective interaction. I will discuss this
need later.

From the above procedure we see that we end up a very simple
model for treating complex projectile reactions. The hope, of
course, is that this procedure will allow us to extract nuclear
structure information from these reactions. We expect that
some difficulties will arise, since analyses of (p,p') have
shown the importance of effects such as core polarization and
exchange. Also we know that a good deal has been swept under
the rug, such as off-energy-shell effects, neglect of the Pauli
exclusion principle, and the fact the general reaction theories
can lead to a complex interaction[4]. However, the procedure
has been to try the simple model and hope for the best. I plan
to indicate where this simple model breaks down, and possible
explanations for its failure.

I first want to discuss only (He^3,t) reactions, since most
of the work on microscopic models has been directed toward this
reaction. Let me first give some indication of why one uses
the (He^3,t) reaction, instead of the more fundamental (p,n) reaction. Of course, the reason for studying the charge exchange
reaction is to obtain spectroscopic information about levels in
odd-odd nuclei, which are made by converting a target neutron
into a proton. For example, a study of the charge exchange
reaction on Ca^{48} immediately identifies the $\pi f7/2$ particle-$\nu f7/2^{-1}$
hole states of Sc^{48}. This feature of the charge exchange reaction
has been beautifully exploited at Argonne using the (He^3,t)
reaction[5]. It is this fact, that the (He^3,t) reaction strongly
populates these particular states, that leads one to hope that a
microscopic model may be applicable.

One of the primary reasons for obtaining a working microscopic model for the (He^3,t) reaction, instead of the (p,n)
reaction, is experimental. A comparison of (p,n)[6] and
(He^3,t)[7] spectra is given in Fig. 1.

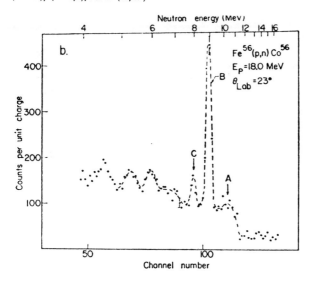

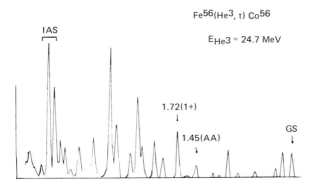

Fig. 1   Comparison of (p,n) (ref. 6) and (He^3,t)(ref. 7) spectra for Fe^{56}. In the (p,n) spectrum A, B, and C indicate the positions of the T$_{lower}$ states, analogue state, and excited analogue state respectively. In the (He^3,t) spectrum the analogue state is indicated by IAS.

One immediately sees that much better energy resolution can be obtained for the (He^3,t) reaction since the final product is charged. Although better energy resolution neutron spectra are available, one typically obtains $\sim$100 kev (FWHM) compared to some results of $\sim$10 kev with the (He^3,t) reaction. The second feature of fig. 1 is the fact that the T$_{lower}$ states are more strongly excited relative to the analogue state, compared to the (p,n) reaction. It seems unlikely (at least in the near future) that one will be able to obtain sufficiently good energy resolution

with the (p,n) reaction to enable one to make detailed studies
of odd-odd nuclei in the medium (or greater) mass region where
the level density is quite large. Thus most of the experimental
data on charge exchange reactions—except for a few isolated levels
—will be obtained using the (He³,t) reaction. Therefore it would
be very desirable to have an accurate microscopic analysis to
extract nuclear structure information from this data.

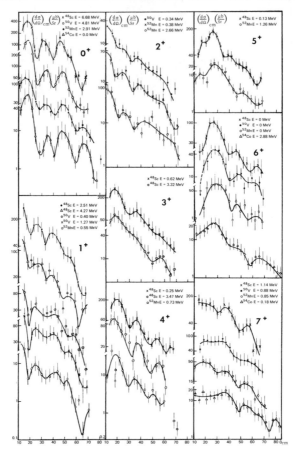

Fig. 2 Taken from ref. 8. Experimental angular distributions for
N = 28 nuclei. The curves are smooth curves drawn through the data.

        Another nice aspect of the (He³,t) reaction is indicated in
Fig. 2, where angular distributions for different J^π states are
shown[8]. Two features stand out in this figure. First, the strong
absorption of the He³ and t gives rise to nice diffraction structure
characteristic of the spin J. The (p,n) angular distributions,
on the other hand, are much smoother and make the identification
of the spin more difficult. Second, you can see that for a given
J^π all of the distributions are essentially the same. Some excep-

tions to this "universal curve" idea have been found, but it is
still very useful in identifying spins and parities. These two
features of the (He^3,t) reaction, allow one to identify the spin
and parity of an unknown state more easily than the (p,n) reaction.

Let me now turn to some applications of the microscopic model
to the (He^3,t) reaction. There is a great deal of experimental
data available and angular distributions are available for many
different orbital angular momentum transfers. As I have mentioned
the (He^3,t) reaction does strongly populate the states expected on
the basis of a microscopic model. Also the dependence on the
neutron excess of the target is essentially in agreement with the
model[8]. Thus, the initial indications are that a microscopic
model might apply to this reaction. However, in carrying out the
analyses a number of problems have been encountered. I would like
to discuss three of these in detail.

First I want to discuss the problem which arises in the
analysis of (He^3,t) transitions to natural parity states. The cases
I shall discuss are ones for which only the term V_{01}(eq.3) enters;
ie, there are no spin flip contributions. Fig. 3 shows some
examples of microscopic fits to natural parity states[9].

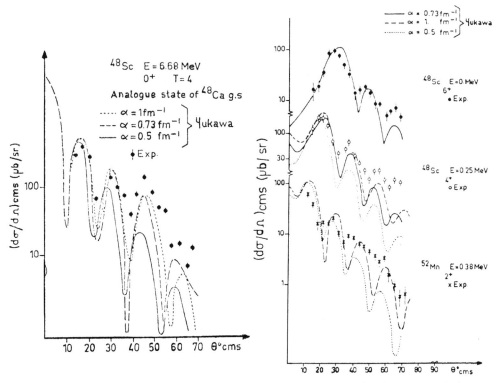

Fig. 3 Taken from ref. 9. Typical microscopic calculations with
a Yukawa interaction ($e^{-\alpha r}/\alpha r$) for different ranges.

Generally one gets good agreement with the experimental angular
distributions, although discrepancies of up to 5° have been observed around the mass 90 region[10]. The shapes of the angular
distributions are not especially sensitive to the range of the
effective interaction, and most analyses have been performed with
a 1 fm range Yukawa interaction. (Recent 30 MeV $Ca^{40}(He^3,t)$ data[11]
does indicate some preference for a 1.4 fm range).

   In order to test the theory, microscopic DWBA calculations
are carried out for each state of a nucleus for which the wave
functions are reasonably well understood. For example, the
$Ca^{48}(He^3,t)Sc^{48}$ reaction to the $(\pi f7/2)(\nu f7/2^{-1})_{0+\leftrightarrow 7+}$ states is
a pretty good calibration nucleus. One then normalizes these
calculations to the data to extract a strength for the He^3-nucleon
interaction. Normalization to the L=0 analogue transition leads
to a strength which is reasonably consistant with the (p,n)
strength-typically 25% to 50% larger. However, the analysis
fails when one considers higher spin states. It is necessary to
increase the He^3-nucleon interaction strength as J is increased.

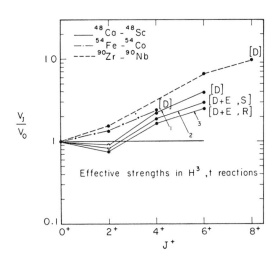

Fig. 4   Taken from ref. 13.   Ratio of effective strengths V_J/V_0
needed to fit the magnitude of the experimental cross sections
using a 1 fm range Yukawa interaction.   [D] is the microscopic
calculation without exchange, and [D+E] the one with exchange.
S and R indicate Serber and Rosenfeld mixtures in the nucleon-
nucleon interaction.

Fig. 4 shows this increase from an analysis by Schaeffer of
several nuclei[12,13] using a 1 fm range Yukawa interaction. The
strength of the 6+ state is roughly 3 1/2 times that for the 0+
transition. This effect is range dependent. For example, if

one increases the range to 1.4 fm the ratio V_{6+}/V_{0+} becomes a factor of ∿10. This is shown in Fig. 7. However, with any reasonable range one cannot avoid this multipole dependence of the strength. There is always a question of configuration mixing, but these nuclei should be reasonably well represented by simple wave functions. In addition it is very difficult to argue with the end points. Analogue transitions are quite insensitive to configuration mixing, and high spin states should have reasonably pure configurations. Thus, one must conclude that the use of the simple microscopic model will lead to predicted cross sections which are factor of 10 to 100 too small for high spin states.

This effect is reminicent of early problems which arose in inelastic proton scattering analyses. In these one also found it necessary to increase the strength of the effective interaction a J became larger. This difficulty in inelastic scattering now seems to be well explained by the inclusion of exchange[14]. Schaeffer[12] has calculated the exchange contribution to the (He^3,t) reaction using a rather simple δ-function interaction. The results of this analysis are shown in Fig. 4 labeled as [D+E]. We see that the exchange contribution is in the right direction and reasonably large, but still does not explain the observed discrepancies--this is particularly true for a 1.4 fm range interaction.

I would like to mention that even with this discrepancy one appears to be able to make use of the microscopic model by parameterizing the strength as a function of J, using Fig. 4. Schaeffer[15] has recently analyzed the Ca^{42}(He^3,t) data of Sherr, et al.[16] using this parameterization, and obtains quite good agreement between experiment and theory when using fairly complete wave functions. This suggests that one might still be able to use the microscopic model predictions to extract structure information by making the strength L dependent.

Before discussing this problem any further I would like to consider a second difficulty which now appears to be related. This problem is the shape of (He^3,t) angular distributions to 0+,T$_{lower}$ states--the so-called anti-analogue states. Fairly recently angular distributions to such states have been measured by Hinrichs, et al.[17] and Goodman and Roos[7,18]. The anti-analogue transition can be treated in a microscopic model as shown by French and MacFarlane[19]. This treatment leads to a DWBA radial form factor which is the difference of two terms. When calculations are carried out with a reasonable He^3-nucleon range, one generally obtains a calculated angular distribution which looks very similar to an L=0 distribution with a magnitude of about 1% of the analogue transition. Fig. 5 shows the experimental angular distributions for these states as well as those for the analogue state. The curves are DWBA calculations for L=0 and L=1.

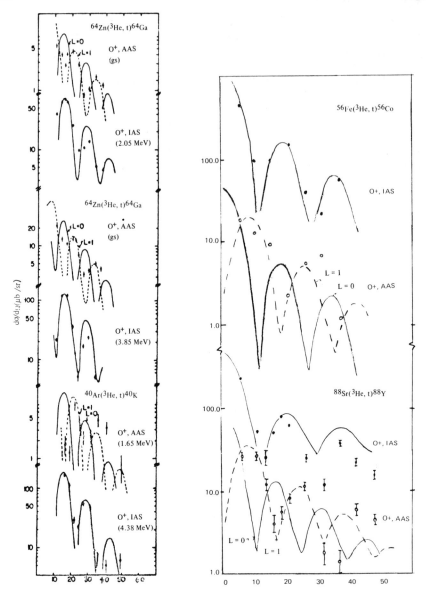

Fig. 5 (He³,t) angular distributions for analogue (IAS) and anti-analogue states (AAS) taken from ref. (7, 17, 18). The curves are DWBA calculations for L=0 (solid curve) and L=1 (dashed curve).

We see that the anti-analogue state experimental distributions are best described by the L=1 curve—or at least that they are out of phase with the L=0 calculation. Actual calculations have been unable to explain this result[20].

There now appears to be an explanation for the two problems I have discussed. Schaeffer and Bertsch[21] have been investigating 2-step contributions to the (He^3,t) reaction. In particular they have considered contributions from (He^3,α) followed by (α,t). The reason for choosing this particular process is primarily that it preserves the same nuclear structure aspects (at least approximately) that are present in the direct microscopic treatment of the (He^3,t) reaction. This has been shown very nicely in Madsen's talk[1]. Also these single nucleon transfer reactions have quite large cross sections. When considering a 2-step process one no longer has the same selection rules that are present for the direct term. The process considered by Schaeffer and Bertsch allows one to have 0+ $\rightarrow$ 0+ transitions with a $\Delta L=0$ or 1 for the projectile. They have performed preliminary calculation (using plane waves with cut-off) and find that L=0 dominates for the analogue transition and L=1 dominates the anti-analogue transition--just as required by the experimental data.

Schaeffer and Bertsch have also applied the same 2-step considerations to the other natural parity states, and find the 2-step process becomes increasingly important, relative to the 1-step charge exchange, as the angular momentum transfer increases. In fact for 6+ states in the f$_{7/2}$ region, they find the 2-step process almost completely dominates the reaction. Thus, it seems probable that inclusion of the 2-step contributions will explain the two discrepancies in the (He^3,t) reaction that I have mentioned. However, one must wait until more exact DWBA calculations are performed. Of course this makes it questionable whether one can extract nuclear structure information using a microscopic model for the (He^3,t) reaction. The hope is that one can include this particular 2-step contribution by means of some effective interaction.

A third problem encountered in analyses of the (He^3,t) reactions was for transitions to unnatural parity states. It was found that angular distributions for unnatural parity states, which have contributions from two possible L-transfers, were generally dominated by the higher L-transfer; eg, 3+ states have very large contributions from L=4. The use of a simple central force, such as that in equation 3, predicts that the lower L should dominate, usually by a large factor. The solution to this problem was found by Rost and Kunz who introduced a tensor force into the He^3-nucleon interaction. The results of their calculations are shown in Fig. 6. We see that when the tensor force is included one obtains large contributions from the higher of the two possible L-transfers. Since these calculations, there have been several refinements of the tensor force and the microscopic theory gives quite reasonable results[13,15]. The strengths extracted for various unnatural parity states in the f$_{7/2}$ region are shown in Fig. 7. We see that the strength is fairly constant as a function of J, unlike

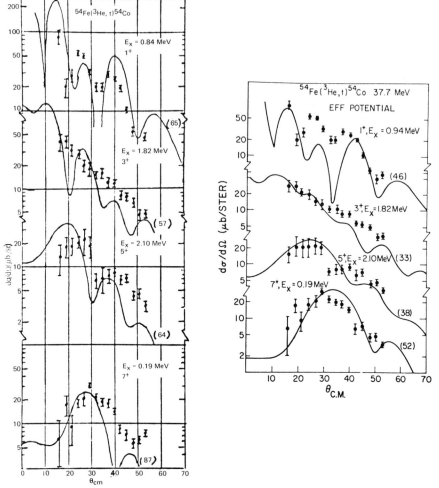

Fig. 6 Taken from ref. 22. The DWBA calculations on the left are done with only a central force. The calculations on the right include a tensor force.

the natural parity states. Thus, there seems to be no difficulties in the theory for these unnatural parity states. Unfortunately, if the 2-step contributions are definitely important, they must also be included in the unnatural parity transitions. This will have the effect of reducing the strength needed for the high angular momentum transfers. Thus again a discrepancy will exist, and one must reinvestigate the role of the tensor force.

The net outcome of the present discussion is that the microscopic effective interaction of eq.(3), even with the tensor force, will not predict the (He^3, t) cross sections. To see if the model can be improved, we will have to wait until better

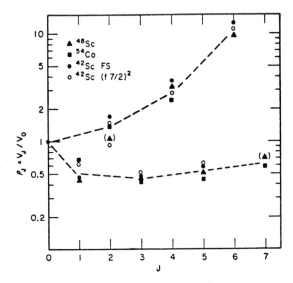

Fig. 7   Taken from ref. 15.   Ratio of effective strengths V_J/V_0
needed to fit the magnitude of the experimental cross sections
using a 1.4 fm range Yukawa interaction.   The unnatural parity
states include the tensor interaction.

calculations for the 2-step process are done.   Before going onto
inelastic scattering I would like to briefly mention several other
considerations which must be applied to the charge exchange reaction.
First, Madsen, et al.[23] have shown the core polarization can be
quite important in the (He^3,t) reaction.   Using the known β-decay
transition rate for the 1+ ground state of Cu^{58}, they have shown
that the Ni^{58}(He^3,t) reaction will be inhibited by about a factor
of about 5, relative to simple two-neutron shell model wave functions.   This work shows the importance of core polarization, and
one must always keep this in mind when trying to obtain information about the effective interaction.   Of course, if an accurate
microscopic model can be found, one can then measure these inhibi-
tions, and obtain information about the nuclear wave functions.
A second consideration is the energy dependence of the effective
interaction.   Fadner, et al.[24] in a study of the (He^3,t) analogue
transitions at several energies find that the strength of the
interaction must be reduced by roughly a factor of 2 from 20 MeV
to 40 MeV.   The reasons for this decrease in strength are not now
understood, and could arise from many different effects; eg, it
could be due to exchange effects, or different contributions
from 2-step processes.   This result does indicate that more information about the (He^3,t) reaction mechanism might be obtained
from a study of the energy dependence.

Now I would like to discuss inelastic scattering for which
relatively few microscopic analyses exist--at least compared to

(He^3,t). First let me briefly mention analyses of elastic
scattering which use this same model. There has recently been
a fair number of analyses of elastic α-scattering[25,26]. The
idea is to fold an α-nucleon interaction, like that of eq.(3),
into the matter distribution to obtain the real part of the
optical potential. One then generally takes the same shape
(or its derivative) for the imaginary potential, and searches
on its strength to fit the data. This model is usually justified
by the fact that the contributions to elastic scattering come
from the tail where the density is small, hopefully reducing many
of the problems with the model. I don't want to discuss the
justification of the model, I only want to say that it works
extremely well. For example, Bernstein[26] and Seidler have taken
the matter distributions of three T=0 nuclei (O^{16},Si^{28},Ca^{40})
from electron scattering (assuming protons and neutrons have
the same distribution), and fit the α-elastic scattering data
quite well using the same α-nucleon strength and range for each.
The strength is within approximately 20% of what you expect on the
basis of the simple microscopic model (eq.3). The hope is that
we can now apply this model to T≠0 nuclei, and by comparing the
results with electron scattering obtain information about the
neutron distribution in the region of low nuclear density. This
is the subject of a contributed paper to this conference by
Bernstein.

There have been some attempts to apply the microscopic
model to inelastic scattering over the past few years. Unfortunately, in almost all cases core polarization was the dominant
effect (often by factors of 10), and almost no information was
obtained about the effective interaction. Therefore, I will confine my discussion primarily to the work of Satchler[2] and
Park and Satchler[27] who have investigated the Ca^{40} and Pb^{208}
3-states for which RPA wave functions are available. Unfortunately,
even these wave functions are not sufficient and miss electromagnetic transition rates by something like 20% for Ca^{40} and about
a factor of 2.5 for Pb^{208}. For Pb^{208} it is assumed that one can
use an isoscalar enhancement for this effect and multiply the
theoretical calculations by approximately a factor of 2. The
effective interaction they use is that given in eq.(3).

The results of their calculations are shown in Figs. 8→11
(Also included is the 5-state for Ca^{40}, but the wave functions are
more uncertain in this case so that I will exclude it from my
discussion). I will assume that no enhancement is necessary for
the Ca^{40}(3-) and that the Pb^{208} cross section must be enhanced of
∿2. There are several interesting results shown in these figures.
First, the cross section for He^3 inelastic scattering is nearly
correct without including exchange (assuming the isoscalar enhancement of Pb^{208}). When exchange is included the theoretical cross

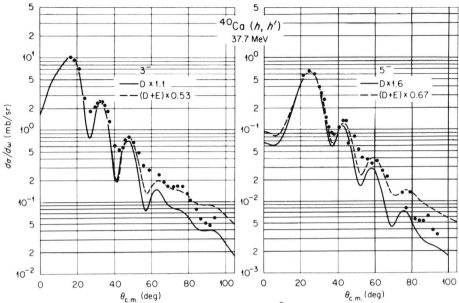

Fig. 8 Microscopic calculations of He^3 inelastic scattering from Ca^{40} (from ref. 27) using the effective interaction given in eqn.3. The labels D and (D+E) indicate calculations without and with exchange included. The theoretical cross sections have been normalized by the numbers appearing in the figure.

section then becomes a factor of 2 too large. Second, the α inelastic scattering is nearly correct when exchange is included. There is a question about the accuracy of the exchange contribution, but one is still left with a discrepancy of about a factor of two between He^3 and α inelastic scattering using the same model. I should point out that including exchange for the Ca^{40}(3-) does seem to improve the fall-off of the theoretical cross section. When one considers only the absolute magnitude a number of questions arise. For example, one can change the theoretical cross section a great deal in Pb^{208} by changing the neutron single particle well. In addition, it would be useful to have better wave functions for these states, so that we would not have to use an isoscalar enhancement factor. Also, it might be best to use a fully microscopic treatment; ie, take the optical potential and the projectile-nucleon strength from a microscopic analysis of the elastic scattering. One will have to investigate these effects. However, there still seems to be a fundamental difference between He^3 and α inelastic scattering (roughly a factor of 2) when treated in the same model.

Satchler[2,27] has also used the microscopic model to investigate the phase in the effective interaction. This can be very neatly done by looking at the interference between coulomb excita-

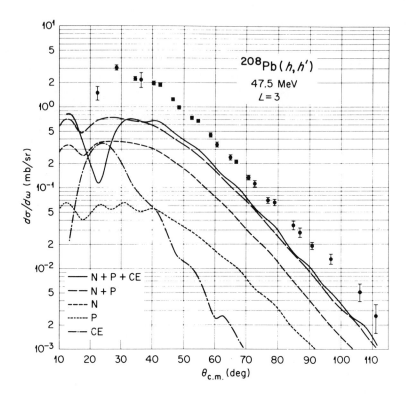

Fig. 9 Microscopic calculations of He^3 inelastic scattering
from Pb^{208} (from ref. 27) using the effective interaction of
eqn.3.   The various curves are the cross sections for the individual components (N=neutron, P=proton, CE=Coulomb excitation).
The inclusion of exchange will increase the theoretical cross section by approximately a factor of 2.

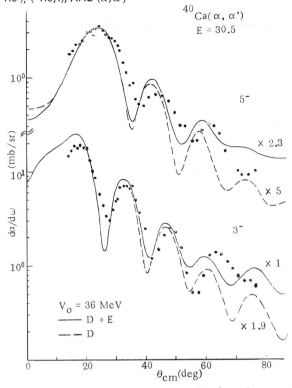

Fig. 10 Microscopic calculations of α inelastic scattering from Ca^{40} (from ref. 2) using the effective interaction of eqn.3. The labels are as in Fig. 8.

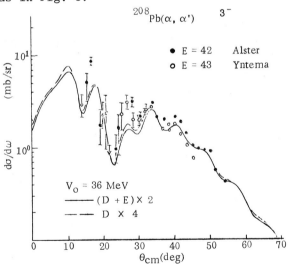

Fig. 11 Microscopic calculations of α inelastic scattering from Pb^{208} (from ref. 2) using the effective interaction of eqn.3. The labels are as in Fig. 8.

tion (repulsive interaction) and the nuclear excitation (attractive).
For Pb^{208} these two interactions are comparable and one sees large
interference effects. Fig. 12 shows results for (He^3,He^3')(27)
where the microscopic form factor has been multiplied by e$^{i\alpha}$.

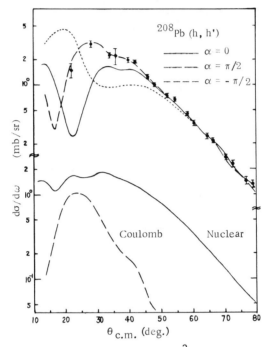

Fig. 12 Microscopic calculations of He^3 inelastic scattering
(from refs. 2,27) using the effective interaction of eqn.3 plus
Coulomb excitation. The effective interaction has been multiplied
by exp(iα).

We can see that there is definite evidence (although more data
would be helpful) that the phase is closest to $\pi/2$, which implies
the microscopic form factor is imaginary and attractive. Satchler
finds similar results for (t,t') on Pb^{208}. Fig. 13 shows the
same calculations for (α,α')(2). In this case one finds the
best results for α=0, or that the interaction is real. Thus
again a difference between He^3 and α inelastic scattering exists,
with the α inelastic scattering microscopic calculations giving
rather good agreement with the data. The reason for the imaginary
interaction is not understood. One can speculate that it might
arise from virtual breakup of the He^3 since it is very loosely
bound when compared to the α-particle. This might explain the
difference between the two.

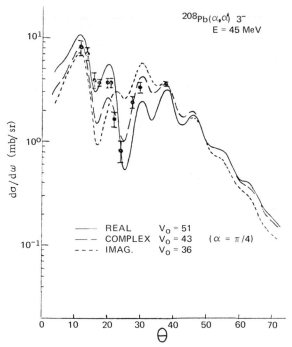

Fig. 13 Microscopic calculations of α inelastic scattering (from ref. 2) using the effective interaction of eqn. 3 plus Coulomb excitation. The effective interaction has been multiplied by $\exp(i\alpha)$. ($\alpha=0$, Real; $\alpha=\frac{\pi}{2}$, Imag.)

    I would also like to show the effect of core polarization in inelastic scattering for a case where the shell model space is very limited. Fig. 14 shows a calculation for $Zr^{90}(t,t')$[27]. As you can see the core polarization completely dominates the cross section, and essentially no information is obtained about the effective interaction.

    This is essentially the stage at which the analysis of inelastic scattering is at the present. There is obviously much to be done. There are refinements obtained from the application of the microscopic model to (He³,t) which must be included or at least be investigated to estimate their importance; for example, inclusion of the tensor force and the 2-step process such as (He³,α), (α,He³) in the He³ inelastic scattering. In considering this last type of process, I found one piece of data which may show this effect in inelastic scattering. Fig. 15 shows the (t,t')[28] results at 20 MeV for the excited 0+ (1.75 MeV) in Zr^{90}. This is again a case in which the direct term arises from the cancellation of two form factors, as in the case of the antianalogue states. The calculation shown with the data is for the microscopic model. Just below this data is shown an L=1 calculation. We can see that if anything the phasing of the data is in

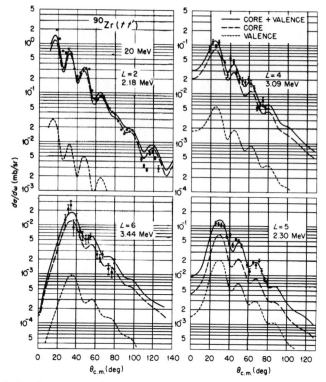

Fig. 14 Calculations of t inelastic scattering including valence
contributions (microscopic model) and core polarization.

better agreement with L=1. Although more data is needed, there
is certainly an indication the 2-step processes might be contributing. In addition, one must understand the reason that the He^3
interaction is complex, and what effects this has on generating
an effective He^3-nucleon interaction.

    Let me emphasize the fact that all of the difficulties with
the effective interaction that I have mentioned are primarily
associated with He^3 reactions. For inelastic α scattering one
has the simplification that the projectile is S=0, T=0, and that
it is very tightly bound. Thus many of the considerations applied
to the He^3 projectile (or t) are no longer necessary (eg, the
tensor force). In fact, the comparisons between experiment and
microscopic theory that I have presented indicate that the α-
nucleon effective interaction of eqn.(3) is quite adequate (providing exchange is included). Of course for other nuclei one
must include core polarization effects which can be dominant.
However, at present inelastic α-scattering appears to be quite
well described by a microscopic model (+ core polarization).

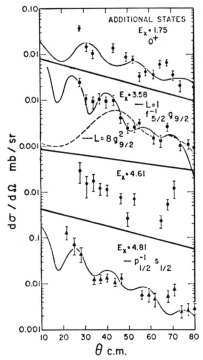

Fig. 15 Experimental angular distributions for $Zr^{90}(t,t')$ (from ref. 28). The curves are DWBA calculations using a microscopic model. Note that the 0+ transition is similar to the L=1 calculation.

Let me conclude this section on inelastic scattering with a brief comment on the topic of why does one bother to work on a microscopic model for He^3's and α's with all the difficulties which are encountered. One can always do (p,p') for which the microscopic analysis now seems to be in rather good shape. In this case one has no experimental justification for using complex particles, except possibly that the strong diffraction patterns make spin assignments easier. This question is most easily answered by assuming that one has a working microscopic model. Then we could compare (e,e'), (p,p'), (He^3,He^3') and (α, α'). This would allow us to obtain information about the transition density of both protons and neutrons, and also allow us to sample it in different regions of the nucleus. Thus the strongly absorbed particles, such as He^3 and α, would measure the transition density far out in the tail of the nucleus.

In conclusion I would like to make several obvious remarks. First, that the very simple microscopic model that I discussed in the first part of my talk is not adequate and will not allow the extraction of structure information. Second, there does seem to be a hope of obtaining an effective interaction (although

possibly quite complicated) which will describe complex projectile
reactions. Last, and most obvious, there is a great deal of theoretical work to be done to explain the discrepancies found in the
present analyses.

References

1)   V.A. Madsen, this conference: Nucl. Phys. 80, 177 (1966)

2)   G.R. Satchler, Invited paper, Growth Point Meeting on
     Nuclear Interactions for Composite Particles; Univ. of
     North Carolina; G.R. Satchler, to be published in Particles
     and Nuclei

3)   J.J. Wesolowski, et al., Phys. Rev. 169, 878 (1968)

4)   G.R. Satchler, Phys. Lett. 35B, 279 (1971)

5)   e.g. J.R. Comfort, et al., Phys. Rev. Lett. 25, 383 (1970);
     R.C. Bearse, et al., Phys. Rev. Lett. 23, 864 (1969)

6)   J.D. Anderson, C. Wong, and J.W. McClure, Phys. Rev. 129,
     2718 (1967)

7)   P.G. Roos and C.D. Goodman in Nuclear Isospin, edited by
     J.D. Anderson, S.D. Bloom, J. Cerney, and W.W. True (Academic,
     New York, 1969) p. 297

8)   G. Bruge, et al., Nucl. Phys. A129, 417 (1969)

9)   P. Kossanyi-Demay, et al., Nucl. Phys. A148, 181 (1970)

10) J.R. Comfort and J.P. Schiffer, to be published

11) J.M. Loiseaux, et al., to be published

12) Richard Schaeffer, Nucl. Phys. A158, 321 (1970)

13) Richard Schaeffer, Proc of the Argonne (He^3,t) Symposium (1970)

14) e.g. J. Atkinson and V.A. Madsen, Phys. Rev. Lett. 21, 295
     (1968); W.G. Love, et al., Phys. Lett. 29B, 478 (1969); R.
     Schaeffer, Nucl. Phys. A132, 186 (1969); F. Petrovich, et al.,
     Phys. Rev. Lett. 22, 895 (1969)

15) Richard Schaeffer, Nucl. Phys. A164, 145 (1971)

16) R. Sherr, et al., to be published

17)   R.A. Hinrichs, et al., Phys. Rev. Lett. <u>25</u>, 829 (1970)

18)   C.D. Goodman and P.G. Roos, BAPS <u>14</u>, 121 (1969)

19)   J.B. French and M.H. Macfarlance, Phys. Lett. <u>2</u>, 255 (1962)

20)   The explanation of Noble (Phys. Rev. Lett. <u>25</u>, 1458 (1970)) is not born out. (E. Rost and R. Schaeffer, Phys. Rev. <u>C3</u>, 2491 (1971))

21)   R. Schaeffer and G. Bertsch, Private Communication.

22)   E. Rost and P.D. Kunz, Phys. Lett. <u>30B</u>, 231 (1969)

23)   V.A. Madsen, et al., Phys. Rev. Lett. <u>26</u>, 454 (1971)

24)   W.L. Fadner, J.J. Kraushaar, and S.I. Hayakawa, to be published

25)   e.g. C.G. Morgan and D.F. Jackson, Phys. Rev. <u>188</u>, 1758 (1969); A.M. Bernstein, Advances in Nucl. Phys. <u>3</u>, 325 (1970); B. Tatischeff and I. Brissaud, Nucl. Phys. <u>A155</u>, 89 (1970)

26)   A.M. Bernstein and W.A. Seidler II, Phys. Lett. <u>34B</u> 569 (1971)

27)   J.Y. Park and G.R. Satchler, Particles and Nuclei <u>1</u>, 233 (1971)

28)   E.R. Flynne, A.G. Blair and D.D. Armstrong, Phys. Rev. <u>170</u>, 1142 (1968)

THE DISTRIBUTION OF NEUTRONS IN THE NUCLEAR SURFACE FROM ELASTIC ALPHA-PARTICLE SCATTERING[*]

Aron M. Bernstein and W. A. Seidler II

Massachusetts Institute of Technology

From electromagnetic measurements, we have a great deal of information about the distribution of protons in nuclei. Although several methods have been proposed to measure the neutron or the total density distributions,[1-3] these methods have not yet achieved accurate results. The purpose of this article is to present an accurate determination of the surface of nuclear density distributions. The measurement of the RMS radius relative to Ca^{40} is achieved to an accuracy of approximately ± 0.1 fm.

Since α particles are strongly absorbed by nuclei, the major contributions to the scattering in the diffraction region comes from grazing distances where the half density points of the α particle and the nucleus do not overlap.[4] This means we need to know the effective α particle, bound-nucleon interaction $V_{eff}(\bar{r}-\bar{r}_\alpha)$, at relatively long distances that are determined primarily from the range of the nucleon-nucleon interaction and the size of the α particle. An estimate using this idea gives V_{eff} as a Gaussian potential whose approximate strength and range are 37 MeV and 2.0 fm.[4] Although the range is the most important parameter, the results presented here are not sensitive to its exact value (10 percent changes would not change our results appreciably).

In this model, we take the optical potential $U(r_\alpha)$ as[4,5]

$$U(r_\alpha) = (\lambda_R + i\lambda_i) \int V_{eff}(\bar{r} - \bar{r}_\alpha)\rho(\bar{r}) \, d\bar{r} \qquad (1)$$

where: $\rho(r)$ is the density distribution of the nucleus whose volume integral is A particles; and λ_R and λ_i are empirically determined functions of E_α (the α-particle energy). Fortunately, the scattering is insensitive to the theoretically undefined imaginary potential; e.g., the derivative form for IM $U(r_\alpha)$ would fit the data equally well.[6] The scattering, particularly the phasing of the differential cross sections, is quite sensitive to Re $U(r_\alpha)$, which contains the information about $\rho(r)$.

For T = 0 nuclei, it can be assumed that $\rho_n(r)$, the neutron distribution, and $\rho_p(r)$, the proton distribution, are approximately equal and can be obtained from electromagnetic experiments. This was checked in detail using Hartree-Fock wave functions.[7,8] By fitting the elastic α-particle scattering to Ca^{40}, the parameters λ_R and λ_i were obtained by varying them until a best fit was obtained.[6] In this paper we shall consider the scattering of 104-MeV α particles,[9] for which the values λ_R = 0.815 and λ_i = 0.46 were previously obtained.[6] Using these values of the parameters, Eq. (1) was tested by successfully predicting the α-particle scattering from the T=0 nuclei, O^{16} and Si^{28} (ref. 6). We conclude that in so far as it is possible we have tested the validity of Eq. (1). For N>Z nuclei, for which there is no accurate experimental evidence for the $\rho_n(r)$, we can therefore use Eq. (1) with fixed λ_R and λ_i to obtain information about $\rho(r)$.

The optical potential was calculated by assuming a Fermi shape, $\rho(r) = \rho_0(1 + \exp(r - c)/a)^{-1}$, for the density distribution. It can be shown that the scattering is not sensitive to the exact value of a. Approximately ±20 percent variations in a can be compensated for by appropriate changes in c.

A typical example of the results is presented in Fig. 1 for Zr^{90}. The goodness of fit parameter Δ (usually called χ^2) is plotted versus the RMS radius, R, and has a deep minimum for which the agreement with experiment is excellent. To estimate the error in R, we have plotted the theoretical cross sections for values of R other than the minimum value of Δ. For $\Delta \geq 2\Delta_{min}$, the fit is unacceptable, as is indicated in Fig. 1.

Table 1 summarizes the results we have obtained. The values of R_p, the RMS radius of the proton distribution, have been obtained from measured charge distributions after the finite size of the proton has been unfolded. The values of R_n are then obtained from the values of R and R_p. From Table 1, it can be seen that R_n-R_p tend to be slightly positive but generally not inconsistent with zero. For Zr^{90} these results are in agreement with Hartree-Fock predictions.[8]

Equation (1) has been previously used to obtain agreement with the scattering of 42-MeV[5] and 166-MeV[10] α particles. In these papers, however, nuclear models were used to evaluate $\rho(r)$; no systematic attempt was made to extract $\rho(r)$ from the data. A new investigation is now under way to extract $\rho(r)$ from the 166-MeV

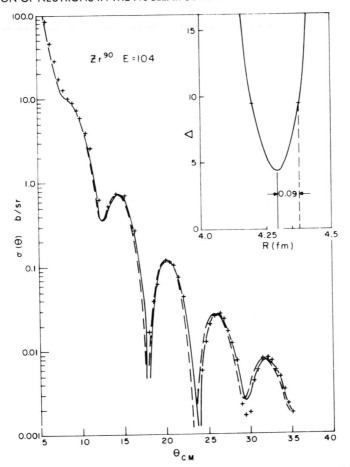

FIGURE 1.--The scattering of 104-MeV α particles from Zr^{90}. The
   insert shows Δ, the difference between theory and experiment
   versus the RMS radius R. The solid curve is calculated for
   Δ = $Δ_{min}$. The dashed curve for Δ = $2Δ_{min}$ when R is 0.09 fm
   larger than its value at $Δ_{min}$.

data using a method similar to the one presented here; the preliminary results are in reasonable agreement with the results
presented herein.

   At the present time we have not yet considered the same nuclei
as have been treated by other methods.[2,3] We shall discuss the
relative accuracies of these methods if time permits.

TABLE 1.--Preliminary Differences of Neutron and Proton RMS Radii.

Element	R_p (rm)	$R_n - R_p$ (fm)
$_{28}Ni^{64}$	3.82 ± 0.02 [a]	0.16 ± 0.16
$_{40}Zr^{90}$	4.16 ± 0.03 [b]	0.23 ± 0.11
$_{50}Sn^{124}$	4.60 ± 0.03 [c]	0.09 ± 0.12

[a] Ni^{64}. J. R. Ficenel, W. P. Trower, J. Heisenberg, and I. Sick, Phys. Lett. 32B, 460 (1970).

[b] Zr^{90}. J. Heisenberg, private communication.

[c] Sn^{124} P. Barreau and J. B. Bellicard, Phys. Lett. 25B, 470 (1967).

It appears that alpha-particle scattering is an accurate method of measuring the density distribution in the nuclear surface. This analysis is being extended to other isotopes and scattering energies. Further experimental work is needed to check that the extracted neutron distributions are independent of energy and to explore further neutron distributions. Theoretical work pertaining to the justification of the method used here and on its accuracy is highly desirable.

REFERENCES

1. Reviews of this subject have been presented by R. Wilson, Comments on Nuclear and Particle Physics, 4, 117 (1970); D. H. Wilkinson, Ibid., 1, 80 (1967) and 1, 112 (1967).

2. J. A. Nolen, Jr. and J. P. Schiffer, Ann. Rev. of Nucl. 19, 471 (1969).

3. G. W. Greenlees et al., Phys. Rev. 2C, 1063 (1970) and 3C, 1560 (1971).

4. A. M. Bernstein, Advances in Nuclear Physics. New York: Plenum Press. M. Baranger and E. Vogt, editors, 3 (1969).

5. D. F. Jackson, Phys. Lett. 32B, 233 (1960); C. G. Morgan and D. F. Jackson, Phys. Rev. 188, 1758 (1969); D. F. Jackson and V. K. Kembhavi, Phys. Rev. 178, 1626 (1969).

6. A. M. Bernstein and W. A. Seidler II, Phys. Lett. 34B, 569
    (1971).

7. D. Vautherin and M. Veneroni, Phys. Lett. 25B, 175 (1967)
    and private communication.

8. J. W. Negele, Phys. Rev. 1C, 1260 (1970).

9. G. Hauser et al., Nucl. Phys. A128, 81 (1969); G. Schatz,
    private communication.

10. B. Tatischeff and I. Brissand, Nucl. Phys. A155, 89 (1970).

11. B. Tatischeff, I. Brissand, and L. Bimbot, private communication.

*This work has been supported in part through funds provided by the
U. S. Atomic Energy Commission under AEC Contract AT(30-1)-2098.

CHARGE-EXCHANGE AND/OR KNOCKOUT SPECTATOR POLES IN THE $D(^3He,tp)p$ REACTION

R. E. Warner[*][†] G. C. Ball, W. G. Davies, A. J. Ferguson

and J. S. Forster, Chalk River Nuclear Laboratories,

Atomic Energy of Canada Limited and Oberlin College

This paper reports, for the first time, intense spectator peaks from the reaction $D(^3He,tp)p$, which may result from quasi-elastic scattering (QES) accompanied by charge exchange (CE). Quasi-elastic scattering was first observed by Kuckes, Wilson and Cooper[1], who found that large peaks (called spectator peaks) are observed in the p-p coincidence cross sections from the reaction $D(p,2p)n$ when momentum is transferred only between the two protons, and the neutron (called the spectator particle) remains nearly at rest in the laboratory. QES from the proton in the deuteron has also been studied[2-4] in the reactions $D(d,dp)n$, $D(^3He,^3He\ p)n$, and $D(\alpha,\alpha p)n$. In CEQES for $D(^3He,tp)p$ reaction (see Fig. 1a), the ^3He and neutron would transfer momentum and exchange charge, emerging as triton and proton, and the proton from the deuteron would remain nearly at rest. Alternatively, a direct knockout (KO) process (see Fig. 1b) might also produce spectator peaks.

A CD_2 target was bombarded with 27 MeV $^3He^{++}$ ions from the Chalk River MP Tandem accelerator. Coincidence events from two ΔE-E counter telescopes, coplanar with and on opposite sides of the beam, were recorded on magnetic tape. Particle identification was achieved during analysis, using a triton range-energy table[5]. Our absolute coincidence cross section data for two geometries appear in Figs. 2 and 3. In Fig. 2a, for instance, the broad peak near E_t = 13 MeV appears to be a spectator peak both because it is near the minimum spectator particle (undetected proton) energy and because of its large size, 30 mb/sr²-MeV; it is half as large as the ordinary QES peak from the $D(^3He,^3He\ p)n$ reaction which was observed simultaneously at this geometry (Fig. 2b). Peaks observed in previous t-p coincidence studies[6,7] resulted from sequential decay of excited

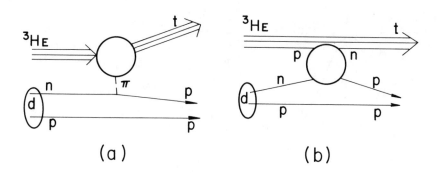

FIGURE 1.--Feynman diagrams for charge-exchange quasi-elastic
    scattering (a) and knockout (b).

states in ^{4}He, but their magnitudes[7] were only a few $mb(sr^{-2})$
(MeV^{-1}).

    To interpret our data we performed a plane wave Born approxi-
mation (PWBA) calculation including both CE and KO processes; our
development is similar to that of Henley et al.   A Hulthen deuteron
wave function[9] with $\alpha = 0.232$ fm^{-1} and $\beta = 1.202$ fm^{-1} and a
Gaussian triton (^{3}He,t) function[8] with $\gamma = 0.36$ fm^{-1} were used.
The transition potential[8] was

$$V(r_{ij}) = v_o [1-B\ P_{\sigma,ij} P_{\tau,ij}]\ \exp(-\delta^2 r_{ij}^2) \tag{1}$$

which contains both ordinary and spin-isospin-exchange interactions,
and causes the matrix elements to separate into spin-isospin and
spatial factors. The range of the force was fixed with $\delta = 0.656$ fm^{-1}.
The matrix elements for the two reactions are:

$$t+p+p:\ \overline{\sum}(M_{fi})^2 = \frac{1}{12}\ (BM_{CE}^+ + M_{KO}^+)^2 + \frac{1}{4}\ (BM_{CE}^- + M_{KO}^-)^2 \tag{2}$$

$$^3He+p+n:\ \overline{\sum}(M_{fi})^2 = \frac{1}{6}(BM_{CE}^- + M_{KO}^-)^2 + M_{KO}^{+2}(6B^2 - 3B + \frac{1}{2})$$

$$+ M_{CE}^{+2}(6-3B+\frac{1}{2}B^2) + M_{KO}^+ M_{CE}^+ (\frac{475}{72}B - \frac{3}{2} - \frac{3}{2}B^2) \tag{3}$$

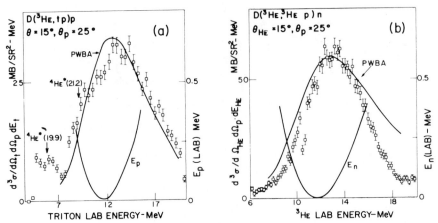

FIGURE 2a.--D(^{3}He,tp)p absolute coincidence cross sections and PWBA
predictions (left-hand scale) and spectator proton lab energy
(right-hand scale) plotted vs triton lab energy at (θ_t=15 , θ_p=25).
Arrows show where enhancements due to sequential decay of the ^{4}He
excited states at 19.9 and 21.2 MeV may occur.

FIGURE 2b.--Similar to 2a except D(^{3}He,^{3}He p)n data, predictions
for the ordinary QES peak, and spectator neutron energy are plotted
vs. ^{3}He lab energy. Both PWBA predictions in Fig. 2 are multiplied
by 0.02.

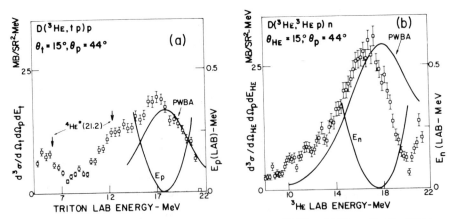

FIGURE 3.--Like Fig. 2, for the geometry (15°, 44°). Both PWBA predictions are multiplied by 0.06. The rise in the ^{3}He spectrum near
21 MeV is attributed to the singlet deuteron final state interaction.

The +(-) superscripts on the M's denote even (odd) parity of the final nucleons.

The shapes of the spectra calculated from these matrix elements are insensitive to B and to whether CE and/or KO is assumed to take place. Renormalized cross sections obtained for CE with B = 2.3 appear in Figs. 2 and 3, and the peaks are generally well fitted by them.

Since we expect the PWBA to predict the ratio R = $d^3\sigma$(t+p+p)/ $d^3\sigma$(^{3}He+p+n) more reliably than the magnitudes of the cross sections, equal but otherwise arbitrary normalization factors were used for the two reactions at each geometry. Acceptable values of R, close to the observed values of 1/2 to 2/3, could be obtained only for pure CE with B = 2.3. Those predicted assuming both CE and KO, or KO alone, were only about 0.05 at B = 0 and decreased monotonically with increasing B. For CE alone they were only 0.08 for B = 1.0, the Serber value[10].

Henley et al.[8] deduced that both pickup and charge exchange are important in the D(^{3}He,tp)p reaction at the highest triton energies, where the p-p final state interaction occurs. The present work provides additional evidence for the existence and strength of charge exchange, but underlines the need for a more sophisticated theoretical approach. The two most striking failures of our model are its inability to give a proper account of the strength of the KO process and, even assuming that only CE takes place, the unrealistic exchange force required to give the observed (t+p+p)/(^{3}He+p+n) cross section ratio. Perhaps these shortcomings could be rectified by straightforward changes within the PWBA framework, such as including tensor forces or other types of exchange forces, completely antisymmetrizing the wave functions for exchange of all 5 nucleons, or considering higher order processes. They may instead require a radically different and more sophisticated approach, such as a solution of the Fadeev equations.

It is a pleasure to thank Professor Robert Hofstadter and Dr. F. C. Khanna for their advice and encouragement.

REFERENCES

1. A. F. Kuckes, R. Wilson and P. F. Cooper, Jr., Annals of Phys. 15, 193 (1961).

2. P. F. Donovan, Rev. Mod. Phys. 37, 501 (1965); B. E. Corey, R. E. Warner, and R. W. Bercaw, Nucl. Phys. to be published.

3. P. A. Assimakopoulos, E. Beardsworth, D. P. Boyd and P. F. Donovan, Nucl. Phys. A144, 272 (1970).

4. R. E. Warner and R. W. Bercaw, Nucl. Phys. A109, 205 (1968).

5. B. Hird and R. W. Ollerhead, Nucl. Instr. and Meth. 71, 231, (1969).

6. P. D. Parker, P. F. Donovan, J. V. Kane and J. F. Mollenauer, Phys. Rev. Lett. 14, 15 (1965).

7. R. E. Warner, B. E. Corey, E. L. Petersen, R. W. Bercaw and J. E. Poth, Nucl. Phys. A148, 503 (1970); G. C. Ball, W. G. Davies, A. J. Ferguson, J. S. Forster and R. E. Warner, to be published.

8. E. M. Henley, F. C. Richards and D. U. L. Yu, Nucl. Phys. A103, 361 (1967).

9. M. J. Moravcsik, Nucl. Phys. 7, 113 (1958).

10. J. M. Blatt and V. F. Weisskopf, Theoretical Nuclear Physics (John Wiley and Sons, New York, 1952) pp. 170-181.

[*]Research supported by the National Science Foundation through Grant No. GP-19269.

[†]Address during 1971-72: Department of Nuclear Physics, Oxford University, Oxford, England.

α--PARTICLE INTERMEDIATE STATE IN THE (h,t) REACTION

G. Bertsch and R. Schaeffer
Michigan State University

We report on preliminary calculations of the (h,t) reaction, assuming that the helion picks up a neutron, propagates some distance as an alpha, and then strips a proton. The following qualitative results are established:

1) The angular distribution to anti-analog states has an L=1 shape while the angular distribution to analog states retains its L=0 shape. This is due to the fact that the transition operator to the analog state,

$$<\text{analog}|\ a_p^+(r_p)a_n(r_n)|\text{parent}> = \theta(r_p,r_n)$$

has an L=0 character when $r_p=r_n$, while for the anti-analog state the operator is very small when $r_p=r_n$ and has an L=1 character when r_p is near r_n.

2) Cross sections fall off much more slowly with increasing angular momentum transfer than first-order calculations predict. The second-order process is similar to an exchange scattering in that the stripped particle need not be the same as the picked up particle. Thus the allowable momentum transfer is characterized more by the single-particle momentum than by the range of the interaction.

SYMPOSIUM SUMMARY

M. Baranger

Massachusetts Institute of Technology, Cambridge,

Massachusetts

I have been asking people what the summary talk is supposed
to be. They told me my job was to unify things. They said "you
are supposed to pick up the diverse material that came to light
during the conference and give it a direction". As a matter of
fact, nuclear physics is very hard to unify. The main unifying
feature is that everything in nuclear physics is about nuclei. It
would be hard to say much more, but that's enough. Nuclear physics
is a huge organism with many complex parts, which are all inter-
related by the fact that they all deal with nuclei.

Some people get impatient with the pragmatism which pervades
much of the work in nuclear physics. They think that if we can't
make our results follow logically from the Schroedinger equation,
then we don't know what we are doing. But to me that is the
description of a dead science. Nuclear physics, on the other hand
is alive, because it contains within it many suborganisms, usually
called models, which co-exist, but not always peacefully; because
new branches of it are being born all the time, while others are
dying; because the sub-units exercise a lot of freedom in their
motions, and if there is a unifying idea which explains the whole
dance, we have not found it yet.

There is a lot of autonomy in the various corners of nuclear
physics. There is no single correct way of working and the nuclear
physicist can do pretty much what he wants as long as he keeps up
the standards. So, without trying to unify that which likes to be
diverse, I shall attempt to pick up one by one some of the themes
which I have felt running through this symposium. Most of these
were already announced by Malcolm Macfarlane in his preview.

First, we talked about the interaction between free nucleons. One theme there, which runs throughout the rest of the symposium also, is that you can either determine the interaction empirically or you can give a microscopic derivation. Both methods have strong drawbacks and only partial success can be claimed for either of them. The empirical determination fails by lack of data and the microscopic derivation because of bad convergence.

In Signell's review of the empirical determination, we learned that progress in understanding and interpreting the on-shell data is being made slowly. The fitting of these data by potentials has recently brought up the possibility of super-soft cores. But these may yet be ruled out by nuclear matter, because it seems that you need a big enough wound in order to saturate at the observed k_F, and the super-soft cores may not give you that big wound. As always, most of the data are on-shell and we have to know the off-shell behavior of the interaction if we want to claim that we know the interaction completely. I shall come back to this later.

As far as the microscopic derivation is concerned, we learned from Gerry Brown that it is being pushed toward smaller inter-nucleon distances. He and his collaborators have made remarkable progress in their derivation of the two-pion exchange term. They start from pion-nucleon scattering at low momenta. They transform it using soft pion theorems and the knowledge of the $\pi-\pi$ phases. What they get is in better agreement with the data than any other theory to date. And there is no need for the σ meson any more. This approach is capable in principle to get the off-shell behavior as well as the on-shell behavior of the interaction.

From the empirical viewpoint, great hopes have been laid in the last half-dozen years on nucleon-nucleon bremsstrahlung as a promising method for finding out about the off-shell behavior of the interaction. We heard about the present situation there from Halbert, Heller and Virginia Brown. The story is rather sad and I am going to try to tell you how I understand it. First, proton-proton bremsstrahlung seems to agree with theory reasonably well. The best theories use the Hamada-Johnston or the Reid potential and both give about the same results. You would need much more experimental accuracy to distinguish between the two. In neutron-proton bremsstrahlung, the data are still extremely uncertain; the various experiments differ sometimes by an order of magnitude, although we learned that there may now exist one good experiment. The most important news is that npγ theory, and to a lesser extent ppγ theory, are hampered by the lack of knowledge of the correct nucleon current. For local potentials we know what current to use, but for non-local potentials and especially for exchange potentials we do not. In the case of short-range non-locality, such as the factor p^2 which appears in many phenomenological or microscopically derived potentials, this lack of knowledge of the current is not

very important. It creates uncertainties of less than 1%. But in
the case of exchange currents, it makes a tremendous difference what
you do. The uncertainty there might give a factor of anywhere from
2 to 5 in npγ. The same uncertainty is present in the exchange
current for ppγ, but it starts to occur only with the two-pion part
of the interaction, and therefore it is less important than in the
case of npγ.

What current should be used? If you use field theory or
diagrams, then you know what to do. Heller gave an example of this
in his talk and the Brown-Franklin contribution is probably the
same kind of thing. But if you use a phenomenological potential,
then you don't know what to do. There exists a specially simple
prescription, called the minimal prescription, but this does not
agree with what you get in a simple field theory example, and in
complicated cases the minimal prescription is not unique. So,
your potential model of the interaction must also include a model
of the current. Given the potential there is lots of arbitrariness
in the current, as Heller discussed in his talk. The conclusion
seems to be that it is impossible to use nucleon-nucleon bremsstrahlung to discriminate between various potential models, all of which
fit on-shell, until these models have been supplemented with current
models.

The situation can also be described in another way. We can
distinguish three kinds of radiation in nucleon-nucleon bremsstrahlung.
First there is external radiation. This is described by diagrams
such as fig. 1a. The photon can issue from any one of the four
external legs of the diagram. External radiation is the simplest
to calculate and it is unique. Next we have rescattering, described
by diagrams such as fig. 1b where the photon is emitted by one of
the internal nucleon legs. Rescattering is small for ppγ under
ordinary conditions and not very big (perhaps of the order of 10%)
for npγ. Finally, we have internal radiation, fig. 1c. Now the
photon line is emitted by a meson line or by any other line or
vertex internal to the nucleon-nucleon blob. In particular, the
photon could be emitted by an exchanged charged meson. Internal
radiation is huge for npγ (this is the factor of 2 to 5 mentioned
above); it is less for ppγ and I don't know whether any estimates
exist there at all.

It seems that nucleon-nucleon bremsstrahlung is not a good way
to learn empirically about the off-shell nucleon-nucleon interaction.
Our best hope for this lies probably in the three-body or four-body
problems, or even in the properties of larger assemblies of nucleons,
namely nuclei.

This brings me to the microscopic calculation of nuclear bulk,
which was discussed by Bethe, and here we have a much happier story.
I shall try to review Bethe's main points. Point #1. Nuclear

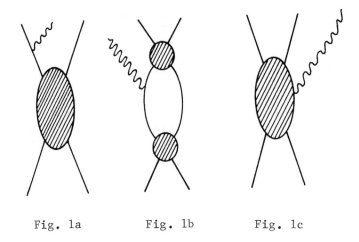

Fig. 1a   Fig. 1b   Fig. 1c

matter theory is in good shape. Any reasonable potential can now
be plugged in and its effect in nuclear matter calculated. The
remaining uncertainties of a few MeV's are due either to the higher-
order corrections or to using the wrong two-body potential. Which
potential is right is another problem, to which I shall come back.
But in any case we can insure agreement with experiment, either by
changing the potential, or by making small phenomenological adjustments as in Negele's thesis. The theory then acquires a predictive
value.

Point #2. By being clever with averages, one can derive a
density-dependent local effective interaction, including a large
exchange term, which reproduces most nuclear matter results and
whose simplicity allows it to be used in Hartree-Fock calculations
of finite nuclei in coordinate space. This was done by Negele with
very good results, after the small adjustments mentioned above.
He calculated things like the total energy, the density distribution,
the root mean square radius, the single particle energies, etc.

Point #3. Another way to treat finite nuclei is to do
Brueckner-Hartree-Fock in a harmonic oscillator basis. This has
best been done by Davies and McCarthy. It is more exact than
Negele's procedure and will presumably replace it ultimately. At
the moment its results do not agree as well with experiment, mostly
because it is not so easy to put in the density-dependent fudge
factor of Negele. This calculation does contain a very important

feature which is not in Negele's: occupation probabilities of less
than unity for the states in the Fermi sea. One strange result of
Davies and McCarthy is a rather prominent density maximum at the
center of the nucleus, which is not in Negele. The wiggles of the
charge distribution are identical in both, and they are also the
same as the wiggles you get by calculating wave functions in a
Woods-Saxon well.

Point #4. A frankly phenomenological way, used very successfully by Brink and Vautherin, takes a very simple potential (delta
functions and derivatives thereof) with density-dependent coefficients
fitted to some crucial data. This is very easy to work with; it
can do deformed nuclei and perhaps even collective motion in heavy
nuclei. The results are mostly excellent, even better than Negele's.
There are indications that the isospin-dependence of the present
force is not quite correct.

Point #5. No theory can predict spin orbit splitting well,
especially when the Fermi surface splits some spin orbit doublets.
This is a major remaining problem. Banerjee thinks that it may help
to take into account the non-Galilean terms in the two-body spin
orbit interaction.

We seem to have reached a contradiction here. On the one
hand, we should be terribly depressed because we have not been able
to learn much about the off-shell nucleon-nucleon interaction. On
the other hand, using only on-shell data, we have derived the Reid
potential which gives remarkably good results for almost everything
after small phenomenological adjustments. If this is true, why worry
about off-shell. Let us ask the question clearly. To what extent
do phase shifts and the properties of the deuteron determine nuclear
structure? This was the subject of Tabakin's talk and of contributions by Haftel and by Pradhan, Sauer, and Vary.

Consider, for example, the work of Elliott and his collaborators
at Sussex. They use only the phase shifts, with reasonable assumptions about smoothness and short range of the potential. Then, by
a series of very clever tricks, they manage to get all the harmonic
oscillator two-body matrix elements that you might need in shell
model calculations. These matrix elements are very good. They work
as well as Tabakin matrix elements or Kuo-Brown matrix elements.
How can you do this when you know only the phase shifts? Isn't it
true that the off-shell part of the interaction contains many more
degrees of freedom than the on-shell part?

My favorite way of representing this is to use the σ function
defined in Tabakin's talk. For a $^1S_0$ partial wave, $\sigma(k, k')$ is a
real symmetric function of two variables, just as the potential
$v(k,k')$ is a real symmetric function of two variables. Thus σ
contains just as much information as v and can be used instead of

v. We can plot the contours of the function σ (divided by the
volume factor k k'), as shown in fig. 2a. The reason that makes

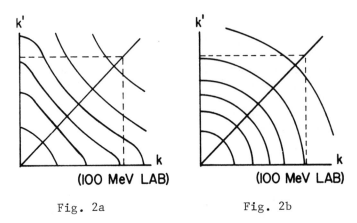

Fig. 2a   Fig. 2b

sigma more convenient to use than a potential is that we know its
value on the diagonal. It is essentially the sine of the phase
shift. So, in order to fit the phase shift, all we need to do is
let sigma have the correct value on the diagonal, and off the
diagonal we can do anything we want. This is much more convenient
than in the case of a potential, where we have to calculate phase
shifts for a whole class of potentials, and then do a complicated
fitting procedure. Here we simply continue σ off the diagonal in
an arbitrary way. Then, once σ is fixed, we have a certain crank
which we turn, a certain collection of formulae and this gives us
all matrix elements of the T matrix off-shell, and also the
potential if we want it. You see on this picture that there is a
lot of space off shell, and we can do what we want with σ there.
Therefore there are many, many σ functions that fit the same phase
shifts, and you would expect naively that the different σ functions
would give different results for the various quantities of nuclear
structure, so that just knowing the phase shifts would be of very
little help. But this is not so as the Sussex work shows, and in
the context of the σ formalism it was discovered by Peter Sauer
when he started to play with these things. He plotted the σ functions from various potentials, all of which give a reasonable fit
to the $^1S_0$ phase shifts, for instance Tabakin and Reid, and to his
surprise he found that these two potentials, which are very very
different from each other, gave essentially the same contours off-

shell as on-shell, all the way up to k or k' corresponding to
about 100 MeV in the lab system. This was quite a surprise, and
so he tried to manufacture some other potentials by forcing the
contours to be different off-shell, like for instance fig. 2b.
Then you can get the potentials by turning the crank which I mentioned earlier. He did this many times and in every instance he
found that, if the contours did not look pretty much like Reid or
Tabakin below 100 MeV, he came out with a potential whose range
was more than 10fm. It seems that the curve giving the phase shift
as a function of energy, which is a very fast varying curve (fig.
3), leads to a very large range, if you interpret the fast variation
of $\delta(k)$ as a direct result of the range. But there is another way
to interpret this variation: it may come from a resonance, and of
course this is the actual case. There exists a resonance at about
50 keV, and the existence of this resonance completely dominates
the behavior of the T-matrix at low energy, both on-shell and off-
shell. We know that this resonance exists simply by looking at the
phase shift curve, because we know that the range must be small,
and then if we insist that there be a resonance we have essentially
no more freedom in the σ function anywhere below 100 MeV. This is
why all the potentials which fit the effective range formula give
the same low energy contours. This is why the effective range
formula itself is valid over a much wider range of energy than it
has any right to be. The situation is presumably the same as in
the $^3S_1$-$^3D_1$ channel, where there is a pole in the T-matrix (the
deuteron) very close to zero energy, and this also dominates the
behavior of σ completely.

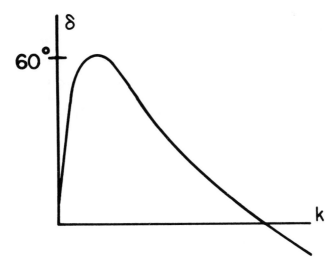

Fig. 3

Therefore the thing to do if you want to vary the off-shell behavior of your interaction is to first subtract from σ its resonating part. The leftover part of σ is very small on-shell up to 100 MeV. Of course, you can still do what you want off-shell, but it is clear that if you do not keep the off-shell part of the same order of magnitude as the on-shell part you are going to introduce some strong unsmooth features in the interaction. These are features for which we have no experimental evidence whatever. They showed up in Tabakin's talk as strange bumps at short distances or zeroes of the wavefunction very close to the origin. I am not saying that I know that such features do not exist in the true nuclear structure. They might very well be there, but we know absolutely nothing about them. Therefore, the natural thing to do until forced otherwise is to invoke the principle of minimal complication and to let the potential and the σ function be reasonably smooth if we can. Then we have very little choice in the σ function below 100 MeV, and this is why all reasonable potentials give approximately the same contours, and this is why the Sussex method works. You do have some choice at higher energies, and there Tabakin and Reid for instance are extremely different from each other. The high energy behavior is important for the saturation of nuclear matter and the Sussex people would not wish to claim that they can reproduce that without further assumptions.

Using the σ method and subtracting the resonance, Sauer has generated families of smooth phase-shift equivalent potentials. He has calculated two-body shell model matrix elements for them, and also the contribution of $^1S_0$ waves to the binding energy of nuclear matter at a fixed k_F, and he has found that most of these quantities do not vary by more than 10 or 20% from one potential to the next.

After dealing with the interaction between free nucleons and the interaction between nucleons in the nuclear bulk, the symposium switched to the effective interaction between nucleons near the Fermi surface. Here again we find our theme of microscopic derivation versus empirical determination, the lack of convergence of the former and the difficulty to obtain data for the latter. The microscopic derivation was reviewed by Barrett. The main results are these: (1) It is possible to derive an effective interaction rigorously from microscopic theory in the form of a series. In fact, it can be done in several ways. (2) This series does not converge, at least for shell model spaces of the usual size. It might converge if you made the shell model space larger, but then you cannot do the shell model calculations. (3) Some valiant attempts are being made to overcome this divergence by partial summations. The most notorious of these attempts are those of Kirson. It is much too early to pass judgment on their success or failure. If they do succeed eventually, a remarkable feat will have been accomplished, a feat comparable in magnitude with the introduction of the G-matrix by Brueckner and Bethe. But let us remember that it was about ten years before G-matrix theory finally stabilized and made sense. The time might be equally long for the shell model interaction, so let's not be too impatient. (4) In the meantime, there is no reliable way of deriving an effective interaction microscopically. A moderately successful interaction like that of Kuo and Brown has to be justified by the fact that it works, not because it came out of some theory. This brings us naturally to the next topic: effective interactions from empirical studies of finite nuclei.

Here we pick up an auxiliary theme. It is possible to concentrate on a few simple processes and treat them very accurately, or you can try to set up a method to explain reasonably well vast masses of data. The first point of view is illustrated by Schiffer's talk. The second by McGrory and by Macfarlane in his over-view. The lessons to be learned from McGrory and Macfarlane seem to be these: (1) It is possible to do large shell model calculations. (2) The features of a two-body interaction that are important to get reasonable fits to large masses of data are rather few. Many interactions give quite similar agreement, presumably because they all contain these few features. It is hard to tell what these features are exactly, but McGrory has some good guesses in terms of relative s-wave radial integrals, and a few others, while Macfarlane extracts them with his error matrix. A random interaction definitely does not work. (3) A parameter-free interaction of Kuo works almost as well as anything else, except for total binding energies, which is saying a lot for Kuo and Brown. (4) The ground state band for each T always works better. (5) The magnetic moments and the quadrupole effects come out very easily, provided you use an effective charge in the latter. (6) The size of the model space can be made quite small without having to change the effective interaction or most one-body operators. All you need to do is normalize the wavefunctions to unity in each case, and you do need to change the effective charge. For instance, around A=20, $d_{5/2}s_{1/2}$ is enough for a model space, but $d_{5/2}$ alone is not enough. Around A=45, $f_{7/2}p_{3/2}$ is enough.

Schiffer finds an interaction which, used in pure particle-particle, hole-hole, or particle-hole configurations, reproduces the multiplet splitting. He has many cases throughout the periodic table which work very well. How do we know that these cases are pure? Because we see all the members of the multiplet, and only them, and the cross section for one-nucleon transfer is proportional to 2J+1 as it should be. Some help is also received from gamma spectroscopy.

The interaction is very simple. It has half a dozen parameters or so and is apparently universal. The problem now is to reconcile this with other kinds of thinking in nuclear physics, but no matter how you do the latter, Schiffer's work stands as a beautiful and concise model encompassing many facts. I do not pretend to be able to do the reconciliation. There is food for a lot of thinking there and a lot is to be learned eventually. The first thing one can ask is how does the interaction compare with theoretical effective interactions such as Kuo-Brown? This is hard to answer because Schiffer's interaction is supposed to be used in only one configuration, for instance proton $h_{9/2}$, neutron hole $f_{5/2}$, in ^{208}Bi, while Kuo-Brown is designed for a much larger model space, namely a major shell for each particle. According to Kuo-Brown, the states in ^{208}Bi are mixed, but Schiffer says they are pure. So you have to reconcile that first. They could both be right, because they are talking about different things. Once again, nuclear physics is not simple and that is what makes its charm. Let us hope that the resolution of the paradox when it comes is as convincing and as elegant as the experimental work itself.

The final topic of the symposium was the effective interaction for reaction calculations, i.e. the interaction between the projectile and the nucleons in the target. All the previous themes are found here again: microscopic derivation vs. empirical determination; looking at a few processes accurately vs. doing systematics. But an additional theme emerges, which I consider the most important theme of the whole symposium: there is a striking parallel between what you do for reactions and what you do for nuclear structure or shell-model calculations. This is new to me and so welcome. At last, reaction theorists and structure theorists can talk to each other; we don't need to have two kinds of theory any more.

The first two talks by Schaeffer and McManus were about the microscopic derivation of the effective interaction for pp' and pn reactions. The story of the recent developments in this field was very nice to hear. Microscopic theory used to be much less reliable than the collective model. The latter works especially well for projectiles which are strongly absorbed like α-particles and for collective excitations, but it works in many other places too, provided you introduce one or two phenomenological parameters. Eventually we shall have to understand why the collective model works so well so often. But for the moment we concentrate on looking at the microscopic model for its own sake. The situation is that, presumably as a result of enormous computing efforts, a sensible microscopic theory has been developed and it works quite well where you would expect it to work. The theory parallels that of the shell model effective interaction but it is much more difficult because of the scattering situation. For instance, you cannot use harmonic oscillator wavefunctions, which is something you always do in nuclear structure, and therefore you cannot work in center-of-mass-relative coordinates and you cannot use Moshinsky brackets. Presumably the scattering effective interaction is affected by the same troubles of non-convergence as the shell-model effective interaction, but

right now it is giving sensible results just as at one time Kuo-
Brown gave sensible results.

The theory contains a two-body interaction which is still
quite simple. It might be possible to get this interaction from
nuclear matter theory. This would be similar to Negele's work but
the conditions would be different and therefore the interaction would
be different. Not too much has been done on this yet. The interaction needs to be local to be usable, and Negele's interaction is
local. The exchange effects are now put in and they are very
important. Core polarization is put in also and is large, as in
Kuo-Brown. Finally, the two-body spin-orbit force is very
important at large angular momenta or large energies.

One is impressed by the enormous amount of structure information which will become available thanks to the good understanding
of these reactions. There was a time not so long ago when a reaction
theorist would be happy if the absolute agreement between his calculation and the data was within a factor of 10. All this is
changed and the theory now claims to get the normalization as well
as the angular distribution. The latter is often not as good as
with the collective model and the reason for this will have to be
understood eventually.

The same effective interaction for pp' or pn reactions can
be determined empirically, as reported by Austin. This was very
much like Schiffer's talk except that reactions cannot claim the
accuracy of energy level data. But qualitatively the amount of
information extracted is about as much. Austin can determine two
parameters in the central part of a simple force; he also has an
estimate of the tensor part. For the spin-orbit part there are
still too many inconsistencies to say much; the fits to polarizations
are never very good. The same question can be asked as in Schiffer's
case: what is the relation of this interaction with the microscopically derived one? This question did not receive much of an
airing at the symposium and it is probably hard to answer, but to
me it is already excellent news that this question can be asked.

In Madsen's and Roos's talks we heard about the microscopic
derivation and the empirical determination of effective interactions to be used in the scattering of complex projectiles. There
is a large number of reactions that come under this category. Some
of them might be considered a luxury which we do not really need to
have in order to understand nuclear phenomena, but others are
essential. An example of the latter is (^{3}He,t) which is a charge
exchange reaction like (p,n), but much easier to measure. You have
to do (^{3}He,t) and therefore you need a theory for it. The theory
is the extension to many nucleons of the pp' or pn theories. If
you compare the study of (pp') or (pn) reactions to Schiffer's study
of the effective interaction between two nucleons outside of a

closed shell, then the theory of complex projectile scattering is
to be compared with McGrory's or Macfarlane's studies. The
developments for reactions follow very closely the developments of
the shell model case. For instance, you can divide your model space
into two or more pieces, with projection operators in and out of
each of them, or else you can use Feynman diagrams just as in the
case of nuclear structure, and the improvement in our intuition and
understanding is the same in both cases. Of course this is the
most difficult of all theories: it is scattering and it is many-
body. The results so far are consistent with pp' and with nuclear
structure, using the same kind of interactions, and the absolute
magnitudes of the cross sections come out correctly, with large
errors of course. Here too, effects like exchange or core polarization can be extremely important. The reactions are direct reactions
which can be classified by their parentage. For example, we have
the reactions whose effect is that of a one-body operator, those
that strip one nucleon or pick up one, etc. . . All reactions
having the same parentage, including beta and gamma processes,
tend to have many features in common. Despite enormous computing
difficulties, we are moving steadily toward a generalized approach
of this kind, a sort of microscopic shell model understanding of
these phenomena. Many problems remain and the understanding may
never be complete, but certainly all nuclear physicists that I have
heard at this symposium seem to be striving toward this end, and
perhaps this is the basic unifying idea of the symposium.

LIST OF PARTICIPANTS

Anantaraman, N., Argonne National Laboratory.
Austin, S. M., Michigan State University.

Banerjee, M. K., University of Maryland.
Baranger, M., Massachusetts Institute of Technology.
Barrett, B. R., University of Arizona.
Benenson, W., Michigan State University.
Bernstein, A., Massachusetts Institute of Technology.
Bernthal, F. M., Michigan State University.
Bertsch, G., Michigan State University.
Bethe, H. A., Cornell University.
Blair, J. S., University of Washington.
Blosser, H. G., Michigan State University.
Borysowicz, J. R., Michigan State University.
Breit, G., State University of New York at Buffalo.
Brown, G. E., Nordita/State University of New York at Stony Brook.
Brown, V. R., Lawrence Radiation Laboratory, Livermore.

Coester, F., Argonne National Laboratory.
Crawley, G. M., Michigan State University.

Davies, W. G., Atomic Energy Commission of Canada, Chalk River.
Doering, R. R., Michigan State University.
Durso, J. W., Nordita/Mount Holyoke College.
Dworzecka, M., Michigan State University.

Eisenstein, R. A., Carnegie-Mellon University.

Fox, S. H., Michigan State University.

Galonsky, A., Michigan State University.
Garvey, G. T., Princeton University.
Gerace, W. J., University of Massachusetts.
Gilbert, K. E., Michigan State University.

Haftel, M. I., United States Naval Research Laboratory.
Halbert, M. L., Oak Ridge National Laboratory.
Hammerstein, G. R., Michigan State University.

Harris, G. I., Wright-Patterson Air Force Base.
Heller, L., Los Alamos Scientific Laboratory.
Hiebert, J. C., Texas A & M University.
Hinrichs, R. A., Michigan State University.
Hodgson, R. J., University of Ottawa.
Holdeman, J. T., Michigan State University.
Hyder, A. K., Wright-Patterson Air Force Base.

Iudice, N. L., University of Toronto.

Kashy, E., Michigan State University.
Kelly, W. H., Michigan State University.
Khanna, F. C., Atomic Energy Commission of Canada, Chalk River.

Lanford, W. A., University of Rochester.
Larson, D. C., Michigan State University.
Larson, N. M., Michigan State University.
Lawson, R. D., Argonne National Laboratory.
Lee, H. T. S., University of Pittsburgh.
Lodhi, M. A. K., Texas Technological University.
Love, W. G., University of Georgia.

Ma, C. W., University of California, Davis.
Macfarlane, M. H., Argonne National Laboratory.
McGrath, R. L., State University of New York at Stony Brook.
McGrory, J. B., Oak Ridge National Laboratory.
McKellar, B. H. J., University of Sydney/Melbourne.
McManus, H., Michigan State University.
Madsen, V. A., Oregon State University.
Mani, G. S., University of Manchester.
Maripuu, S., Wright-Patterson Air Force Base.
Miller, P. S., Michigan State University.
Moszkowski, S. A., University of California, Los Angeles.
Müller-Arnke, A., Darmstadt.

Noble, C. J., Bartol Research Foundation.
Nolen, J. A., Michigan State University.

Ophel, T. R., Australian National University.

Parish, L. J., University of Minnesota.
Petrovich, F. L., Michigan State University.

Reid, N. E., University of Manitoba.
Reid, R. V., Jr., University of California, Davis.
Rice, J. A., Michigan State University.
Richards, K. C., Bartol Research Foundation.
Robertson, R. G. H., Michigan State University.
Roos, P. G., University of Maryland.

Schaeffer, R., Saclay/Michigan State University.
Schiffer, J. P., Argonne National Laboratory.
Serduke, F. J., Argonne National Laboratory.
Shamu, R. E., Western Michigan University.
Signell, P. S., Michigan State University.
Sprung, D. W. L., McMaster University/Orsay.
Svenne, J. P., University of Manitoba.

Tabakin, F., University of Pittsburgh.
True, W. W., University of California, Davis/Los Alamos Scientific
                                                        Laboratory.

Vary, J., Massachusetts Institute of Technology.

Wagner, W. T., Michigan State University.
Warner, R. A., Michigan State University.
Wildenthal, B. H., Michigan State University.
Woods, T. J., University of Minnesota.

Zamick, L., Rutgers University.

AUTHOR INDEX

Austin, S. M. 285

Ball, G. C. 360
Baranger, M. 367
Barnes, P. D. 244
Barrett, B. R. 155, 175, 234
Bernstein, A. M. 355
Bertsch, G. F. 243, 365
Bethe, H. A. 133
Brown, G. E. 29, 174
Brown, V. R. 123

Canada, T. R. 244
Chemtob, M. 48

Davies, W. G. 360
deTourreil, R. 43
Durso, J. W. 48
Dworzecka, M. 243

Eisenstein, R. A. 244
Ellegaard, C. 244
Epstein, G. N. 51

Feldmaier, H. 241
Ferguson, A. J. 360
Forster, J. S. 360
Franklin, J. 123

Haftel, M. I. 132
Halbert, M. L. 53
Harvey, M. 229
Heller, L. 77, 79
Hodgson, R. J. W. 177

Jopko, A. 229

Khanna, F. 229, 234

Lanford, W. A. 241

Macfarlane, M. H. 1
Madsen, V. A. 315
Manakos, P. 241
Mani, G. S. 313
McGrory, J. B. 183
McKellar, B. H. J. 51
McManus, H. 243, 265
Müller-Arnke, A. 242

Noble, C. J. 51

Petrovich, F. 242
Pradhan, H. C. 126

Richards, K. C. 51
Riska, D. O. 48
Roos, P. G. 333

Sauer, P. U. 126
Schaeffer, R. 242, 245, 365
Schiffer, J. P. 205
Seidler, W. A. 355
Signell, P. 9
Sprung, D. W. L. 43, 229

Tabakin, F. 101
Trilling, R. 242

Vary, J. P. 126

Warner, R. E. 360
Woods, T. J. 307
Wolff, T. 241

Zamick, L. 235, 242